T0226300

Lecture Notes in Artificial Intelligence 10642

Subseries of Lecture Notes in Computer Science

More information about this series at http://www.springer.com/series/1244

Gita Sukthankar · Juan A. Rodriguez-Aguilar (Eds.)

Autonomous Agents and Multiagent Systems

AAMAS 2017 Workshops, Best Papers
São Paulo, Brazil, May 8–12, 2017
Revised Selected Papers

 Springer

Editors
Gita Sukthankar ⓘ
University of Central Florida
Orlando, FL
USA

Juan A. Rodriguez-Aguilar ⓘ
IIIA-CSIC
Bellaterra
Spain

ISSN 0302-9743   ISSN 1611-3349   (electronic)
Lecture Notes in Artificial Intelligence
ISBN 978-3-319-71681-7   ISBN 978-3-319-71682-4   (eBook)
https://doi.org/10.1007/978-3-319-71682-4

Library of Congress Control Number: 2017960858

LNCS Sublibrary: SL7 – Artificial Intelligence

Printed on acid-free paper

This Springer imprint is published by Springer Nature
The registered company is Springer International Publishing AG
The registered company address is: Gewerbestrasse 11, 6330 Cham, Switzerland

Preface

AAMAS is the leading scientific conference for research in autonomous agents and multiagent systems, which is annually organized by the non-profit organization, the International Foundation for Autonomous Agents and Multiagent Systems (IFAAMAS). The AAMAS conference series was initiated in 2002 by merging three highly respected meetings: the International Conference on Multi-Agent Systems (ICMAS); the International Workshop on Agent Theories, Architectures, and Languages (ATAL); and the International Conference on Autonomous Agents (AA).

Besides the main program, AAMAS hosts a number of workshops, which aim at stimulating and facilitating discussion, interaction, and comparison of approaches, methods, and ideas related to specific topics, both theoretical and applied, in the general area of Autonomous Agents and Multiagent Systems. The AAMAS workshops provide an informal setting where participants have the opportunity to discuss specific technical topics in an atmosphere that fosters the active exchange of ideas.

This book compiles the best papers of the AAMAS 2017 workshops. In total, AAMAS 2017 ran 18 workshops. To select the best papers, the organizers of each workshop were asked to nominate up to two papers from their workshop and send those papers, along with the reviews they received during their workshop's review process, to the AAMAS 2017 workshop co-chairs. The AAMAS 2017 workshop co-chairs then studied each paper carefully, in order to assess its quality and whether it was suitable to be selected for this book. Notice that not all workshops were able to contribute to this volume. The result is a compilation of 17 papers selected from 13 workshops, which we list below.

- The 22nd International Workshop on Coordination, Organization, Institutions, and Norms in Agent Systems (COIN 2017)
- The 18th International Workshop on Multi-Agent-Based Simulation (MABS 2017)
- The 8th International Workshop on Optimisation in Multiagent Systems (OptMAS 2017)
- The 10th International Workshop on Agents Applied in Health Care (A2HC 2017)
- The 19th International Workshop on Trust in Agent Societies (Trust 2017)
- The 17th International Workshop on Adaptive Learning Agents (ALA 2017)
- The 1st International Workshop on Teams in Multi-Agent Systems (TEAMAS 2017)
- The 10th International Workshop on Agent-Based Complex Automated Negotiations (ACAN 2017)
- The 1st International Workshop on Transfer in Reinforcement Learning (TiRL 2017)
- The 5th International Workshop on Engineering Multi-Agent Systems (EMAS 2017)
- The 4th International Workshop on Multiagent Interaction without Prior Coordination (MIPC 2017)

- The 8th International Workshop on Cooperative Games and Multiagent Systems (CoopMAS 2017)
- The 2nd International Workshop on Agent-Based Modelling of Urban Systems (ABMUS 2017).

We note that a similar process was carried out to select the most visionary papers of the AAMAS 2017 workshops. While best papers follow the style of more traditional papers, visionary papers are papers with novel ideas that propose a change in the way research is currently carried out. The selected most visionary papers may be found in the Springer book LNCS 10643.

The AAMAS 2017 workshops are the second AAMAS workshop series to publish their (selected) papers in the form of a collective book. We hope that this book can better disseminate the most notable results of these workshops and encourage authors to submit top-quality research work to the AAMAS workshops.

August 2017   Gita Sukthankar
                                                          Juan A. Rodriguez-Aguilar

Organization

AAMAS 2017 Workshop Co-chairs

Gita Sukthankar   University of Central Florida, USA
Juan A. Rodriguez-Aguilar   Artificial Intelligence Research Institute, Spain

AAMAS 2017 Workshop Organizers

COIN 2017

Felipe Meneguzzi   Pontifical Catholic University of Rio Grande do Sul,
                                Brazil
Wamberto Vasconcelos   University of Aberdeen, UK

MABS 2017

Graçaliz Pereira Dimuro   Universidade Federal do Rio Grande - FURG, Brazil
Luis Antunes   University of Lisbon, Portugal

OptMAS 2017

Archie Chapman   University of Sydney, Australia
Sebastian Stein   University of Southampton, UK
Long Tran-Thanh   University of Southampton, UK
William Yeoh   New Mexico State University, USA
Roie Zivan   Ben-Gurion University of the Negev, Israel

A2HC 2017

Sara Montagna   University of Bologna, Italy
Eloisa Vargiu   Fundació Eurecat - eHealth Unit, Spain
Marcia Ito   BM Brazil/Faculty of Technology of São Paulo, Brazil
Daniel Castro Silva   University of Porto, Portugal
Pedro Henriques Abreu   University of Coimbra, Portugal
Michael Ignaz Schumacher   University of Applied Sciences Western Switzerland,
                                Switzerland

Trust 2017

Jie Zhang   Nanyang Technological University, Singapore
Murat Sensoy   Ozyegin University, Turkey
Rino Falcone   ISTC-CNR, Rome, Italy

ALA 2017

Tim Brys	Vrije Universiteit Brussels, Belgium
Anna Harutyunyan	Vrije Universiteit Brussels, Belgium
Patrick Mannion	Galway-Mayo Institute of Technology, Ireland
Kaushik Subramanian	Georgia Institute of Technology, USA

TEAMAS 2017

Ewa Andrejczuk	IIIA-CSIC, Spain
Juan M. Alberola	Universitat Politecnica de Valencia, Spain
Leandro Soriano Marcolino	Lancaster University, UK
Mehdi Farhangian	University of Otago, New Zealand

ACAN 2017

Susel Fernandez Melian	Nagoya Institute of Technology, Japan
Katsuhide Fujita	Tokyo University of Agriculture & Technology, Japan
Naoki Fukuta	Shizuoka University, Japan
Takayuki Ito	Nagoya Institute of Technology, Japan
Minjie Zhang	University of Wollongong, Australia
Quan Bai	Auckland University of Technology, New Zealand
Fenghui Ren	University of Wollongong, Australia
Miguel Angel Lopez Carmona	University of Alcala, Spain
Ivan Marsa Maestre	University of Alcala, Spain
Tim Baarslag	Centrum Wiskunde & Informatica, The Netherlands
Reyhan Aydogan	Ozyegin University, Turkey

TiRL 2017

Anna Helena Reali Costa	Universidade de São Paulo, Brazil
Doina Precup	McGill University, Montreal, Canada
Manuela Veloso	Carnegie Mellon University, USA
Matthew Taylor	Washington State University, USA
Felipe Leno da Silva	Universidade de São Paulo, Brazil
Ruben Glatt	Universidade de São Paulo, Brazil

EMAS 2017

Amal El Fallah-Seghrouchni	LIP6 - University Pierre and Marie Curie, France
Alessandro Ricci	DISI, University of Bologna, Italy
Son Tran Cao	New Mexico State University, USA

MIPC 2017

Tathagata Chakraborti	Arizona State University, USA
Katie Genter	University of Texas at Austin, USA
Trevor Santarra	University of California Santa Cruz, USA

COOPMAS 2017

Edith Elkind	University of Oxford, UK
Tomasz Michalak	University of Oxford, UK
Yair Zick	National University of Singapore, Singapore

ABMUS 2017

Koen H. van Dam	Imperial College London, UK
Jason Thompson	University of Melbourne, Australia
Pascal Perez	University of Wollongong, Australia
Nick Malleson	University of Leeds, UK
Alison Heppenstall	University of Leeds, UK
Andrew T. Crooks	George Mason University, USA
Claudia Pelizaro	University of Melbourne, Australia

Contents

Elastic & Load-Spike Proof One-to-Many Negotiation to Improve the Service Acceptability of an Open SaaS Provider

Amro Najjar$^{(\boxtimes)}$, Olivier Boissier, and Gauthier Picard

Hubert Curien Laboratory, Saint-Etienne, France
{amro.najjar,olivier.boissier,gauthier.picard}@emse.fr

Abstract. Service acceptability rate and user satisfaction are becoming key factors to avoid client churn and secure the success of any Software as a Service (SaaS) provider. Nevertheless, the provider must also accommodate fluctuating workloads and minimize the cost it pays to rent resources from the cloud. To address these contradicting concerns, most of existing works carry out resource management unilaterally by the provider. Consequently, end-user preferences and her subjective acceptability of the service are mostly ignored. In order to assess user satisfaction and service acceptability recent studies in the domain of Quality of Experience (QoE) recommend providers to use quantiles and percentile to gauge user service acceptability precisely. In this article we propose an elastic, load-spike proof, and adaptive one-to-many negotiation mechanism to improve the service acceptability of an open SaaS provider. Based on quantile estimation of service acceptability rate and a learned model of the user negotiation strategy, this mechanism adjusts the provider negotiation process in order to guarantee the desired service acceptability rate while meeting the budget limits of the provider and accommodating workload fluctuations. The proposed mechanism is implemented and its results are examined and analyzed.

Keywords: Adaptive one-to-many negotiations · Acceptability rate SaaS · Cloud computing elasticity

1 Introduction

Client churn is one of the most negative indicators affecting any online SaaS provider [1]. Yet, user satisfaction and client churn are not the only challenges providers need to overcome in order to remain in the business. A recent Cisco report predicted that busy-hour Internet traffic will grow twice more rapidly than the average Internet traffic between 2015 and 2020 [10]. Therefore, tomorrow's market is shaped by intensive but fluctuating demand and high user expectations. In order to cope with this rapid evolution, Application Service Providers (ASP) are increasingly migrating to the cloud to manage their resources in elastic

© Springer International Publishing AG 2017
G. Sukthankar and J. A. Rodriguez-Aguilar (Eds.): AAMAS 2017 Best Papers,
LNAI 10642, pp. 1–20, 2017.
https://doi.org/10.1007/978-3-319-71682-4_1

manner thereby accommodating the fluctuating workload as well as minimizing their operational costs.

Thus, the ASP or the SaaS provider has to balance two concerns: minimizing client churn while meeting its budget constraints. In the context of cloud computing, this issue is known as elasticity management or auto-scaling [22] and it has received considerable attention in the recent years. However, most of the works tackling this issue adopt a centralized approach where the ASP takes the resource decision unilaterally [22]. Consequently, the end-user preferences are mostly overlooked and it is often presumed that their acceptability threshold tolerates the best-effort service proposed by the provider.

Multi-agent systems have been outlined as a platform to distribute resource allocation and coordination in the cloud computing ecosystem [35]. Furthermore, recent works used mult-agent system to account for end-user expectations and satisfaction [17,23]. One-to-many multi-agent negotiation provides a potential platform to involve the end-user into the elasticity management process. Nevertheless, in contrast to the case of a SaaS provider where the goal is to maximize the acceptability rate by reaching as much agreements as possible, the majority of existing works in the literature addresses a scenario where the seller seeks to find one *atomic* agreement reached with one of the concurrent buyers. Furthermore, most of these works assume a *closed* set of participants in which the users are assumed to be known in advance before the outset of the negotiation process. This assumption does not hold in today's online and cloud ecosystem where thousands of users may enter the negotiation process every minute.

In this article we present AQUAMan, a novel adaptive, elastic and open multi-issue negotiation and coordination mechanism[1] allowing the provider to achieve a targeted service acceptability rate while satisfying its budget constraints. Users can decide whether to accept or reject the proposed service depending on their expectations and subjective estimation of the service quality [17]. Based on its measurements of the portion of users finding the service unacceptable, the provider adjusts its negotiation strategy in order to restore the acceptability rate to its predefined goals. Using the proposed mechanism, the SaaS provider can *(i)* ensure a precise predefined service acceptability rate while meeting its budget constraints. *(ii)* accommodate the dynamic and open nature of the cloud ecosystem where the workload (*i.e.* the number of users entering the system) is variable and is subject to sudden load-spikes. The proposed negotiation and coordination mechanisms are developed in the EMan architecture [17,18,23].

The rest of this article is organized as follows: Sect. 2 argues how one-to-many negotiation can provide a potential solution allowing the provider to satisfy its business goals and integrate the user into the decision process. Section 3 reviews the EMan architecture, its agents, its negotiation and coordination protocols. Section 4 details AQUAMan, the adaptive negotiation mechanism, and explains the underlying opponent learning and modeling approach. Section 5

---

[1] The terms "AQUAMan" and "the adaptive mechanism" may be used interchangeably.

details the evaluation process and discusses the results. Section 6, discusses the related works. Finally, Sect. 7 concludes this paper and points out future research perspectives.

2   Motivation

Elasticity and cloud resource management have received considerable attention since the emergence of cloud computing. Several works in the literature have formulated elasticity management as a constraint satisfaction problem whose goal is to maximize a global utility function defined by the provider while meeting the budget constraints [4]. However, these problems are inherently NP-hard. Moreover, they presuppose that the user's requests and the workload applied to the system are known in advance. Consequently, several heuristic policies have been proposed to overcome these limitations (c.f. [4] and the references therein). Nevertheless, in most of these works the end-user preferences and their subjective service acceptability are either overlooked or assumed to be known in advance by the provider. Hence, the elasticity management process is done unilaterally by the provider.

To understand end-users' satisfaction and acceptability threshold, we will rely on Quality of Experience (QoE). The latter is a metric that appeared in 2000's to assess the end-user satisfaction. QoE is defined as the service quality as perceived subjectively by the user [16] and it is known to be a key determinant of the user's decision to accept or reject a service [8,16]. QoE is targeted to provide a practical measure allowing to quantify user satisfaction and acceptance of the service [16]. In particular, QoE-management emerged as a process aiming at maximizing QoE while optimizing the used resources [32]. However, QoE-management literature largely relies on the Mean Opinion Score (MOS) to assess satisfaction and service acceptability. Nevertheless, since it is an average of users' opinions, MOS hides important information about user diversity and their personal preferences [7]. For this reason, other measures have been proposed to allow the provider to understand the end-user satisfaction and estimate client churn [8]. For instance, percentiles have been proposed as a measurement tool allowing the provider to ascertain that, say, 95% of its users find the service to be acceptable or better [8].

By definition, an agent is usually self-interested and is bound to an individual perspective [36]. This makes agents potential candidates to represent the subjectivity of users' opinions and acceptance of a given service [23]. Furthermore, the principles of negotiation behavior discussed in [26] provide a useful tool to represent end-user expectations. The latter are key determinants of user satisfaction and service acceptability [30,37] (for more about this discussion please refer to [17]).

One-to-many negotiation is a sub-type of multi-agent negotiation [13] where one agent (e.g. a seller) negotiates simultaneously with multiple agents (e.g. potential buyers). In multi-agent negotiation literature, several approaches have been proposed to tackle this type of negotiation. Even though they address different applications, the majority of these solutions (e.g. [2,14,27]) share a common architectural blueprint. Yet, the negotiation and coordination strategies used in each solution depends on its purpose [18].

One-to-many multi-agent negotiation is a potential solution to allow the provider to negotiate simultaneously with several end-users and integrate their subjective service acceptability into the elasticity management process. Nevertheless, in contrast to the case of a SaaS or online provider where the goal is to maximize the acceptability rate by reaching as much agreements as possible, the majority of existing works in the literature of one-to-many negotiations addresses a scenario where the seller seeks to find one *atomic* agreement reached with one of the concurrent buyers. Furthermore, most of these works assume a *closed* set of participants in which the opponents of the single agent are known in advance before the outset of the negotiation process. Moreover, in most of the existing works, offers are sent and received in a synchronous manner. These assumptions do not hold in today's open cloud ecosystem where thousands of users may enter/leave the system every minute and where the negotiation sessions are not synchronized. Section 6 provides further discussions about the related works in one-to-many negotiations and their limitations.

3   The Elasticity Management Architecture (EMan)

In this article we present a novel one-to-many negotiation mechanism dubbed as AQUAMan (Adaptive QUality of experience Aware elasticity Management). AQUAMan is implemented in the EMan architecture [17,18]. The latter is multi-agent architecture for SaaS elasticity management. EMan (Fig. 1) follows the same architectural blueprint discussed in the previous section and shared by most of existing one-to-many solutions in the literature. In the earlier version of EMan [17,18], users are represented by autonomous agents whose goal is to maximize the QoE of their respective users. However, these earlier versions do not include the adaptive mechanism proposed in this article.

This section introduces the types of agents involved in the EMan architecture (Sect. 3.1) and discusses the role assumed by the coordinator (Sect. 3.2).

3.1   Agents

The EMan architecture depicted in Fig. 1 models the negotiation taking place between a SaaS provider and its end-users or Service Users (SU). The negotiation between the SaaS providers and the cloud providers is considered beyond the scope of this article. The EMan architecture contains three types of agents: service user agents (denoted as sa_i), delegate agents (denoted as da_i) and a single coordinator (denoted ca). The latter two types represent the provider.

In the EMan architecture, da_i and sa_i do not have access to the preferences and negotiation strategies of other agents. Therefore, their negotiation behavior follows the *negotiation decision function* [6] and it is determined by the utility function and negotiation strategy. In order to accommodate the open and

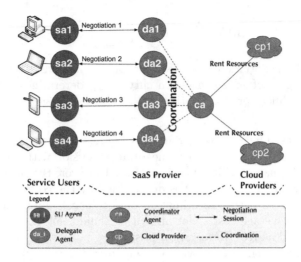

Fig. 1. The EMan architecture deployed in the cloud ecosystem.

dynamic nature of the cloud ecosystem, negotiation sessions in the EMan architecture are non-synchronous *i.e.* some sessions will be already terminated while other sessions will be still active or have not started yet. For further information about the negotiation protocol and the acceptance strategy please refer to [17,18].

The following sections present the agents of the EMan architecture.

Service User Agents. A service user agent participates in the negotiation process on behalf of a service user. A sa_i has a utility function M_{sa_i} that encodes its preferences. M_{sa_i} is used at each cycle t to assess the utility of offers $o^t_{da_i}$ received from the corresponding delegate. The utility function of users depends on the user's personal expectations, its expertise and preferences. Formulating user utility function is beyond the scope of this article. For further information please refer to [17,18] where rely on evidence from the literature of QoE and Psychophysics (*i.e.* the Weber-Fechner Law and the logarithmic hypothesis [28]).

In order to make an accept/reject decision, a sa_i relies on its utility function and on its current Aspiration Rate (AR). $AR^t_{sa_i}$ expresses how much utility sa_i expects to obtain in this negotiation cycle t. $AR^t_i \in [0, 1]$. When $AR^t_i = 1.0$, this means that a_i expects obtaining an ideal service. On the other hand, $AR^t_i = 0.0$ indicates that the agent is ready to accept the worst possible offer. In order to reach agreements, sa_i makes concessions by reducing its AR. All the sa_i follow a Time-Based Concession strategy (TBC) [6]. This assumption is quite common in the literature notably in works whose goals is to construct a model of the opponent behavior (c.f. [3] and the references therein). Therefore, $\Delta AR^t_{sa_i}$, the concession made by sa_i for the cycle t, depends on the time left before reaching the deadline. It is computed as follows [34]:

$$\Delta AR_{sa_i}^t = AR_{sa_i}^{t-1} \cdot \left(\frac{t}{T_{sa_i}}\right)^{\lambda_{sa_i}} \tag{1}$$

where T_{sa_i} is the negotiation time deadline. λ_{sa_i} is a parameter that controls the convexity degree of sa_i's concession curve. λ_{sa_i} determines the behavior of sa_i (conciliatory, linear or conservative) [34].

Delegate Agents. Once a new user agent sa_i enters the system, a new delegate da_i is created and it enters a bilateral negotiation session with sa_i.

Like sa_i, a da_i has a utility function M_{da_i}. However, the utility of an offer $o_{sa_i}^t$ from a delegate da_i standpoint is determined by the cost required to serve the offer. In order to ensure meeting the budget constraints, the coordinator imposes that the average cost spent on a user should not exceed a predefined parameter denoted as RC (Reservation Cost).

In order to reach agreements with sa_i, da_i may use two types of concession strategies:

- *Time-Based Concession (TBC)* computed as follows:

$$\Delta AR_{da_i}^t = \frac{1}{T_{da_i}} \tag{2}$$

  Where T_{da_i} is the time deadline of the delegate. When da_i reaches T_{da_i}, it stops making more concessions.
- *Behavioral-Based Concession (BBC)* or *tit-for-tat*: Since da_i does not have access to sa_i's real concession, it relies on its own utility function to assess sa_i's concession, which is defined as follows:

$$\Delta AR_{da_i}^t = M_{da_i}(o_{sa_i}^t) - M_{da_i}(o_{sa_i}^{t-1}). \tag{3}$$

  Where $(o_{sa_i}^t)$ and $(o_{sa_i}^{t-1})$ are the previous couple of offers received from sa_i.

Note that the results and conclusions drawn in this article are valid independently of $da_i's$ negotiation strategy. Therefore, comparing the different da_i negotiation strategies is beyond the scope of this article. For such a comparison please refer to [21].

3.2 Coordination Strategies

The coordinator, denoted as ca, is defined by the following tuple:

$$ca = \langle \Pi, Cost, \Omega_{initial}, AcceptRate, Surplus \rangle \tag{4}$$

Where Π and $Cost$ are the performance model and the cost function. The former estimates the amount of resources needed from the cloud to satisfy the user's request whereas the latter calculates the cost needed to rent these resources. The provider builds them based on its business knowledge. $\Omega_{initial}$ is the delegate initial negotiation strategy and it is used to initialize delegates when they are spawned by the *ca*. *AcceptRate* is the acceptance rate *i.e.* the percentage of users who accepted the service to all users who entered the system up till a given moment. *Surplus* is a variable where the ASP accumulates all the surplus obtained from successful sessions.

As long as the ASP business goals are not violated (a coordinator intervention case detailed in Sect. 4), delegates assume the bilateral negotiation with the sa_i in an autonomous manner. Yet, delegates solicit the coordinator when a negotiation session is terminated either successfully or unsuccessfully. The following sections list these cases.

Successful Negotiation Session. When a delegate da_i reaches agreement with the corresponding sa_i, it notifies the coordinator. In order to estimate the acceptance rate, the coordinator keeps track of successful negotiation sessions. Furthermore, when a session is declared successful, the coordinator updates its *Surplus* variable as follows: $Surplus = Surplus + surplus_i$, where $surplus_i$ is the surplus obtained from the successful negotiation session i. $surplus_i$ is computed as follows:

$$surplus_i = RC - Cost(\Pi(\hat{o}_i)) \tag{5}$$

Where RC is the reservation cost of the delegates. $\hat{o}_i$ is the accepted offer, and $Cost(\Pi(\hat{o}_i))$ is the cost to be paid by the ASP to satisfy this offer. Thus, the difference between the actual cost and the reservation (maximum) cost is considered as surplus.

Failed Negotiation Session. Whenever a delegate da_i observes that the corresponding sa_i has left the negotiation session before reaching an agreement, da_i notifies the coordinator. The latter updates the *AcceptRate* variable in order to ascertain that the targeted acceptance rate (e.g. 95% of users) is met. Otherwise, the coordinator triggers the adaptation process. This process is detailed in the next section where we detail the adaptive algorithm proposed in this current work.

4   Adaptive Negotiation Strategy

This section details the adaptive algorithm. In order to estimate the current acceptance rate at any given time t, the provider relies on a quantile estimation algorithm presented in Sect. 4.1. If the targeted acceptance rate is violated, the coordinator activates the adaptive mode. With this mode active, a delegate da_i has to analyze the behavior of its opponent sa_i, learn a model of its negotiation strategy in order to estimate its negotiation time deadline $\bar{T}_{sa_i}$, and send this estimation to the coordinator (explained in Sect. 4.2). The latter chooses the high-priority sessions (based on their time deadlines) and adjust their negotiation strategies while respecting its budget constraints (Sect. 4.3).

4.1 Triggering the Adaptation Mechanism

Quantiles and percentiles are values that partition a finite set of values into q subsets of (nearly) equal sizes. As discussed in Sect. 2, the literature on QoE and user satisfaction recommend providers to rely on quantiles and percentiles as more accurate measures (compared with MOS) to gauge user acceptability of the service [8].

Whenever session i is terminated, the coordinator is notified by da_i about the outcome of this session. Using these data, the coordinator runs a quantile estimation algorithm to detect the current service acceptability rate.

Let Q the quantile/percantile estimation function. Let R be the dataset containing the outcomes of the terminated sessions. R can contain either 0's, for failed sessions, or 1's for successful sessions. If the coordinator seeks to ensure that β percent of users who requested the service so far have had a successful negotiation session (hereby accepting the proposed service quality), the coordinator needs to verify that the $(100 - \beta + 1)^{th}$ percentile equals 1:

$$Q(R, 100 - \beta + 1) = 1 \tag{6}$$

As long as this condition holds, the coordinator has no need to intervene into the negotiation process. To calculate Q we rely on quantile/percentile estimation algorithm.

Once the condition in Eq. 6 is violated, the coordinator triggers the adaptation mechanism by commanding all working delegates to activate their adaptive mode. Furthermore, if the coordinator has already activated the adaptive mode, when it spawns new delegates they will have this mode active as well.

Note that the coordinator continues to evaluate Eq. 6 even after the activation of the adaptive mode. Next section details the delegate's role after the activation of the adaptive mode.

4.2 Opponent Learning and Modeling Algorithm

Opponent Concession Estimation. Even when the adaptive mode is not active, when a delegate da_i receives an offer $o^t_{sa_i}$ at cycle t from the corresponding sa_i, it estimates the concession made by sa_i by comparing $o^t_{sa_i}$ with $o^{t-1}_{sa_i}$ the previous offer made by sa_i. Since da_i does not have access to sa_i preferences or utility function M_{sa_i}, it cannot calculate the real concession made by sa_i. Instead, it relies on its own utility function to estimate the concession made by sa_i by assuming that a concession made by sa_i is synonymous with a utility gain for da_i. Thus, $c^t_{sa_i}$, the estimated concession made by sa_i at the negotiation cycle t is defined as:

$$c^t_{sa_i} = M_{da_i}(o^t_{sa_i}) - M_{da_i}(o^{t-1}_{sa_i}) \tag{7}$$

To learn sa_i concession behavior, da_i keeps track of sa_i concessions during the negotiation session. Figure 2a compares $\Delta AR^t_{sa_i}$ the real concessions made by sa_i (in blue) with $c^t_{sa_i}$, da_i's estimation of the same concession (in red).

In this example, $T_{sa_i} = 80$ cycles and $\lambda_{sa_i} = 3.0$. As can be seen from the figure, although the red curve is noisy (because it is estimated using da_i's utility function and then normalized), it seems to provide an adequate estimation of sa_i concession since the main goal is to learn T_{sa_i}, the negotiation time deadline of sa_i.

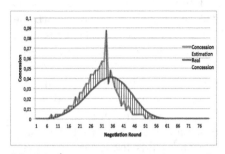

(a) $\Delta AR^t_{sa_i}$, the real concession made by sa_i (in blue) compared with $c^t_{sa_i}$ da_i's estimation of the same concession (in red).

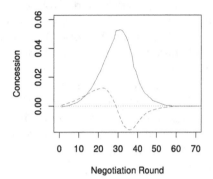

(b) A smoothed version $c^t_{sa_i}$ (solid line) and its derivative (dashed line).

Fig. 2. Learning the user time-deadline T_{sa_i} (Color figure online)

Opponent Model Learning. If the adaptation mode is active, the concession data collected in Sect. 4.2 should be used to establish a model of sa_i concession behavior and infer T_{sa_i}, its negotiation time deadline. This can be achieved, as has been shown in the literature [3], using non-linear regression assuming that all users follow a time-based concession defined by Eq. 1. Yet, unlike the predominantly one-to-one existing works in the literature [3] where learning the negotiation model of the opponent is done and corrected each cycle once a new concession is made (c.f. Sect. 6), in the one-to-many negotiation settings addressed by this article such an approach is not practical for scalability reasons since hundreds or thousands of users may be negotiating with the provider simultaneously. Furthermore, if the provider runs delegates' learning algorithms on resources rented from the cloud, this may increase the costs considerably.

For this reason, in the proposed solution a delegate can run the non-linear regression algorithm only once. The next decision is to determine when a delegate should do so. On the one hand, if a delegate launches this process too early in the negotiation session, the regression process will have only few input points. On the other hand, if the delegate waits too long, the corresponding user risks to reach its time deadline and quit the negotiation process.

All sa_i follow time-based concession strategy in which the rate of concession sa_i slows down when sa_i has already made most of the concessions it was going to make. Therefore, regardless of the type of sa_i (conservative or linear, c.f. Eq. 1) and of its actual time deadline T_{sa_i}, when the rate of change (or the derivative) of da_i's estimation of sa_i concession decays significantly into negative values, this means that sa_i has made most of its concessions and that, for the rest of the negotiation session before sa_i reaches its T_{sa_i}, sa_i will stop making considerable concessions. Thus, if da_i launches its non-linear regression at this stage, it will have the most of useful data points.

Figure 2b superimposes da_i's estimate of the concession made by sa_i (red solid line) with its first derivative (red dashed line). The sa_i in this figure is exactly the same of Fig. 2a, but we used a Savitzky-Golay filter [31] to smooth the curve and filter out the noise. To calculate the rate of change of the concessions, we calculate the derivative and we smooth the result using the a variant of the same filter.

The non-linear regression algorithm takes the negotiation cycle (t) and the estimated concession ($c_{sa_i}^t$) as input variables and outputs an estimated values of the $\bar{T}_{sa_i}$ and λ_{sa_i} parameters. Then, based on $\bar{T}_{sa_i}$, da_i computes $\bar{r}_i = \bar{T}_{sa_i} - t$ the estimated number of cycles remaining in session i. As long as the session i is not terminated, da_i updates $\bar{r}_i$ every cycle and sends it repeatedly to the coordinator.

4.3   Negotiation Adaptation

As was explained in the previous section, the coordinator receives $\bar{r}_i$ from all negotiation sessions i whose delegate da_i considers that sa_i is approaching its time deadline and deserves to be prioritized. These estimations $\bar{r}_i$ are stored in the *priority list* which is continuously sorted in ascending manner (a session whose time deadline is estimated to be sooner will be in the top of the list).

Note that since negotiation sessions are non-synchronized, the coordinator (ca) has to repeat the sorting whenever it receives a new estimation from a delegate da_i. Furthermore, whenever a session is terminated either successfully or not, ca removes its record from the priority list.

Thus, in the priority list, sessions are sorted from highest to lowest priorities. Now ca must decide how many of these priority sessions should receive a preferential treatment. To do so, the coordinator undertakes this process as follows:

1. Estimate the current acceptability rate (using the quantile estimation function as discussed in Sect. 4.1). Then compute the difference between the desired acceptability rate and its actual current value.
2. Estimate p the number of successful sessions required to restore the desired rate and choose the first p sessions from the priority list and adjust their negotiation strategies to encourage them accept the service.

When the negotiation strategy of a delegate da_i is adjusted, da_i retains a time-based concession strategy defined in Sect. 3.1. Yet, with the following modifications.

First, the negotiation time deadline of da_i becomes $\bar{r}_i$ instead of T_{da_i}. Second, the reservation cost of da_i is increased by value denoted as $RcPrio$. $RcPrio$ is computed by dividing the surplus available in *Surplus* among all the prioritized sessions. Note that delegates cannot get more than a $RcPrio_{max}$ considered as the maximum value for prioritized sessions. Thus, the provider ensures the satisfaction of its budget constraints. Third, the preferred cost of da_i is changed to the cost of the last offer it has made. $AR^t_{da_i}$ is restored to 1.

This section presented the opponent modeling and the negotiation adaptation processes. Next section presents the experimental evaluation.

5 Evaluation

To evaluate AQUAMan, the adaptive mechanism, we implement it in the EMan architecture ([17,18]). The latter is implemented using Repast Simphony [25], a multi-agent simulation environment. The evaluation is organized into three experiments. The first experiment (Sect. 5.2) aims at evaluating the adaptation algorithm. The second experiment (Sect. 5.3) examines the impact of the workload applied to the provider (i.e. the number of user entering the system per minute) on the acceptability rate. The third experiment (Sect. 5.4), evaluates the overhead of the negotiation/coordination mechanisms. Before discussing these experiments, next section presents their parameters.

5.1 Experiments Parameters

The total number of users entering the system is denoted as $|SU|$. In the following experiments, $|SU| = 10000$ users. Users enter the simulation following a Poisson random process whose mean value is A per minute. The service in the experimental scenario involves two attributes: one is the service delivery time while the other represents the service quality. The user profiles (reservation, preferred values and weights for each attributes) are generated randomly. The negotiation time deadlines of sa_i are generated randomly $T_{sa_i} \in [40 : 120]$, and $\lambda_{sa_i} \in [1 : 8]$ (*i.e.* users can be linear or conservative). The cost of services acceptable by users ranges from 0.1\$ to 0.9\$. RC, delegate reservation cost (the maximum cost allocated to a non-prioritized user) is set to 0.60\$. This parameter represents the provider's budget constraints. Therefore, without the adaptation mode, approximately a quarter of users will not accept the service since RC cannot satisfy their least expected service. $RcPrio_{max} = RC/2$. Therefore, when a user sa_i is prioritized, the RC of the respective delegate (da_i) is at most increased by $RC/2$. *Goal* is the percentage of users that the provider seeks to satisfy. Its impact is evaluated in the next subsection. Note that the results discussed below remain valid even when the values of the parameters above (e.g. RC, $RcPrio_{max}$, etc.) are changed. Please refer to our previous work [21] for a complete evaluation and analysis of the impact of these cost-related parameters. Furthermore, we obtained similar results when user attribute utility function is a logarithmic function derived from the logarithmic hypothesis [17,28]. However, these results were not included due to space constraints.

5.2 Evaluating the Adaptation Mechanism

Figure 3 shows the results of this experiment. The blue curve draws the acceptance rate when the $Goal = 95\%$. As can be seen from the figure, at the outset of the simulation, the value of the acceptance rate oscillated before getting stabilized on the goal value ($Goal = 95\%$). The red and the brown curves, that plot the acceptance rate when $Goal = 90\%$ and $Goal = 85\%$, show similar results. To compare the results of the adaptation mechanism, the black curve plots the acceptance rate when the mechanism is deactivated. Note that the results of Fig. 3 where obtained with $A = 160$ users per minute.

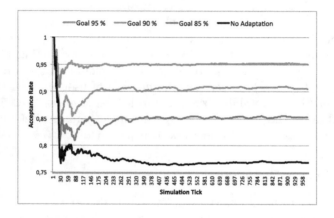

Fig. 3. Acceptance Rate with $A = 160$. (Color figure online)

As can be seen from the results, the adaptation mechanism managed to restore the acceptance rate to the predefined *goal* in a precise manner. This is explained by the fact that the prediction algorithm discussed in Sect. 4.2 managed to predict the user negotiation deadline T_{sa_i} with acceptable precision. The average error rate between the prediction $\bar{T}_{sa_i}$ and the real value T_{sa_i} is about 6.0 cycles which means a delegate da_i overestimates/underestimates T_{sa_i} by three negotiation cycles on average.

Intuitively, the service acceptability rate achieved by the prediction algorithm comes with more cost invested per user. However, this increase does not violate the provider budget constraint: the average costs per user were 0.55$, 0.52$, 0.5$ and 0.44$ for $Goal = 95\%$, $Goal = 90\%$, $Goal = 85\%$ and non-adaptive respectively. Thus, the budget constraint of the provider (*i.e.* average cost per user does not exceed RC=0.6$) was satisfied because the provider relies on *Surplus* to serve prioritized users. Note that studying the cost-efficiency of the adaptive mechanism and its limits under given cost constraints is beyond the scope of this article since these issues have been addressed in our previous work [21].

5.3   The Impact of the Workload A

This experiment studies the impact of A, the workload applied to the system, on the acceptance rate in order to evaluate the elasticity of the adaptation mechanism. The aim of this experiment is twofold: First, we study the systems response to high but constant workloads. Second, to evaluate the load-spike proofness of the system, we will apply a sudden load-spike.

Table 1. The impact of the workload parameter A.

$A =$	20	40	80	160	320	640	1280	2500
Adaptation	95%	95%	95%	95%	95%	94.9%	94.7%	93.3%
No-Adaptation	77%	77%	77%	77%	77%	77%	77%	77%

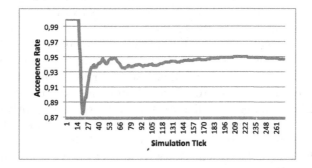

Fig. 4. Acceptance rate ($Goal$=95% and $A = 1280$).

Intense Workload A. Table 1 shows the acceptance rate, at the end of the simulation, when the $Goal = 95\%$ [2] and with A increasing from 20 to 2500 users per minute. As can be seen from the table, the mechanism proved to be highly elastic. When $A \in [20 : 320]$ the acceptance rate does not change as A increases. However, when A increases to higher values $[640 : 2500]$ the acceptance rate witnesses a slight decrease. However, even with very high A, the results achieved by the adaptation mechanism remain very close to the goal. This decrease is explained as follows: the adaptation algorithm is reactive and the coordinator can only assess the acceptance rate each time a session terminates successfully or unsuccessfully. Thus, when A is too high, a significant number of sessions may fail in the same simulation tick and the adaptation algorithm executed by the coordinator suffers from a slight delay and becomes a bit short from meeting

---

[2] We conducted the same experiment with different values of $Goal \in \{92\%, 90\%, \text{etc.}\}$ and obtained similar results.

the predefined goal ($Goal = 95\%$) as can be seen from the table. Yet, this result should be taken relatively for the following reason:

$|SU|$ the total number of users in the simulation influence the elasticity threshold of the adaptation mechanism. For instance, when $|SU| = 10000$ and $A = 2500$, users enter the architecture in 4 massive waves each carrying roughly a quarter of the total number of users. Thus, the coordinator does not have enough time restore the acceptance rate to ($Goal = 95\%$). Figure 4 shows how the adaptation mechanism reacts to a drop in acceptance rate when $A = 1280$. The coordinator manages to restore the value of the predefined goal ($Goal = 95\%$). Yet, a new sudden drop occurs with the new wave of users. Note that in real life scenario, when $|SU| = \infty$, the coordinator will be always able to restore the acceptance rate into its goal value even with a significant A.

Load-Spike Proofness. To assess the impact of load-spikes, in this experiment A, the arrival rate, will undergo a sudden increase (*i.e.* a load-spike). On average 80 users per minute will enter the system. Yet, for four minutes, a load-spike occurs increasing A to 800 per minute. Thus, four waves of users will enter the system each wave containing 800 users on average.

Figure 5a plots the acceptance rate achieved by the adaptive mechanism with a workload of $A = 80$ users per minutes on average and a load-spike of $A = 800$ taking place around the 45th minute and lasting for four minutes[3].

As can be seen from the figure, after a drop in the acceptability rate at the beginning of the experiment (around the 10th minute), the adaptive mechanism is activated and it manages to restore the acceptability rate defined by the provider ($Goal = 95\%$ plotted by the blue dashed line on the figure). More importantly, the load-spike does not have an impact on the achieved acceptance rate. The $Goal = 95\%$ was attained before the load-spike and it remains satisfied during and after the load-spike.

Figure 5b plots the value of the acceptance rate but this time a uniform workload of $A = 80$ users per minute on average was applied. Comparing the two figures shows that the main difference between the two is that the simulation in Fig. 5a ended earlier (around the 90th minute) because of the load-spike whereas in Fig. 5b the simulation lasted about 30 min longer. Despite this minor difference, the load-spike did not have a significant impact on the acceptance rate achieved by the adaptive mechanism and in both cases (with and without load-spikes), the predefined $Goal$ was achieved in a precise manner.

To understand this load-spike proofness, Fig. 5c shows the impact of this load-spike on the surplus collected by the surplus redistribution mechanism (c.f. Sect. 3.2). The red curve plots the content of the *Surplus* used to serve prioritized sessions. As can be seen from the figure, the amount available in the *Surplus* rises significantly during the load-spike and then drops down to its average values after the end of the load-spike. This is explained as follows: the load-spike raises the number of users entering the systems. In less than four minutes about 2400 users enter the system in four massive waves. Some of users arriving in the load-spike

---

[3] In this experiment a *minute* is equivalent to 30 simulation ticks.

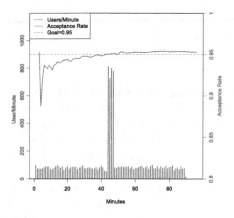

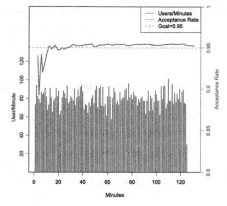

(a) The impact of the load-spike on the acceptance rate (the blue curve).

(b) The acceptance rate (the blue curve) with a uniform workload $A = 80$.

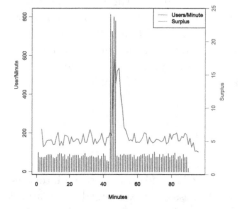

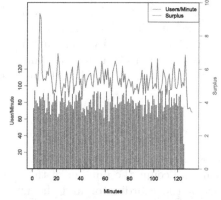

(c) The impact of the load-spike on the *Surplus* (the red curve).

(d) The *Surplus* (the red curve) with a uniform workload $A = 80$.

Fig. 5. The gray vertical lines plot the number of users entering the system per minute. The blue and the red curves respectively plot The acceptance rate and the content of the *Surplus* with a load-spike (left side) and without a load-spike (right side). (Color figure online)

can be satisfied quickly and easily due to their relatively less costly expectations. Therefore, the number of successful sessions will increase thereby raising the total amount of surplus collected by the coordinator. This surplus will be then used by surplus redistribution mechanism to be spent on intransigent users in order to retain the *Goal* value. This is why the content of the *Surplus* will drop down to its pre-spike levels when the load-spike is over.

Finally, Fig. 5d plots the content of the *Surplus* when no load-spike is applied. As can be seen from the figure, after an initial increase of the surplus, the adaptive mechanism is activated and the *Surplus* is used to serve prioritized sessions. Yet, since no load-spike is applied, unlike Fig. 5c, *Surplus* in Fig. 5d does not witness any significant increase.

Discussion. This experiment evaluated the impact of workload on the acceptance rate achieved by the adaptive mechanism. In the first part of this experiment (Table 1 and Fig. 4) we showed that the acceptance rate achieved by the adaptive mechanism remained unchanged with relatively intense workloads (up to $A = 1280$ users per minute). However, intense but uniform workload is not often the case. Instead, in today's cloud computing market, SaaS providers are often exposed to load-spikes or heavy workloads in rush hours, week-ends, or holiday seasons. For this reason, in the second part of this experiment, we applied a load-spike which increased the workload tenfold in less than five minutes. The results of this experiment (Fig. 5) proved that the adaptive mechanism is load-spike proof. This makes it capable to handle the dynamic nature of the cloud ecosystem where thousands of users may rush into the service portal in few minutes. In addition, we explained how this load-spike proofness feature is realized by the surplus redistribution mechanism. In particular, the surplus redistribution mechanism absorbs the surplus obtained from successful negotiation sessions and use it to serve intransigent users. Thus, the *Surplus* acts like a buffer allowing to stabilize the acceptability rate (c.f. Figure 5a) despite the load-spike.

5.4   Coordination and Negotiation Overhead

To assess the overhead introduced by the coordination and the negotiation mechanisms we count the number of the messages exchanged to undertake interventions of the coordinator and the number of messages exchanged in bilateral negotiation sessions respectively.

Without the adaptive mechanism, the coordinator is solicited only once per session to get the result of the session (success/failure). With the adaptive mechanism, the number of messages exchanged between delegates and the coordinator increases about 50%. We consider this increase to be not significant for the following couple of reasons. First, even with the adaptive mechanism, the tasks assumed by the coordinator are lightweight as has been discussed in Sects. 3.2, 4.1 and 4.3. Second, since it is likely that the SaaS runs both the delegates and the coordinator on the same cloud data-center, the cost of communication between them is negligible.

As for the number of negotiation messages, the result show that they witness a slight decrease ($\approx 7\%$) when the adaptive mechanism is active since the latter helps reaching agreements faster.

6   Related Works

This section discusses related works addressing adaptation and learning in multi-agent negotiation (Sect. 6.1) as well as related works in the domain of one-to-many negotiation (Sect. 6.2).

6.1   Learning Opponent Negotiation Model

Learning and modeling negotiation behavior of the opponent is a mature body of research [3]. The model of the opponent typically used to help an agent adapt to its opponent behavior, reach win-win settlements, or minimize the negotiation cost. Various learning techniques are used in the literature. Some works learn the acceptance strategy of the opponent or its preference profile. More importantly, learning the opponent's time deadline has received considerable attention. In particular, some of these works use Bayesian learning techniques [12], while others use non-linear regression [9] or use both. Nevertheless, all these works address bilateral negotiation, a choice not applicable for elasticity management where the SaaS provider seeks to maximize the service acceptability rate.

6.2   One-to-many Negotiation

Despite the relatively rich literature of one-to-many negotiations, the goal of most of these works, typically modeling negotiations place between a buyer and numerous competing sellers, is to find a single (*i.e.* atomic) agreement that maximizes the buyer's utility whereas other, less beneficial, sessions are aborted (e.g. [14,27]). Furthermore, existing negotiation and coordination mechanisms (e.g. [14,24,27]) are predominantly *(i)* closed: participants are identified before the outset of the negotiation process, and *(ii)* synchronous: the coordinator needs to receive all the offers from buyers before making its analysis and commanding delegates to send new offers. Therefore, these solutions cannot accommodate the agile nature of the cloud ecosystem where thousands of SUs may surge the service portal.

Recent works in the domain of service composition use one-to-many negotiation to reach *composite* agreements with more than one (maybe all) sellers of atomic services (e.g. [29] and [15]) for the sake of bundling a composite service. In [29], the authors propose a one-to-many negotiation mechanism that redistributes the surplus obtained from successful negotiation sessions to ongoing negotiation sessions to increase their chances of success. In [15] Mansour *et al.* develop a more sophisticated approach that adapts delegates reservation values for attributes shared among multiple sessions. Nevertheless, these works assume *closed* set of atomic sellers all known before the outset of the negotiation.

One-to-many negotiation has been used in the domain of cloud computing. The authors in [2] develop a negotiation-based approach to handle resource allocation in cloud computing. However, a provider accepts an offer if and only if it can gain some immediate payoff by accepting the offer. For this reason, user acceptability rate is not taken into account and no adaptive mechanism is proposed. Siebenhaar *et al.* [33] develop a mechanism for concurrent SLA negotiation in cloud-based systems. This work supports composite negotiation. Yet, the work's main contribution seems to be focused on the protocol. Therefore, it does not provide a mechanism to represent user acceptability nor it offers a solution to adjust the delegate negotiation behaviors to account for the objectives of the provider.

In our earlier works [19,20] we presented an initial version of the adaptive mechanism. In this initial version, workload and load-spike proofness aspects were not studied. Furthermore, these works contained only partial experimental evaluation.

7   Conclusions and Future Works

This paper presented an adaptive one-to-many negotiation mechanism designed to improve the acceptability rate of a SaaS provider while accommodating load-spikes and meeting its budget constraints. The proposed approach endows the provider with a fine-grained control of the desired acceptability rate. Furthermore, as has been shown in Sect. 5.3 the results, with the proposed solution the SaaS provider is capable to cope with the dynamic & open nature of the cloud ecosystem and to respond to load spikes in an adequate manner.

Our future research work will be directed towards giving the provider a finer-grained control over the level of user satisfaction it seeks to attain. In particular, the provider should be able to ensure that a predefined percentage of users consider the service to be *Good* or *Better* [8]. Several reports of the ITU [11] and the ETSI [5] recommend moving in this direction (c.f. [8]). However, most of the existing works adopt a provider-centric approach based on the MOS [8]. Our future research works will address this issue. This requires upgrades in the user modeling algorithm and the resulting adaptation mechanism. Another work in progress is allowing service user agents sa_i to use negotiation strategies other than time-based concessions (TBC).

References

1. Accenture: Accenture 2013 Global Consumer Pulse Survey Global & U.S. Key Findings (2013)
2. An, B., Lesser, V., Irwin, D., Zink, M.: Automated negotiation with decommitment for dynamic resource allocation in cloud computing. In: Proceedings of the 9th International Conference on Autonomous Agents and Multiagent Systems, vol. 1. pp. 981–988. International Foundation for Autonomous Agents and Multiagent Systems (2010)
3. Baarslag, T., Hendrikx, M.J., Hindriks, K.V., Jonker, C.M.: Learning about the opponent in automated bilateral negotiation: a comprehensive survey of opponent modeling techniques. In: Autonomous Agents and Multi-Agent Systems, pp. 1–50 (2015)
4. Casalicchio, E., Silvestri, L.: Mechanisms for SLA provisioning in cloud-based service providers. Comput. Netw. **57**(3), 795–810 (2013)

5. ETSI: European telecommunications standards institute, http://www.etsi.org/
6. Faratin, P., Sierra, C., Jennings, N.R.: Negotiation decision functions for autonomous agents. Robot. Auton. Syst. **24**(3), 159–182 (1998)
7. Hobfeld, T., Schatz, R., Egger, S.: Sos: The mos is not enough! In: 2011 Third International Workshop on Quality of Multimedia Experience (QoMEX), pp. 131–136. IEEE (2011)
8. Hoßfeld, T., Heegaard, P.E., Varela, M., Möller, S.: QoE beyond the MOS: an in-depth look at QoE via better metrics and their relation to MOS. Qual. User Exp. **1**(1), 2 (2016)
9. Hou, C.: Predicting agents tactics in automated negotiation. In: Proceedings. IEEE/WIC/ACM International Conference on Intelligent Agent Technology, IAT 2004, pp. 127–133. IEEE (2004)
10. Cisco Visual Networking Index: Cisco VNI Forecast and Methodology, 2015–2020. Cisco white paper, 1 June 2016 (2016)
11. ITU: International telecommunications union, https://www.itu.int/
12. Ji, S.J., Zhang, C.J., Sim, K.M., Leung, H.F.: A one-shot bargaining strategy for dealing with multifarious opponents. Appl. Intell. **40**(4), 557–574 (2014)
13. Lomuscio, A.R., Wooldridge, M., Jennings, N.R.: A classification scheme for negotiation in electronic commerce. Group Decis. Negot. **12**(1), 31–56 (2003)
14. Mansour, K., Kowalczyk, R.: A meta-strategy for coordinating of one-to-many negotiation over multiple issues. In: Wang, Y., Li, T. (eds.) Foundations of Intelligent Systems. AISC, vol. 122, pp. 343–353. Springer, Heidelberg (2012). https://doi.org/10.1007/978-3-642-25664-6_40
15. Mansour, K., Kowalczyk, R.: On dynamic negotiation strategy for concurrent negotiation over distinct objects. In: Marsa-Maestre, I., Lopez-Carmona, M.A., Ito, T., Zhang, M., Bai, Q., Fujita, K. (eds.) Novel Insights in Agent-based Complex Automated Negotiation. SCI, vol. 535, pp. 109–124. Springer, Tokyo (2014). https://doi.org/10.1007/978-4-431-54758-7_6
16. Möller, S., Raake, A.: Quality of Experience. Springer, Cham (2014)
17. Najjar, A., Gravier, C., Serpaggi, X., Boissier, O.: Modeling user expectations satisfaction for SaaS applications using multi-agent negotiation. In: 2016 IEEE/WIC/ACM International Conference on Web Intelligence (WI), pp. 399–406, October 2016
18. Najjar, A.: Multi-Agent Negotiation for QoE-Aware Cloud Elasticity Management. Ph.D. thesis, École nationale supérieure des mines de Saint-Étienne (2015)
19. Najjar, A., Boissier, O., Picard, G.: An adaptive one-to-many negotiation to improve the service acceptability of an open SaaS provider. In: International Workshop on Agent-based Complex Automated Negotiations (ACAN) (2017)
20. Najjar, A., Boissier, O., Picard, G.: Aquaman: an adaptive QoE-aware negotiation mechanism for SaaS elasticity management (extended abstract). In: International Conference on Autonomous Agents and Multi-Agent Systems, AAMAS 2017, 1655–1657 May 2017
21. Najjar, A., Mualla, Y., Boissier, O., Picard, G.: Aquaman: QoE-driven cost-aware mechanism for SaaS acceptability rate adaptation. In: 2017 IEEE/WIC/ACM International Conference on Web Intelligence (WI), August 2017
22. Najjar, A., Serpaggi, X., Gravier, C., Boissier, O.: Survey of elasticity management solutions in cloud computing. In: Mahmood, Z. (ed.) Continued Rise of the Cloud. CCN, pp. 235–263. Springer, London (2014). https://doi.org/10.1007/978-1-4471-6452-4_10

23. Najjar, A., Serpaggi, X., Gravier, C., Boissier, O.: Multi-agent systems for personalized QoE-management. In: 2016 28th International Teletraffic Congress (ITC 28), vol. 3, pp. 1–6. IEEE (2016)
24. Nguyen, T.D., Jennings, N.R.: Coordinating multiple concurrent negotiations. In: Proceedings of the Third International Joint Conference on Autonomous Agents and Multiagent Systems, vol. 3. pp. 1064–1071. IEEE Computer Society (2004)
25. North, M.J., Howe, T.R., Collier, N.T., Vos, J.: The repast simphony runtime system. In: Agent 2005 Conference on Generative Social Processes, Models, and Mechanisms. Argonne National Laboratory. Citeseer, Argonne, Illinois, USA (2005)
26. Pruitt, D.G.: Negotiation Behavior. Academic Press, New York (2013)
27. Rahwan, I., Kowalczyk, R., Pham, H.H.: Intelligent agents for automated one-to-many e-commerce negotiation. In: Australian Computer Science Communications. vol. 24, pp. 197–204. Australian Computer Society, Inc. (2002)
28. Reichl, P., Egger, S., Schatz, R., D'Alconzo, A.: The logarithmic nature of QoE and the role of the weber-fechner law in QoE assessment. In: 2010 IEEE International Conference on Communications (ICC), pp. 1–5. IEEE (2010)
29. Richter, J., Baruwal Chhetri, M., Kowalczyk, R., Bao Vo, Q.: Establishing composite slas through concurrent QoS negotiation with surplus redistribution. Concurrency Comput. Pract. Experience **24**(9), 938–955 (2012)
30. Sackl, A., Schatz, R.: Evaluating the impact of expectations on end-user quality perception. In: Proceedings of International Workshop Perceptual Quality of Systems (PQS), pp. 122–128 (2013)
31. Savitzky, A., Golay, M.J.: Smoothing and differentiation of data by simplified least squares procedures. Anal. Chem. **36**(8), 1627–1639 (1964)
32. Schatz, R., Fiedler, M., Skorin-Kapov, L.: QoE-based network and application management. In: Möller, S., Raake, A. (eds.) Quality of Experience. TSTS, pp. 411–426. Springer, Cham (2014). https://doi.org/10.1007/978-3-319-02681-7_28
33. Siebenhaar, M., Nguyen, T.A.B., Lampe, U., Schuller, D., Steinmetz, R.: Concurrent negotiations in cloud-based systems. In: Vanmechelen, K., Altmann, J., Rana, O.F. (eds.) GECON 2011. LNCS, vol. 7150, pp. 17–31. Springer, Heidelberg (2012). https://doi.org/10.1007/978-3-642-28675-9_2
34. Son, S., Sim, K.M.: A price-and-time-slot-negotiation mechanism for cloud service reservations. IEEE Trans. Syst. Man Cybern. B (Cybern.) **42**(3), 713–728 (2012)
35. Talia, D.: Clouds meet agents: toward intelligent cloud services. IEEE Internet Comput. **2**, 78–81 (2012)
36. Wooldridge, M.: An Introduction to Multiagent Systems. Wiley, New York (2009)
37. Zeithaml, V.A., Berry, L.L., Parasuraman, A.: The nature and determinants of customer expectations of service. J. Acad. Mark. Sci. **21**(1), 1–12 (1993)

Opponent Modeling with Information Adaptation (OMIA) in Automated Negotiations

Yuchen Wang[✉], Fenghui Ren, and Minjie Zhang

School of Computing and Information Technology,
University of Wollongong, Wollongong, NSW 2522, Australia
yw808@uowmail.edu.au, {fren,minjie}@uow.edu.au

Abstract. Opponent modeling is an important technique in automated negotiations. Many of the existing opponent modeling methods are focusing on predicting the opponent's private information to improve the agent's benefits. However, these modeling methods overlook an ability to improve the negotiation outcomes by adapting to different types of private information about the opponent when they are available beforehand. This availability may be provided by some prediction algorithms, or be prior knowledge of the agent. In this paper, we name the above ability as Information Adaptation, and propose a novel Opponent Modeling method with Information Adaptation (OMIA). Specifically, the future concessions of the opponent will firstly be learned based on the opponent's historical offers. Then, an expected utility calculation function is introduced to adaptively guide the agent's negotiation strategy by considering the availability and value of the opponent's private information. The experimental results show that OMIA can adapt to different types of information, helping the agent reach agreements with the opponent and achieve higher utility values comparing to those which lack the information adaptation ability.

Keywords: Automated negotiations · Opponent modeling
Information adaptation

1 Introduction

Negotiation is an important activity between people or parties who discuss issues intending to reach agreement. Negotiation often involves significant cost in human resources and time. As a result of this, automated negotiation techniques have attracted increasing attentions during the last two decades [10]. The benefits include resource-saving in manpower and time [4], avoiding social confrontation [3] and automatic bargains in e-markets [15].

One key challenge in automated negotiations is to reach beneficial negotiation results when private information about the opponent is unknown. The private information is in contrast to public information known by all negotiators.

© Springer International Publishing AG 2017
G. Sukthankar and J. A. Rodriguez-Aguilar (Eds.): AAMAS 2017 Best Papers,
LNAI 10642, pp. 21–35, 2017.
https://doi.org/10.1007/978-3-319-71682-4_2

For instance, the public information includes the maximum negotiation time, the historical offers exchanged by negotiators, etc., while the private information contains a negotiator's personal reservation value, deadline, etc. Obviously, sharing this private information between negotiators is not applicable. To tackle this problem, researchers have put effort into opponent modeling techniques [2]. These techniques mostly give agents an ability of exploring the opponent's private information and adapting to the opponent's negotiation behavior in order to achieve satisfactory negotiation outcomes.

In current literature, a number of opponent modeling methods have been developed by employing different learning methods, such as Bayesian learning [18], Non-linear regression [14], Kernel density estimation [5], and Artificial neural networks [12]. Among these methods, four types of opponent's private information are commonly selected as their learning goals, which are reservation value [17], negotiation deadline [9], bidding strategy [16] and offer acceptance possibility [13]. However, these approaches lack an ability of information adaptation.

The information adaptation indicates an ability to make better negotiation decisions when some types of the opponent's private information are available. This availability may be provided by some prediction algorithms, or be prior knowledge of the agent. For example, when the opponent's reservation value is available, the agent should adapt its negotiation strategy and try to make agreements close to this reservation value. Lacking the ability of information adaptation means losing a potential behavior guidance toward different types of available information, thus the negotiation outcomes would be negatively affected.

In order to give agents such an information adaptation ability, in this paper, we propose a novel opponent modeling method called Opponent Modeling with Information Adaptation (OMIA). Traditionally, an opponent modeling method could guide the agent's negotiation strategy adapting to the historical offers of the opponent. In this paper, OMIA should not only adapt to the historical offers, but also adapt to different types of private information both when they are available and unavailable. The types of private information considered in this paper are reservation value, deadline, bidding strategy and acceptance possibility due to their importance in automated negotiations [1].

To establish OMIA, the major challenge is how to create one opponent model that can simultaneously adapt to the five types of information based on their availability (historical offers are always available) and values. Our idea is to establish OMIA from a probability point of view. In particular, these probability distributions are going to be estimated: (i) the probability distributions of the opponent offers' utility in future time; (ii) the probability that the opponent accepts a particular offer; (iii) the probability that the opponent quits in particular future time. The process of OMIA is as follows. First, OMIA predicts the future concessions (i) of the opponent using the historical offers. Then, an expected utility calculation function is introduced to estimate (ii) and (iii). Then, based on (i), (ii) and (iii), this function calculates the expected utility the agent will gain when it takes different concession strategies. The maximum expected utility determines the best concession strategy that the agent should take. The availability and values of the private information will influence the results of all probability distributions, affecting the value of the expected utility, and thus guiding the concession strategy of the agent.

The merits of OMIA are: (a) OMIA could adapt to the behavior of the opponent only using the historical offers. (b) OMIA could adapt to different types of private information based on their availability and values, and the agent can choose the types of information to adapt to. (c) OMIA makes little assumptions about the opponent (e.g. bidding strategy, utility function, etc.), making itself a highly robust model.

The remainder of this paper is organized as follows. Section 2 describes the general negotiation setting. Section 3 introduces the proposed OMIA. Section 4 describes how OMIA adapts to various types of information. Section 5 demonstrates the experimental results. Related work is presented in Sect. 6, and Sect. 7 makes a conclusion.

2   Negotiation Setting

In this paper, we study the bilateral single-issue automated negotiation, which consists of two agents negotiating over a single issue. The alternating offers protocol [11] is employed in this paper where two agents exchange offers in turns until one agent accepts an offer or reaches its deadline. The time can either be a continuous or discrete variable. In this paper, we use the discrete time setting as we are focusing on adapting to various types of input information and currently do not take computational cost into account. A monotonic concession process is assumed where there are no decommitment behaviors during the negotiation, i.e., the agent will not regret its compromise and ask for more benefits from the opponent in newly generated offers. In bilateral single-issue automated negotiation, the utility of the opponent's offers is monotonically increasing.

The bidding strategy of the opponent denotes how it provides its offers during a negotiation. A common one is called the time-dependent strategy where the offer is given based on time [7]. The utility function is given by:

$$u_t = u^{min} + (u^{max} - u^{min})(\frac{t_c}{t_{max}})^\beta, \tag{1}$$

where u^{min} and u^{max} denote the minimum and maximum utility of the opponent's offers respectively. t_c and t_{max} mean the current and the maximum negotiation time respectively. β is the concession parameter. $0 < \beta < 1$, $\beta = 1$ and $\beta > 1$ represent three types of concession strategies, which are called Conceder, Linear and Boulware respectively. The Conceder concedes dramatically toward u^{max} at the early stage of a negotiation, while the Boulware only makes significant concession when the time is close to t_{max}.

3   Opponent Modeling with Information Adaptation

3.1   Basic Notation

Before introducing OMIA, we first establish the notation for future use. We use t to denote the time of a negotiation, with different subscripts indicating specific time. For example, t_1 is the first time when the agent exchanges an offer with the opponent, t_c the current time, and t_{max} the maximum negotiation time. We use a discrete time setting so that the time is measured based on the negotiation rounds elapsed.

The offers received from the opponent are measured as utility values, which represent the agent's preferences over them. Higher utility values are preferred than lower ones. These utility values are the only information that the agent could obtain and make use of to build its opponent model when no other information is available. Also, the agent does not know the opponent's utility function. Thus, all offers exchanged by both sides will be measured in the agent's utility space. The utility that the agent receives from and offers to the opponent at time t could be respectively written as:

$$u_t^{oppo}, u_t^{my} \in [0, 1] \tag{2}$$

The utility is quantified between 0 and 1. Here we do not specify the types of utility function that maps offers to utility values because this is dependent on particular negotiation scenarios and should be determined accordingly.

As this paper is focusing on modeling the opponent, to make the formulas neater, we simplify the notation of utility received from the opponent by removing the superscript:

$$u_t \Leftrightarrow u_t^{oppo} \tag{3}$$

At time t_c, the agent will have received c offers from the opponent. These are called the *historical offers* and we use U_{t_c} to indicate it:

$$U_{t_c} = \{u_{t_i} | i = 1, ..., c\} \tag{4}$$

These historical offers are the only information that the agent has at the current time when no other sources of information are available.

All the notations used in this paper are listed in Table 1.

3.2   Predicting Future Concessions

The process of OMIA is first to construct the probability distributions $P(u_{t_f})$ of utility u_{t_f} for every future time t_f based on the historical offers. We will introduce this step in this Subsect. (3.2). Then, an expected utility calculation function is introduced to incorporate $P(u_{t_f})$, the probability that the opponent accepts or rejects the agent's offer $P(a_t = \{1, 0\})$, and the probability that the opponent quits or does not quit the negotiation at time t $P(q_t = \{1, 0\})$. This step will be described in the next Subsect. (3.3).

Table 1. Notation

Notation	Meaning
t_i	Negotiation time[a]
t_c	Current time
t_f	Future time
u_t	Received offer's utility at time t
u_t^{my}	Utility of the agent's counter-offer at time t
U_{t_c}	Historical offers at current time
$Learn(\cdot)$	Learning algorithm
$GP(\cdot)$	Gaussian process
$P(u_t)$	Probability distribution for u_t
$f_n(u_t; \mu_t, \sigma_t, 0, 1)$	Normalized and truncated probability density function for u_t
$F_n(u_t; \mu_t, \sigma_t, 0, 1)$	Normalized and truncated cumulative distribution function for u_t
E_{t_f}	Expected utility the agent will gain when trying to reach agreement at time t_f
t_m	Time between t_c and t_f
$u_{t_m}^{my}(t_f)$	Counter-offer corresponding to E_{t_f} provided at time t_m
$E_{t_f}(u_{t_m}^{my}(t_f))$	Expected utility from the agent's offers $u_{t_m}^{my}$ accepted by the opponent between time t_c and t_f
$E_{t_f}(u_{t_f})$	Expected utility from accepting the opponent's offer u_{t_f}
E^*	Maximum expected utility among E_{t_f}
t^*	Time corresponding to E^*
$P(a_t = \{1, 0\})$	Probability that the opponent accepts or rejects the agent's offer u_{t-1}^{my}
$P(q_t = \{1, 0\})$	Probability that the opponent quits or does not quit the negotiation at time t

[a]We use a positive integer subscript i to denote the negotiation time in accordance with the discrete time setting

First, we are going to predict the probability distributions $P(u_{t_f})$ using the historical offers. These distributions can be regarded as an estimation of the opponent's future concessions. Thus, the agent can exploit the concessions of the opponent by adaptively providing counter-offers. Predicting $P(u_{t_f})$ naturally requests a learning algorithm to analyze the historical offers and provide predictions for any future time:

$$Learn(U_{t_c}) \Rightarrow < P(u_{t_{c+1}}), ..., P(u_{t_{max}}) >, \tag{5}$$

where $Learn(\cdot)$ is a learning function that can satisfy this requirement. Any applicable learning approaches can be applied here. This allows the users of OMIA to flexibly choose the learning approaches.

In this paper, we choose the Gaussian process technique based on three considerations: (1) Gaussian process is a powerful non-linear interpolation tool and has been applied to address various learning tasks [6,8]. (2) Gaussian process provides both an estimation and its uncertainty (essentially a Gaussian distribution) for an unseen utility at any time, which is highly in accordance with our aims. (3) Gaussian process can work with variables with a continuous domain, which gives potential to expand OMIA to real-time negotiation. For the parameters, we use Matérn covariance function and linear mean function due to their robustness [16].

The Gaussian process will predict new probability distributions $P(u_{t_f})$ using the historical offers U_{t_c} at time t_c:

$$GP(U_{t_c}) \Rightarrow < P(u_{t_{c+1}}), ..., P(u_{t_{max}}) > \tag{6}$$

The output of the Gaussian process for any time t is a Gaussian distribution given by:

$$P(u_t) = f(u_t; \mu_t, \sigma_t) = \frac{1}{\sigma_t \sqrt{2\pi}} exp(\frac{-(u_t - \mu_t)^2}{2\sigma_t^2}), \tag{7}$$

where μ_t is the mean, i.e. the most likely value of u_t, and σ_t is the standard deviation.

The utility space is bounded in the range of $[0, 1]$, and the expected mean μ_t may exceed this bound. Thus, a normalization is needed to make the mean between 0 and 1:

$$f_n(u_t; \mu_t, \sigma_t) = normalize(f(u_t; \mu_t, \sigma_t)) \tag{8}$$

Furthermore, a truncated normal distribution should also be used to make the distribution in the range of $[0, 1]$. Its probability density function is given by:

$$P(u_t) = f_n(u_t; \mu_t, \sigma_t, 0, 1)$$
$$= \frac{f_n(u_t; \mu_t, \sigma_t)}{F_n(1; \mu_t, \sigma_t) - F_n(0; \mu_t, \sigma_t)} \tag{9}$$

where $f_n(u_t; \mu_t, \sigma_t)$ is the normalized probability density function of opponent's utility at time t and $F_n(u_t; \mu_t, \sigma_t)$ is its cumulative distribution function.

When a negotiation goes to time t_c, Eqs. (6), (7), (8) and (9) will be performed to get probability distributions $P(u_{t_{c+1}}), ..., P(u_{t_{max}})$.

3.3   Making Concessions by Calculating Expected Utility

We assume that the opponent provides its offer first, so at time t_c, the opponent's latest offer is u_{t_c} and the agent's latest offer is $u_{t_c-1}^{my}$. After constructing the probability distributions for every future time after t_c, the agent needs to make a decision about whether to accept u_{t_c} or provide a counter-offer $u_{t_c}^{my}$.

Our idea is to exploit the future concessions of the opponent by calculating the expected utility E_{t_f} when the agent tries to reach agreement with the opponent at particular future time t_f. Here we regard reaching agreement at future time t_f as conceding toward μ_{t_f} and accepting the opponent's offer u_{t_f} if the agent's offers before time t_f are not accepted by the opponent. The μ_{t_f} is the estimated mean value of u_{t_f}. Let $u_{t_c}^{my}(t_f)$ be the agent's counter-offer at time t_c when conceding toward μ_{t_f}. Let E^* be the maximum expected utility among all E_{t_f}, and the corresponding time is t^*. The $u_{t_c}^{my}(t^*)$ will be set as the agent's final counter-offer $u_{t_c}^{my}$ ready to be provided to the opponent.

We choose μ_{t_f} as concession targets because of these two considerations:

(1) The μ_{t_f} is a moderate value which is not too high. As the agent does not know how the opponent will react to its concessions, avoiding being extreme is a reasonable way. Also, it is not too low so the agent would gain sufficient utility when conceding toward it.
(2) The μ_{t_f} indicates the most possible value that the opponent will offer at time t_f. Thus, conceding toward it would have a high chance of reaching agreement with the opponent.

In terms of determining how the agent concedes toward μ_{t_f}. Again, avoiding being extreme is a reasonable way. To be more exact, the agent should not keep its offers unchanged until time t_f nor concede immediately to μ_{t_f}. So we choose to concede linearly toward every μ_{t_f}.

When trying to reach agreement at future time t_f, the expected utility E_{t_f} consists of following two parts.

First, the opponent has a chance to accept one of the counter-offers $u_{t_m}^{my}(t_f)$ provided by the agent from time t_c to t_{f-1}. We use $E_{t_f}(u_{t_m}^{my}(t_f))$ to represent this part of expected utility. The $u_{t_m}^{my}(t_f)$ is calculated by performing linear interpolation:

$$u_{t_m}^{my}(t_f) = u_{t_c-1}^{my} + (\mu_{t_f} - u_{t_c-1}^{my})\frac{t_m - t_{c-1}}{t_f - t_{c-1}} \tag{10}$$

Each $u_{t_m}^{my}(t_f)$ is accepted when the opponent rejects all previous offers, does not quit, and accepts this one. The $E_{t_f}(u_{t_m}^{my}(t_f))$ is defined as:

$$E_{t_f}(u_{t_m}^{my}(t_f)) = (\prod_{n=c+1}^{m} P(a_{t_n} = 0)P(q_{t_n} = 0))$$

$$\times P(a_{t_{m+1}} = 1)u_{t_m}^{my}(t_f) \tag{11}$$

When no extra private information is available, the $P(q_{t_n} = 0)$ could be specified by some prior assumptions, e.g. the opponent will not quit.

Second, the opponent has a chance to reject all offers from time t_c to t_{f-1}, does not quit, and gives u_{t_f} at time t_f. Then the agent will accept u_{t_f}, which constitutes the second part of the expected utility. We use $E_{t_f}(u_{t_f})$ to denote this part and it can be calculated as:

$$E_{t_f}(u_{t_f}) = (\prod_{n=c+1}^{f} P(a_{t_n} = 0)P(q_{t_n} = 0))$$

$$\times \int_0^{u_{t_f-1}^{my}(t_f)} u_{t_f}P(u_{t_f})du_{t_f}, \tag{12}$$

where $\int_0^{u_{t_f-1}^{my}(t_f)} u_{t_f} P(u_{t_f}) du_{t_f}$ is the expected utility that the opponent will give when it rejects offer $u_{t_f-1}^{my}(t_f)$.

In terms of determining the acceptance possibility $P(a_t = 1)$ when no extra private information is available, a common assumption for the opponent to accept an offer is that the offer will give extra benefit to the opponent, i.e., the agent provides less utility than what the opponent is going to give:

$$
\begin{aligned}
P(a_t = 1) &= P(u_{t-1}^{my} \le u_t) \\
&= 1 - P(u_t < u_{t-1}^{my}) \\
&= 1 - F_n(u_{t-1}^{my}; \mu_t, \sigma_t, 0, 1),
\end{aligned}
\tag{13}
$$

where $F_n(u_t^{my}; \mu_t, \sigma_t, 0, 1)$ is the normalized and truncated cumulative probability distribution of the opponent offers' utility.

The final expected utility E_{t_f}, which the agent is expected to reach at time t_f, is given by:

$$
\begin{aligned}
E_{t_f} &= \sum_{m=c}^{f-1} E_{t_f}(u_{t_m}^{my}) + E_{t_f}(u_{t_f}) \\
&= \sum_{m=c}^{f-1} (\prod_{n=c+1}^{m} P(a_{t_n} = 0)P(q_{t_n} = 0))P(a_{t_{m+1}} = 1)u_{t_m}^{my} \\
&\quad + (\prod_{n=c+1}^{f} P(a_{t_n} = 0)P(q_{t_n} = 0)) \int_0^{u_{t_f-1}^{my}} u_{t_f} P(u_{t_f}) du_{t_f}
\end{aligned}
\tag{14}
$$

For each future time t_f, we follow the same process to calculate the expected utility. The maximum expected utility E^* and the corresponding time t^* can be obtained by:

$$
E^* = max(E_{t_f}) \quad t_f \in [t_{c+1}, t_{max}]
\tag{15}
$$

$$
t^* = \text{argmax}_{t_f \in [t_{c+1}, t_{max}]} E_{t_f}
\tag{16}
$$

Finally, we compare E^* with u_{t_c}. As u_t is monotonically increasing, u_{t_c} is the maximum utility that the agent has received at current time. If $E^* > u_{t_c}$, it means that conceding toward μ_{t_f} may be more valuable so that the agent will reject u_{t_c} and choose to concede toward μ_{t^*} by setting $u_{t_c}^{my}$ as $u_{t_c}^{my}(t^*)$. Otherwise, if $E^* \le u_{t_c}$, the agent will accept u_{t_c}. This decision procedure at time t_c can be formed as:

$$
Decision(t_c) = \begin{cases} \text{accept } u_{t_c}, & E^* \le u_{t_c} \\ \text{reject } u_{t_c} \text{ and provide } u_{t_c}^{my}(t^*), & E^* > u_{t_c} \end{cases}
\tag{17}
$$

4 Adaptation to Four Types of Private Information

In this section, we demonstrate how to use the proposed OMIA to adapt to four types of private information, which are (1) reservation value, (2) deadline, (3) bidding strategy and (4) acceptance probability.

Reservation value. The opponent will never provide an offer exceeding its reservation value. To adapt to a given reservation value u^r, the $P(q_{t_n} = 0)$ in Eq. 14 will be influenced by the opponent offers' utility. Given an u^r, the probability of not quitting $P(q_t = 0)$ could be specified by:

$$P(q_t = 0) = \begin{cases} 0, & u_t > u^r \\ 1, & u_t \leq u^r \end{cases} \tag{18}$$

Deadline. Similar to the reservation value, we also adapt to the deadline by specifying the $P(q_t = 0)$. Given an deadline t^d, $P(q_t = 0)$ is set by:

$$P(q_t = 0) = \begin{cases} 0, & t > t^d \\ 1, & t \leq t^d \end{cases} \tag{19}$$

Bidding strategy. Knowing the information of the opponent's bidding strategy means the agent knows what the opponent will offer at specific time. That is, pairs of time and utility (t_i, u_{t_i}) are given beforehand. OMIA utilizes (t_i, u_{t_i}) together with the historical offers U_{t_c} to train the Gaussian process models and predict u_{t_f}.

$$GP(U_{t_c}, (t_i, u_{t_i})) \Rightarrow < P(u_{t_{c+1}}), ..., P(u_{t_{max}}) > \tag{20}$$

Acceptance possibility. OMIA models the acceptance possibility $P(a_t = 1)$ from a probability point of view only with historical offers of the opponent. If there are other kinds of methods giving more precise estimations, OMIA can directly use them in Eq. 14 so that the expected utility is computed with adaptation to these acceptance possibility calculation methods.

5   Experimental Results

5.1   Experimental Setting

In this experiment, an agent and an opponent negotiate over a single issue. The utility of the opponent's offers follows a time-dependent strategy, i.e. Conceder, Linear or Boulware. The concession parameter is denoted as β. The utility of both sides is quantified in the agent's utility space in $[0, 1]$. The negotiation time is set from 1 to 50.

We design a series of experiments to show the information adaptation ability of OMIA. Table 2 shows the detailed experimental setting, including the experiment type, the information adapting to, opponent strategy, assumed estimated private information of the opponent, and assumptions for the opponent's decision for quit. The estimations of these types of private information may be provided by some prediction algorithms, or be prior knowledge of the agent. In this experiment, these estimations are set as prior knowledge and are assumed to be true.

We conduct 2 types of experiments. One is a case study in which the agent negotiates with a Boulware opponent with β being 4. A non-adaptive behavior is

Table 2. Experimental setting

Experiment type	Adapting to	Opponent strategy	Assumed estimated private information	Quit of opponent
Case study	No adaptation	Boulware, $\beta = 4$	/	Assume not quit
	Historical offers		/	
	Historical offers and bidding strategy		An offer worth utility 0.1296 will be given by the opponent at time 30	
	Historical offers and acceptance possibility		The acceptance possibility will become 0% after time 30	
Empirical analysis	Historical offers and reservation value	Conceder, Linear and Boulware	Three experiments with reservation value being 0.6, 0.7 and 0.8	Quit after its offer's utility > the estimated reservation value
	Historical offers and deadline	Uniformly select 20 β for each strategy type in the range of $[0.3, 0.8]$, $[0.9, 1.1]$, and $[2, 4]$ respectively	Three experiments with deadline being 30, 35 and 40	Quit after time > the estimated deadline

firstly studied. Then, we study three types of information to adapt to, which are the historical offers, bidding strategy and acceptance possibility. The historical offers are public information and can be used directly. For bidding strategy, we select a pair of time and utility $(40, 0.4096)$ on the curve of opponent's offers as its assumed estimated value, and use it along with the historical offers as the input of Gaussian process. This means that the agent knows the opponent will offer 0.4096 at time 40. Thus, we can compare the behavior of the agent with and without this estimated pair of time and utility. For acceptance possibility, we assume that the acceptance possibility will become 0 after time 30. We can then compare how will the agent behave with and without the estimation for acceptance possibility.

The other type of experiment is empirical analysis for the private information of reservation value and deadline. For each one, we have three experiments using three assumed estimated values, which are 0.6, 0.7 and 0.8 for reservation values, and 30, 35 and 40 for deadlines. In every experiment, different numbers of β are selected to cover three types of opponent (Conceder, Linear and Boulware) in order to get average results. We uniformly select 20 β for each strategy type in the range of $[0.3, 0.8]$, $[0.9, 1.1]$, and $[2, 4]$, respectively. Totally, we get 60 results for an experiment with particular assumed estimated value.

5.2 Results Analysis

The experimental results for the case studies are presented in Figs. 1 and 2. The results for the empirical analysis are showed in Tables 3 and 4.

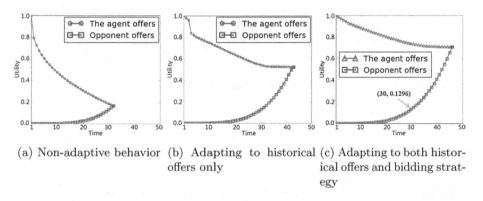

(a) Non-adaptive behavior   (b) Adapting to historical offers only   (c) Adapting to both historical offers and bidding strategy

Fig. 1. Negotiation process of the agent with non-adaptive behavior, adaptation to historical offers, and adaptation to both historical offers and bidding strategy

Figure 1(a) shows the negotiation process of the agent with a non-adaptive Conceder negotiation strategy. Figure 1(b) shows the process when the agent applies OMIA, adapting to the opponent's historical offers U_{t_c}. Figure 1(c) shows the process when the agent adapts to both U_{t_c} and the bidding strategy of the opponent. It can be seen that the agent with non-adaptive strategy achieves a low utility value, while the agent with adaptation to U_{t_c} achieves a higher utility value than the non-adaptive agent. When the opponent's bidding strategy (represented by the assumed estimated time-utility pair $(40, 0.4096)$) is given, we can see that the agent has more confidence about the future concessions of the opponent, makes fewer concessions during the early stage of the negotiation process, and finally achieves a higher utility value than the agent with adaptation only to U_{t_c}.

Figure 2(a) and (b) illustrate the negotiation process without and with adaptation to the assumed estimated acceptance possibility. The acceptance possibility will become 0 after time 30. The agent, receiving this information but not adapting to it, fails to reach agreement with the opponent. By contrast, the agent, adapting to this information, makes more concessions and finally reaches agreement at time 30. This is because the given acceptance possibility makes the expected utility when conceding toward time 30 maximum.

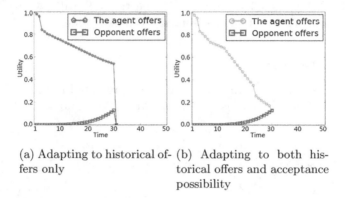

(a) Adapting to historical offers only

(b) Adapting to both historical offers and acceptance possibility

Fig. 2. Negotiation process of the agent without and with adaptation to the assumed estimated acceptance possibility

Table 3. Average utility achieved without and with adaptation to estimated reservation value

Estimated reservation value u^r	Average utility based on historical offers only	Average utility based on historical offers and estimated reservation value
0.6	0.22 (±0.008)	0.54 (±0.002)
0.7	0.36 (±0.006)	0.63 (±0.005)
0.8	0.54 (±0.009)	0.72 (±0.006)

Table 3 shows the average utility achieved without and with adaptation to estimated reservation value u^r. 95% confidence intervals are listed in the parentheses. We can see that the agent without adaptation to u^r achieves a lower average utility value than that with adaptation. This is caused by a high chance of failing to reach agreement with the opponent for the agent with a Boulware strategy and without adaptation to u^r. In addition, for the agent with adaption to u^r, the average utility achieved is always slightly lower than u^r. This shows that the agent makes agreement when the opponent offers' utility is close to u^r in order to gain as much utility as possible.

Table 4 shows the average utility achieved without and with adaptation to estimated deadline t^d. Similar to the result of the reservation value, the agent without adaptation to t^d achieves significantly lower average utility than that with adaptation. As there is a high chance of failing to reach agreement when the opponent approaches its deadline, but the agent does not adapt to it.

Table 4. Average utility achieved without and with adaptation to estimated deadline

Estimated deadline t^d	Average utility based on historical offers only	Average utility based on historical offers and estimated deadline
30	0.11 (±0.012)	0.51 (±0.014)
35	0.19 (±0.008)	0.60 (±0.004)
40	0.32 (±0.009)	0.70 (±0.009)

In summary, the experimental results show that OMIA could adapt to the historical offers by exploiting the future concessions of the opponent. Also, OMIA can adaptively guide the agent's negotiation behaviors by utilizing the availability and values of the opponent's private information in an expected utility measurement. As a result, The agent is able to successfully reach agreement with the opponent and achieve higher utility values comparing to those which lack the information adaptation ability.

6   Related Work

A lot of opponent modeling methods with different learning goals have been developed. Yu et al. [17] apply non-linear regression and Bayesian learning to estimate the opponent's reservation value and deadline. They introduce a concept named detecting region to estimated the lower and upper boundary of the reservation value and deadline. Historical offers are used to make the prediction more accurate during the process of the negotiation. Williams et al. [16] use the Gaussian process to estimate the future behavior of the opponent. The concession rate of the agent is then adaptively set based on the predictions during a single negotiation session. Time-based discounts and a risk function are applied in their model to handle the uncertainty. Oshrat et al. [13] create a negotiator called *KBAgent*. It can use the past negotiation results as a knowledge base to compute the acceptance probability for unseen offers. This approach allows agents to negotiate with people and can gain more utility than humans.

However, these opponent modeling methods are focusing on the use of the information they have. These methods overlook the fact that there may exist potentially available information. Having the ability of utilizing the potential information will help negotiation agents to reach agreement with the opponent and gain more utility values.

7   Conclusion

In this paper, we proposed a novel opponent modeling method called OMIA. The proposed method can not only adapt to the historical offers of the opponent, but also simultaneously adapt to four types of commonly used information in automated negotiations. Agents using OMIA can flexibly choose which types of information to adapt to. Also, OMIA makes little assumptions about the opponent, making itself a highly robust model. The experimental results showed that OMIA could exploit the future concessions of the opponent only based on the historical offers. When extra estimated information is given, OMIA could adaptively give guidance to the agent's behaviors, and the agent is able to gain as much utility from the opponent as possible under the estimated information.

Acknowledgments. This research is supported by a DECRA Project (DP140100007) from Australia Research Council (ARC), a UPA and an IPTA scholarships from University of Wollongong, Australia.

References

1. Baarslag, T., Hendrikx, M.J., Hindriks, K.V., Jonker, C.M.: Learning about the opponent in automated bilateral negotiation: a comprehensive survey of opponent modeling techniques. Auton. Agents Multi-Agent Syst. **30**(5), 849–898 (2016)
2. Baarslag, T., Hendrikx, M.J., Hindriks, K.V., Jonker, C.M.: A survey of opponent modeling techniques in automated negotiation. In: Proceedings of the 2016 International Conference on Autonomous Agents and Multiagent Systems, pp. 575–576. International Foundation for Autonomous Agents and Multiagent Systems (2016)
3. Broekens, J., Jonker, C.M., Meyer, J.J.C.: Affective negotiation support systems. J. Ambient Intell. Smart Environ. **2**(2), 121–144 (2010)
4. Carbonneau, R., Kersten, G.E., Vahidov, R.: Predicting opponent's moves in electronic negotiations using neural networks. Expert Syst. Appl. **34**(2), 1266–1273 (2008)
5. Coehoorn, R.M., Jennings, N.R.: Learning on opponent's preferences to make effective multi-issue negotiation trade-offs. In: Proceedings of the 6th International Conference on Electronic Commerce, pp. 59–68. ACM (2004)
6. Deisenroth, M.P., Fox, D., Rasmussen, C.E.: Gaussian processes for data-efficient learning in robotics and control. IEEE Trans. Pattern Anal. Mach. Intell. **37**(2), 408–423 (2015)
7. Fatima, S.S., Wooldridge, M., Jennings, N.R.: Multi-issue negotiation under time constraints. In: Proceedings of the First International Joint Conference on Autonomous Agents and Multiagent Systems: Part 1, pp. 143–150. ACM (2002)
8. Gal, Y., van der Wilk, M., Rasmussen, C.E.: Distributed variational inference in sparse Gaussian process regression and latent variable models. In: Advances in Neural Information Processing Systems, pp. 3257–3265 (2014)
9. Ji, S.J., Zhang, C.J., Sim, K.M., Leung, H.F.: A one-shot bargaining strategy for dealing with multifarious opponents. Appl. Intell. **40**(4), 557–574 (2014)
10. Kersten, G.E., Lai, H.: Negotiation support and E-negotiation systems: an overview. Group Decis. Negot. **16**(6), 553–586 (2007)

11. Lin, R., Kraus, S., Baarslag, T., Tykhonov, D., Hindriks, K., Jonker, C.M.: Genius: an integrated environment for supporting the design of generic automated negotiators. Computat. Intell. **30**(1), 48–70 (2014)
12. Moosmayer, D.C., Chong, A.Y.L., Liu, M.J., Schuppar, B.: A neural network approach to predicting price negotiation outcomes in business-to-business contexts. Expert Syst. Appl. **40**(8), 3028–3035 (2013)
13. Oshrat, Y., Lin, R., Kraus, S.: Facing the challenge of human-agent negotiations via effective general opponent modeling. In: Proceedings of the 8th International Conference on Autonomous Agents and Multiagent Systems-Volume 1, pp. 377–384. International Foundation for Autonomous Agents and Multiagent Systems (2009)
14. Ren, F., Zhang, M.: Predicting partners' behaviors in negotiation by using regression analysis. In: Zhang, Z., Siekmann, J. (eds.) KSEM 2007. LNCS (LNAI), vol. 4798, pp. 165–176. Springer, Heidelberg (2007). https://doi.org/10.1007/978-3-540-76719-0_19
15. Ren, F., Zhang, M.: A single issue negotiation model for agents bargaining in dynamic electronic markets. Decis. Support Syst. **60**, 55–67 (2014)
16. Williams, C.R., Robu, V., Gerding, E.H., Jennings, N.R.: Using Gaussian processes to optimise concession in complex negotiations against unknown opponents (2011)
17. Yu, C., Ren, F., Zhang, M.: An adaptive bilateral negotiation model based on Bayesian learning. In: Ito, T., Zhang, M., Robu, V., Matsuo, T. (eds.) Complex Automated Negotiations: Theories, Models, and Software Competitions. SCI, vol. 435, pp. 75–93. Springer, Heidelberg (2013). https://doi.org/10.1007/978-3-642-30737-9_5
18. Zhang, J., Ren, F., Zhang, M.: Bayesian-based preference prediction in bilateral multi-issue negotiation between intelligent agents. Knowl. Based Syst. **84**, 108–120 (2015)

Uncertainty Assessment in Agent-Based Simulation: An Exploratory Study

Carolina G. Abreu$^{(\boxtimes)}$ and Célia G. Ralha

Department of Computer Science, University of Brasília (UnB), Brasília, DF, Brazil
{carolabreu,ghedini}@unb.br

Abstract. This paper presents an overview of uncertainty assessment in agent-based simulations, mainly related to land use and cover change. Almost every multiagent-based simulation review has expressed the need for statistical methods to evaluate the certainty of the results. Yet these problems continue to be underestimated and often neglected. This work aims to review how uncertainty is being portrayed in agent-based simulation and to perform an exploratory study to use statistical methods to estimate uncertainty. MASE, a Multi-Agent System for Environmental simulation, is the system under study. We first identified the most sensitive parameters using Morris One-at-a-Time sensitivity analysis. The efforts to assess agent-based simulation through statistical methods are paramount to corroborate and improve the level of confidence of the research that has been made in land use simulation.

1 Introduction

Land use and cover change (LUCC) investigation are of importance to promote insightful management of Earth's land use to refrain environmental damage. Moreover, LUCC is a complex process that relates the interaction between environmental, economic and social systems at different temporal and spatial scales. Computational frameworks are the most used technique to simulate LUCC models for its ability to cope with its complexity.

Agent-based model (ABM) has been incorporated into LUCC models, and many other real-world problems, to explicitly simulate the effects of human decisions in complex situations. They are based on the multi-agent system paradigm that features autonomous entities that interact and communicate in a shared environment. These entities perceive the environment, reason about it and act on it to achieve an internal objective. Therefore, ABM can capture emergent phenomena and provide an original description of the modeled system.

The Multi-Agent System for Environmental simulation (MASE) is a freeware software developed at the University of Brasilia. MASE is a tool for exploring potential impacts of land use policies that implement a land use agent-based model [28]. Considering the purpose and reliance upon external data, MASE may be characterized as a predictor-type agent-based simulation (ABS) model [12]:

© Springer International Publishing AG 2017
G. Sukthankar and J. A. Rodriguez-Aguilar (Eds.): AAMAS 2017 Best Papers,
LNAI 10642, pp. 36–50, 2017.
https://doi.org/10.1007/978-3-319-71682-4_3

a data-driven model with the overall goal of performing medium to long term predictions. MASE simulations were calibrated to match available GIS data [4]. Simulation results were validated according to a standard methodology for spatially explicit simulations [27] and then compared to similar frameworks [29]. MASE performance was found to be higher than other 13 LUCC modeling applications with nine different traditional peer-reviewed LUCC models according to [27]. Despite this fact, the lack of uncertainty assessment and sound experimentation is the main reason for criticism and questioning about the real contribution of frameworks to decision support for LUCC.

According to [3], any ABS has levels of uncertainty and errors associated with it. ABS continues to harbor subjectivity and hence degrees of freedom in the structure and intensity of agent's interactions, learning, and adaptation [18]. There are significant chances of finding results which may be the consequence of biases. Furthermore, almost every ABS review have expressed the need for statistical methods to validate models and evaluate the results to improve the transparency, replicability and general confidence in results derived from ABS. These problems continue to be underestimated and often neglected. Some authors [12], likewise, argued that validation is one of the most important aspects of a model building because it is the only means that provides some evidence that a model can be used for a particular purpose. However, at least 65% of the models in their survey were incompletely validated. Of the models validated in some way, surprisingly less than 5% used statistical validation techniques. Traditionally, ABS types of systems are difficult to analyze given their non-linear behavior and size [6].

Treatment of uncertainty is particularly important and usually difficult to deal with in the case of ABM's stochastic models. While acknowledging the differences in data sources and the causes of inconsistencies, there is still need to develop methods to optimally extract information from the data, to document the uncertainties and to assess common methodological challenges. To look away could reinforce inconsistent results and damage the integrity and quality of simulation results.

This work aims to briefly discuss how uncertainty is being portrayed in ABS and to perform an exploratory study to use statistical methods to estimate uncertainty in a LUCC agent-based prediction simulation tool. The MASE system will be the simulator under study. The Cerrado case study simulations [29] will be the basis for the analysis. As a first investigation step, we assessed the uncertainty within the inputs and configuration parameters of the simulation. Our final goal would be to document, quantification and to foresee its propagation impacts in the results. A particular challenge in performing measurements is coming up with appropriate metrics. The thorough experimentation and repeatability would, therefore, improve our understanding of the uncertainty and relations among the variables that characterize a simulation. The remainder of the paper is structured as follows. In Sect. 2, we present some background on uncertainty and in Sect. 3 some related work. In Sect. 4, we summarize the MASE characteristics and case study. We also present the methodology for the exploratory study.

In Sect. 5 we show results together with discussions. In Sect. 6 we conclude with a summary.

2 Overview of Uncertainty in ABS

The relevance of the treatment of uncertainty is dependent of the modeling objective. Requirements regarding model uncertainty may be less critical for social learning models, where communication and interaction among stakeholders would be of more significance. Conversely, parameters, measurements, and conditions used for model runs influence much more data-based predictions of future states. Projection, forecasting and prediction models are usually very affected by the variation of a system output from observed models.

Also, there are different sources of uncertainty that can influence the prediction of a simulation model. It can arise from simulation variability in stochastic simulation models or from structural uncertainty within assumptions of a model. We will emphasize input uncertainty, what McKay [24] defined as incomplete knowledge of 'correct' values of model inputs, including model parameters. If the inputs of a model are uncertain, there is an inherent variability associated with the output of that model. Therefore it is crucial to communicate it effectively to stakeholders and technical audiences when outputting model predictions.

Uncertainty in environmental prediction simulations may limit the reliability of predicted changes. This issue is one of the recurrent conclusions of the Intergovernmental Panel on Climate Change (IPCC). Back at 1995, IPCC stated that "uncertainties in the simulation of changes in the physical properties have a major impact on confidence in projections of future regional climate change" [13] and that was necessary to reduce uncertainties to increase future model capabilities and improve climate change estimates. Since 2010, IPCC dedicates an integral feature of its reports to the communication of the degree of certainty within IPCC assessment findings [23]. In the most recent report, IPCC assesses a substantially larger knowledge base of scientific, technical and socio-economic literature to reduce uncertainty and uses a large number of methods and formalization [7]. Especially for future predictions, validating a model's predictive accuracy is not straightforward due to a lack of appropriate data and methods for 'validation' [15]. That is another reason why applications, frameworks, and methods of formalization in this research area are relevant and should be promoted.

Regarding the type of modeling, there are approaches such as Bayesian networks, able to explicitly deal with uncertainty in the interpretation of data, measurements or conditions. In contrast, other approaches such as ABMs require the development of comprehensive or compelling analysis of output data and a lot of resource-intensive attention [18]. The level of testing required to develop this understanding is rarely carried out, mainly due to time and other resource constraints [15].

Indeed, uncertainty assessment in ABM can be a hard task for even relatively small models. Due to their inherent complexity, ABS are often seen as black boxes, where there is no purpose in explaining why the agents acted as they did, as long as the modeler presents some form of validation (i.e., shows a good fit). According to Marks [22], ABMs simulations can prove existence, but not in general necessity. Despite that, there is a research effort to make ABS more transparent and to demonstrate that the simulations behave as intended through efforts in standardization in simulation model analysis and result sharing [21]. Besides from verification, uncertainty assessment aims to increase understanding, to improve the reliability of the predicted changes and to inform the degree of certainty of key findings. To achieve this effort, some techniques and methods such as uncertainty and sensitivity analysis should be part of the modeling process.

Uncertainty Quantification is defined as the identification, characterization, propagation, analysis and reduction of uncertainties. Sensitivity analysis (SA) is defined as the study of how uncertainty in the output of a model can be apportioned to different sources of uncertainty in the model input [30] and is a method to assess propagation of uncertainties. SA responds the question of which inputs are responsible for the variability of outputs. Local SA explores the output changes by varying one parameter at a time, keeping all the others constant. Although it is a useful and straightforward approach, it may be location dependent. Global SA gives a better estimate of uncertainty by varying all parameters at the same time by using probability density functions to express the uncertainty of model parameters. Uncertainty analysis is a related broader uncertainty propagation practice to SA. It focuses rather on quantifying uncertainty in model output, addressing the variability of results. Ideally, uncertainty and SA should be run in tandem.

3   Related Work

There are a growing number of attempts to assess uncertainty in ABS. However, there is a lack of specific guidance on effective presentation and analysis of the simulation output data. There is a variety of approaches to quantifying or reduce uncertainty. The work of [18] offers an overview of the state-of-the-art methods on the social simulation area, in particular examining the issues around variance stability, SA and spatiotemporal analysis. Because of our interest in LUCC simulations, we choose to review how those approaches are being applied and communicated on spatially-explicit simulations.

In [1], the authors propose an algorithm as an alternative to goodness-of-fit traditional validation to answer if the agents in a simulation are behaving as expected. To them, the key for effective interaction in multi-agent applications is to reason explicitly about the behavior of other agents, in the form of a hypothesized behavior. This approach would allow an agent to contemplate the correctness of a hypothesis. In the form of a frequentist hypothesis test, the algorithm allows for multiple metrics in the construction of the test statistic and learns its distribution during the interaction process. It is an interesting approach to addressing the uncertainties within the model and agents behavior. We believe it would be even more effective if coupled with an uncertainty quantification technique.

The work of [26] assesses uncertainty that is characteristic of spatially explicit models and simulations. The authors propose a benchmarking scheme of LUCC modeling tools by various validation techniques and error analysis. The authors investigate LUCC tools that are based on map comparisons to analyze the accuracy of LUCC models in terms of quantity, pixel by pixel correctness and LUCC components such as persistence and change. Also, they investigated the map outputs of these simulations to test the fidelity of spatial patterns and the congruency of the simulation maps from different modeling tools. Although the variability of LUCC models does not allow strict comparisons, there is still room for improvements in methodologies, validation and uncertainty quantification.

The work of [8] assesses model output analysis through a global SA, a commonly used approach for identifying critical parameters that dominate model behaviors. They use the Problem Solving environment for Uncertainty Analysis and Design Exploration (PSUADE) software, to evaluate the effectiveness and efficiency of widely used qualitative and quantitative SA methods. Each method is tested using a variety of sampling techniques to screen out the most relevant parameters from the insensitive ones. The Sacramento Soil Moisture Accounting (SAC-SMA) model, which has thirteen tunable parameters, is used for illustration. The South Branch Potomac River basin near Springfield, West Virginia in the U.S. is chosen as the study area. The authors show how different sampling methods and SA measurements can indicate different sensitive and insensitive parameters and that a comprehensive SA is paramount to avoid misleading results.

The work of [20] also performed a global SA to show which model parameters are critical to the performance of land surface models. The authors considered 40 adjustable parameters in The Common Land Model and therefore compare different SA methods and sampling. The size of each sample would vary as well. The sampling techniques and SA measures that were considered optimal were distinct from the results found by [8], meaning that not all LUCC ABS propagate uncertainty the same way.

Another approach was performed by [17], also in a LUCC model. They use the method of independent replication. In the case study, the authors replicated the simulation 12 times for each mechanism and computed the mean values of the impact indicators and their confidence intervals (CI) at a reliability of 95%. They used uncertainty quantification to define a minimum certainty threshold in the simulation outputs.

All these authors used several indicators to measure the variability of model results based on changing input parameters. Table 1 illustrates a brief comparison among those works. MASE exploratory uncertainty assessment will be described in the next sections. A large panel of statistical tools exist to help with the accuracy of the predictions such as Dakota[1], PSUADE [32], UQ-PyL[2], MEME Suite[3] and MC2MABS [2]. There are initiatives to apply the potential of classic Design of Experiments (DOE) for ABS [16,21]. ABS field of research would benefit from a systematic empirical research with standardized procedures, but ABS idiosyncrasies in model output turn the task even harder. Researchers so far failed to reach consensus and to determine sound methodological guidelines. Hence the studies are still mostly investigative and exploratory.

Table 1. Overview of the general characteristics of each related work

Reference	Model	Uncertainty Methods
[1]	Generic ABS	Correctness Hypothesis test and runtime statistical verification in the agent's behavior
[26]	Land use models	Image statistical comparison of pixel/maps and error analysis to find uncertainty drivers
[8]	SAC-SMA hydrological model	Global SA with 15 sampling techniques, 9 different sample sizes and 12 SA methods
[20]	Land surface model	Local SA and 4 Global SA methods with 3 sampling techniques, and 6 sample sizes
[17]	LUDAS: land use ABS	Independent Replications and Confidence Intervals to assess output variation
MASE	MASE: land use ABS	Global SA with different sample configurations, independent replications, and Confidence Intervals

4   MASE Exploratory Study

The MASE Project[4] objective is to define and implement a multi-agent tool for simulating environmental change. MASE enables modeling and simulations of LUCC dynamics using a configurable user model. The multi-agent architecture is composed of three hierarchical layers (from top to bottom) [29]: a User Interface (UI), a Pre-processing and an Agent layer. In the agent layer, there are cell agents representing land units hosting natural processes, such as crop/forest grow, and

---

[1] https://dakota.sandia.gov/.
[2] http://www.uq-pyl.com/.
[3] http://meme-suite.org/.
[4] Software Availability: http://mase.cic.unb.br/.

there are transformation agents, representing human agents and their behavior as farmers or cattle rancher.

The Cerrado-LUCC model of MASE is used as a test problem. The simulations depict the land use and cover changes of the most endangered biome in Brazil. The Cerrado is the second largest biome in South America and harbors significant endemism and biodiversity. The landscape has been undergoing severe transformation due to the advance of cattle ranching and soy production. To promote transparency and replicability, the Cerrado-LUCC simulation model was documented and described employing the standard ODD-protocol (Overview, Design concepts, and Details) [10,11]. We also applied empirically grounding ABM mechanisms for the characterization of agent behaviors and attributes in socio-ecological systems [31]. In this article, we provide some core information of MASE and the Cerrado-LUCC Model, mainly about the parameters and outputs. Readers who are interested in the details of this model and the implementation of MASE multi-agent system should refer to [29] and [28], respectively.

The input of the simulation is a couple of grid raster maps consisting of the land cover of the region, from two different time periods (an initial and a final map). Also, each simulation carries a set of maps to describe the physical characteristics of the environment, such as water courses, water bodies, slope, buildings, highways, environmental protected areas, and territorial zoning maps.

The simulations are calibrated from the two time-steps and project the land use and cover change for future steps. The result of a MASE simulation is a couple of predicted maps (Fig. 1), with the allocation of change and a set of metrics calculated during runtime. The resulting image is submitted to a goodness-of-fit measurement and the quality and errors of the quantity of change and allocation of land use change are calculated.

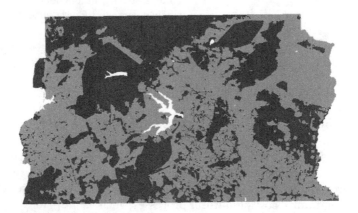

Fig. 1. A land cover predicted map of the Cerrado in Federal District, Brazil

Methodology

The objective is to perform an exploratory analysis, based on classical statistics, to reduce uncertainty and to understand how the model behave. MASE LUCC model is under input uncertainty investigation, to calculate their influence in the simulation output. For exploratory purposes, we want insight on the parameters that affects the multi-agent system implementation, so we selected a subset of Cerrado-LUCC model inputs for this demonstration. The subset of input parameters of the multi-agent system are displayed in Table 2: TA-Number of Transformation Agents, TG- Number of Group Transformation Agents, IE-Potential of Individual Exploration and GE- Potential of Group Exploration. These parameters characterize the instantiation of MASE agents and therefore, should be analyzed regarding uncertainty.

Table 2. MASE multi-agent input configuration parameters

ID	Parameter	Description	Range
I1	TA	Number of Transformation Agents	[1, 100]
I2	TG	Number of Group Transformation Agents	[10, 100]
I3	IE	Potential of Individual Exploration	[1, 500]
I4	GE	Potential of Group Exploration	[1, 1500]

The number of transformation agents is a parameter that reflects the number of computational agents (in the multi-agent system paradigm) instantiated in a simulation run. In this study case, one agent does not represent one single individual. The Cerrado-LUCC model was formulated based on an empirical characterisation of agent behaviors, proposed by [31], with two basic steps: the development of behavioral categories and the scaling to the whole population of agents. TA was derived from the Brazilian Agricultural Census of 2006 and comprises a set of Producer legal status. The range of 1 to 100 is an abstraction to the 3407 register producers in the region that may be active or inactive in a given period. The details of this agent characterization are thoroughly illustrated in [29]. Likewise, a particular type of agent is GT, which represent not an individual but an organization, cooperative, business or so. The range is an abstraction of the 548 group producers, 10 of which have permanent exploration licenses.

The potential of exploration, individual or of a group, represent the impact an agent can produce in the natural vegetation cover of a cell during a step. In the Cerrado LUCC Model, considering the deforestation process, the potential of exploration is again an abstraction for the amount of m^3 of wood that can be obtained from a particular grid cell, until a nominal limit that represents resource depletion.

In addition to the final LUCC maps, the simulation generates a set of metrics as results, mainly spatial analysis measurements, which includes pixel by pixel comparison, a quantitative and an allocation agreement. Those measurements are certain statistical LUCC indices to determine the produced map accuracy, proposed by [27]. It includes an objective function called the figure of merit (FoM), a ratio between correct predicted changes and the sum of observed and predicted changes. To evaluate the response of the model to the different parameters, the experiments considered the outputs described in Table 3 and tried to identify and quantify the influence of the simulation input configurations on the model outputs. The identification (ID) of each of the outputs follows the numbering of its generation in the file .csv produced by MASE at the end of each simulation.

Table 3. MASE output parameters

ID	Output	Description
O1	TM	Total time of the simulation
O4	FoM	Figure of Merit
O5	PA	Image Producer's Accuracy
O6	UA	Image User's Accuracy
O7	WC	*Pixel's Wrong Change*: observed change predicted as persistence
O8	RC	*Pixel's Right Change*: observed change predicted as change
O9	WP	*Pixel's Wrong Persistence*: observed change predicted as persistence

To identify and analyze these uncertainties we performed a method of elementary effects (EE) of global SA on the MASE LUCC model. For this calculation, we used the software package developed by Tong [32] called PSUADE, containing various methods for parameter study, numerical optimization, uncertainty analysis and SA.

Screening methods are based on a discretization of the inputs in levels, allowing a fast exploration of the system behavior [14]. The aim of this type of method is to identify the non-influential inputs with a small number of model calls. The most used screening method is based on the one-parameter-at-a-time (OAT) design, where each input is varied while fixing the others. The simplicity is one of OAT's advantages, but there are drawbacks when applying to ABM. For one, it does not consider parameter interactions and may cover a slight fraction of the input space.

The EE method we chose to apply is the Morris method (MOAT) proposed by [25] and refined by [5], an expansion of the OAT approach that forsakes the strict OAT baseline. It means that a change in one input is maintained when examing a switch to the next input and the parameter set is multiply repeated while randomly selecting the initial parameters settings. EE is suited for spatially explicit simulations, usually computationally expensive models with large input sets.

MOAT allows classifying the inputs into three groups: inputs having a negligible effect, inputs having large linear effects without interactions and inputs having significant non-linear and interaction effects. In overall effect and interaction effect of each parameter can be approximated by the mean μ and standard deviation σ of the gradients of each parameter sampled from r.

The MOAT sampling technique was designed for the particular MOAT method. The work of [8] details how the MOAT sampling works: the range of each parameter is partitioned into $p - 1$ equal intervals. Thus the parameter space is an n-dimension p-level orthogonal grid, where each parameter can take on values from these p determined values.

First, r points are randomly generated from the orthogonal grid; and then, for each of the r points, other sample points are generated by perturbing one dimension at a time. Therefore, sample size will be $(n+1) \cdot r$. For the sampling size, [19] report that one needs at least $10 \cdot n$ samples to identify key factors among the parameters.

To avoid the effect size on the sample, we determining a minimum sample size of $800 (= 20 \cdot 4)$, for four inputs. For MOAT sampling we used 160 replications, resulting in sample size of 800 $(= (4 + 1) \cdot 160)$.

Moreover, as in other stochastic models, it is not advisable to draw conclusions from a single MASE simulation run. For an initial uncertainty assessment, we applied the method of independent replications proposed by [9]. We run the model approximately eighty-five thousand times (an arbitrary choice to explore all the input parameter space) and randomly clustered the results into five independent replication groups. We computed the mean values of the outputs and their confidence intervals (CI) at a reliability of 95%. Another approach to estimating the uncertainty of the model output is to study the variance in the model outputs by using the Coefficient of Variation (CV) (the ratio of the standard deviation σ of a sample to its mean μ), to compare the variance of different frequency distributions.

5   Results

In the current work, we analyzed four input parameters, displayed in Table 2, regarding the multi-agent configuration of MASE LUCC model. First, we present the results of the SA. Figure 2 presents the EE of CERRADO-LUCC model parameters. Figure 2 (left) illustrates the modified means of MOAT gradients and also their spreads based on bootstrapping. The results show that GE and TA are the most sensitive parameters in term of having the largest average median (26.466 and 25.205, respectively). The other two parameters have median sensitivities close to zero, denoting the impact of these parameters on the simulation output is minimal.

Figure 2 (right) is a MOAT diagram that shows a consensus view among mean μ and standard deviation σ of the gradients of each parameter sampled from r. The more sensitive the parameter, the closer it is to the upper right corner of the graph. These results show a positive correlation between input and output uncertainties. Since GE and TA describe the amount of land transformation in a simulation, high values of these parameters will increase the model output.

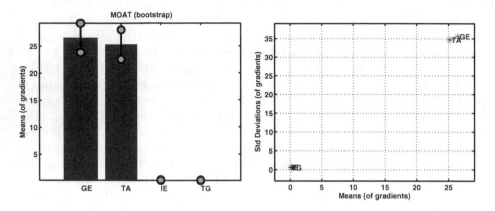

Fig. 2. Parameter sensitivity rankings of MOAT method

GE is the most sensitive parameter, followed by TA. To understand and to reduce uncertainty within this two variables will, therefore, reduce the uncertainty of the simulation as a whole.

GE represents the amount of land cover that is transformed by a group of human agents in a cell of the map. GE is a sensitive value for it indicates the voracity and velocity of the current land exploitation, what will directly affect the result of the simulation. GE is probably sensitive because the socio-economic groups responsible for large-scale cattle ranching and permanent agriculture are the principal driver of deforestation in Cerrado. Their rates of land change are more significance than the number of groups, what explain TG as an insensitive parameter to the output. As for TA, the more agents one instantiates in a simulation, more land cover will be affected, higher will be the land use transformation rates. Conversely, the potential of exploration of a single individual is less determinant than the number of single individuals acting on the land, with SA indicating TA a sensitive and IE as an insensitive parameter.

To investigate MOAT sensitivity results, we used different replications times r and different levels p to know for sure the relevance of the parameters as displayed in Fig. 3. It is possible to see that even within the same method, results may vary. The results for four replications are not very consistent with the other replication results, mainly with the mean. The results with $r = 56$, $r = 108$ and $r = 160$ present minor variations. We can infer that four replications are not enough to identify the parameters sensitivity in the MASE model successfully and therefore the number of replications should be higher to be effective.

Table 4 is a summary of the Basic Output Statistics of the MASE LUCC model. Each replication is assigned by $i = [1..5]$, the sample mean from the coefficient variation by CV_i, and the mean of all replications by $\bar{Z}$. We performed independent replications to verify the variation of the indicators, and for an initial analysis, we consider this variation as noise (uncertainty). Any impact conclusions in predictions can only be drawn if the changes in standards are

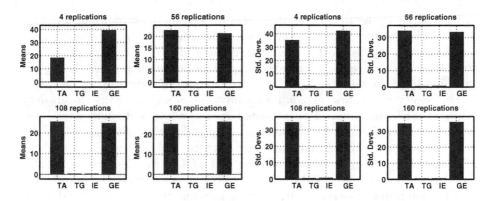

Fig. 3. Sensitivity of parameters at different replication times r

greater than the uncertainty rate. Therefore, we have a first threshold to define if some result is valid, compared to the simulations behavior.

We also estimated the expected average FoM for simulations, using the five replication grouped results ($b = 5$). Considering the $\bar{Z}_{FoM} = 43.87$ and the estimated Variance $\hat{V}_R = 100.99$, we have an approximately $100(1 - \alpha)\%$ two-sided CI for θ, according to the formalization proposed by [9]. For level $\alpha = 0.05$, we have $t_{0.025,4} = 2.78$, and gives $[31.39, 56.34]$ as a 95% CI for the expected FoM for MASE simulations.

Table 4. Coefficient of variation for MASE outputs

Output	CV_1	CV_2	CV_3	CV_4	CV_5	$\bar{Z}$
Time	0.300	0.130	0.250	0.260	0.200	**0.230**
Figure of Merit	0.015	0.011	0.008	0.007	0.090	**0.100**
Producer's Accuracy	0.015	0.011	0.008	0.007	0.009	**0.010**
User's Accuracy	0.006	0.005	0.004	0.004	0.003	**0.004**
Wrong Change	0.030	0.030	0.030	0.030	0.020	**0.030**
Wrong Persistence	0.007	0.007	0.008	0.008	0.013	**0.009**
Right Change	0.015	0.011	0.008	0.008	0.009	**0.010**

6   Conclusions

In this study, we first identified the most sensitive parameters for the MASE LUCC model using MOAT SA. We investigated some proper sampling design and sample size needed for MOAT screening the parameters effectively. Although these conclusions are model-specific, it corroborates possible variation among sampling techniques and SA methods.

This paper is the first exploratory study towards quantifying uncertainty within MASE simulations. Following experiments must be done to promote more standardization to this effort through the application of Design of Experiments. We look forward to investigating further on the model parameters, analyzing the remaining inputs besides the agent's quantities and their impacts.

This paper is the first exploratory study towards quantifying uncertainty within MASE simulations. The presented results allow us to understand the uncertainty when defining the parameters of the simulation of the LUCC model under study. Our feeling is that the uncertainty is very high which means that either model need to dramatically improve or LUCC policy need to be reevaluated. Most simulation tools fail to validate models and to state the uncertainty in simulation results. Consequently, policymakers and the general public develop opinions based on misleading research that fails to give them the appropriate interpretations required to make informed decisions. The efforts to assess ABMs through statistical methods are paramount to corroborate and improve the level of confidence of the research that has been made in LUCC simulation.

References

1. Albrecht, S.V., Ramamoorthy, S.: Are you doing what i think you are doing? criticising uncertain agent models. In: Proceedings of the 31st Conference on Uncertainty in Artificial Intelligence, Amsterdam, Netherlands, p. 10 (2015)
2. Herd, B., Miles, S., McBurney, P., Luck, M.: MC^{2}MABS: a monte carlo model checker for multiagent-based simulations. In: Gaudou, B., Sichman, J.S. (eds.) MABS 2015. LNCS (LNAI), vol. 9568, pp. 37–54. Springer, Cham (2016). https://doi.org/10.1007/978-3-319-31447-1_3
3. Bommel, P.: Foreword. In: Adamatti, D.F. (ed.) Multi-Agent Based Simulations Applied to Biological and Environmental Systems, pp. xv-xviii. IGI Global, Hershey (2017)
4. Coelho, C.C.G., Abreu, C.G., Ramos, R.M., Mendes, A.H.D., Teodoro, G., Ralha, C.G.: MASE-BDI: agent-based simulator for environmental land change with efficient and parallel auto-tuning. Appl. Intell. **45**(3), 904–922 (2016)
5. Campolongo, F., Braddock, R.: The use of graph theory in the sensitivity analysis of the model output: a second order screening method. Reliab. Eng. Syst. Saf. **64**(1), 1–12 (1999), https://doi.org/10.1016/S0951-8320(98)00008-8, http://linkinghub.elsevier.com/retrieve/pii/S0951832098000088
6. Casti, J.L.: Complexification: Explaining a Paradoxical World through the Science of Surprise, reprint edn. HarperCollins (1995)
7. Intergovernmental Panel on Climate Change: Climate Change 2013 - The Physical Science Basis: Working Group I Contribution to the Fifth Assessment Report of the Intergovernmental Panel on Climate Change. Cambridge University Press, Cambridge (2014)
8. Gan, Y., Duan, Q., Gong, W., Tong, C., Sun, Y., Chu, W., Ye, A., Miao, C., Di, Z.: A comprehensive evaluation of various sensitivity analysis methods: a case study with a hydrological model. Environ. Model. Softw. **51**, 269–285 (2014)
9. Goldsman, D., Tokol, G.: Output analysis procedures for computer simulations. In: Joines, J., Barton, R.R., Kang, K., Fishwick, P. (eds.) Proceedings of the 2000 Winter Simulation Conference, pp. 39–45 (2000)

10. Grimm, V., Berger, U., Bastiansen, F., Eliassen, S., Ginot, V., Giske, J., Goss-Custard, J., Grand, T., Heinz, S.K., Huse, G.: A standard protocol for describing individual-based and agent-based models. Ecol. Model. **198**(1–2), 115–126 (2006), http://linkinghub.elsevier.com/retrieve/pii/S0304380006002043
11. Grimm, V., Berger, U., DeAngelis, D.L., Polhill, J.G., Giske, J., Railsback, S.F.: The ODD protocol: a review and first update. Ecol. Model. **221**(23), 2760–2768 (2010), http://linkinghub.elsevier.com/retrieve/pii/S030438001000414X
12. Heath, B., Hill, R., Ciarallo, F.: A survey of agent-based modeling practices (January 1998 to July 2008). JASSS **12**(4), 1–49 (2009)
13. Houghton, J., Filho, L.M., Callander, B., Harris, N., Kattenberg, A., Maskell, K. (eds.): Climate Change 1995 The Science of Climate Change. The Intergovernmental Panel on Climate Change (1996)
14. Iooss, B., Lemaître, P.: A review on global sensitivity analysis methods. In: Dellino, G., Meloni, C. (eds.) Uncertainty Management in Simulation-Optimization of Complex Systems. ORSIS, vol. 59, pp. 101–122. Springer, Boston (2015). https://doi.org/10.1007/978-1-4899-7547-8_5
15. Kelly (Letcher), R.A., Jakeman, A.J., Barreteau, O., Borsuk, M.E., ElSawah, S., Hamilton, S.H., Henriksen, H.J., Kuikka, S., Maier, H.R., Rizzoli, A.E., van Delden, H., Voinov, A.A.: Selecting among five common modelling approaches for integrated environmental assessment and management. Environ. Model. Softw. **47**, 159–181 (2013)
16. Kleijnen, J.P., Sanchez, S.M., Lucas, T.W., Cioppa, T.M.: A user's guide to the brave new world of designing simulation experiments. INFORMS J. Comput. **17**(3), 263–289 (2005), https://harvest.nps.edu/papers/UserGuideSimExpts.pdf
17. Le, Q.B., Seidl, R., Scholz, R.W.: Feedback loops and types of adaptation in the modelling of land-use decisions in an agent-based simulation. Environ. Model. Softw. **27–28**, 83–96 (2012)
18. Lee, J.S., Filatova, T., Ligmann-Zielinska, A., Hassani-Mahmooei, B., Stonedahl, F., Lorscheid, I., Voinov, A., Polhill, G., Sun, Z., Parker, D.C.: The complexities of agent-based modeling output analysis. JASSS **18**(4), 1–25 (2015)
19. Levy, S., Steinberg, D.M.: Computer experiments: a review. AStA Adv. Stat. Anal. **94**(4), 311–324 (2010)
20. Li, J.D., Duan, Q.Y., Gong, W., Ye, A.Z., Dai, Y.J., Miao, C.Y., Di, Z.H., Tong, C., Sun, Y.W.: Assessing parameter importance of the common land model based on qualitative and quantitative sensitivity analysis. Hydrol. Earth Syst. Sci. Discuss. **10**(2), 2243–2286 (2013)
21. Lorscheid, I., Heine, B.O., Meyer, M.: Opening the 'black box' of simulations: increased transparency and effective communication through the systematic design of experiments. Comput. Math. Organ. Theory **18**(1), 22–62 (2012)
22. Marks, R.E.: Validating simulation models: a general framework and four applied examples. Comput. Econ. **30**(3), 265–290 (2007)
23. Mastrandrea, M.D., Field, C.B., Stocker, T.F., Edenhofer, O., Ebi, K.L., Frame, D.J., Held, H., Kriegler, E., Mach, K.J., Matschoss, P.R., Plattner, G.K., Yohe, G.W., Zwiers, F.W.: Guidance note for lead authors of the ipcc fifth assessment report on consistent treatment of uncertainties. In: Intergovernmental Panel on Climate Change (IPCC), pp. 1–7 (2010)
24. McKay, M.D., Morrison, J.D., Upton, S.C.: Evaluating prediction uncertainty in simulation models. Comput. Phys. Commun. **117**(1–2), 44–51 (1999)
25. Morris, M.D.: Factorial sampling plans for preliminary computational experiments. Technometrics **33**(2), 161–174 (1991)

26. Paegelow, M., Camacho Olmedo, M.T., Mas, J.F., Houet, T.: Benchmarking of LUCC modelling tools by various validation techniques and error analysis. Cybergeo **701**(online), 29 (2014)
27. Pontius, R.G., Boersma, W., Castella, J.C., Clarke, K., Nijs, T., Dietzel, C., Duan, Z., Fotsing, E., Goldstein, N., Kok, K., Koomen, E., Lippitt, C.D., McConnell, W., Mohd Sood, A., Pijanowski, B., Pithadia, S., Sweeney, S., Trung, T.N., Veldkamp, A.T., Verburg, P.H.: Comparing the input, output, and validation maps for several models of land change. Ann. Reg. Sci. **42**(1), 11–37 (2008)
28. Ralha, C.G., Abreu, C.G.: A multi-agent-based environmental simulator. In: Adamatti, D.F. (ed.) Multi-Agent Based Simulations Applied to Biological and Environmental Systems, chap. 5, pp. 106–127. IGI Global, Hershey (2017)
29. Ralha, C.G., Abreu, C.G., Coelho, C.G., Zaghetto, A., Macchiavello, B., Machado, R.B.: A multi-agent model system for land-use change simulation. Environ. Model. Softw. **42**, 30–46 (2013)
30. Saltelli, A., Ratto, M., Andres, T., Campolongo, F., Cariboni, J., Gatelli, D., Saisana, M., Tarantola, S.: Global Sensitivity Analysis: The Primer. Wiley (2008)
31. Smajgl, A., Brown, D.G., Valbuena, D., Huigen, M.G.A.: Environmental modelling & software empirical characterisation of agent behaviours in socio-ecological systems. Environ. Model. Softw. **26**(7), 837–844 (2011), https://doi.org/10.1016/j.envsoft.2011.02.011
32. Tong, C.: PSUADE Short Manual (Version 1.7). Lawrence Livermore National Laboratory (LLNL), Livermore, CA (2015)

Developing Multi-agent-based Thought Experiments: A Case Study on the Evolution of Gamete Dimorphism

Umit Aslan[✉], Sugat Dabholkar, and Uri Wilensky

Northwestern University, Evanston, IL, USA
{umitaslan,sugatdabholkar2020}@u.northwestern.edu, uri@northwestern.edu

Abstract. Multi-agent modeling is a computational approach to model behavior of complex systems in terms of simple micro level agent rules that result in macro level patterns and regularities. It has been argued that complex systems approaches provide distinct advantages over traditional equation-based mathematical modeling approaches in the process of scientific inquiry. We present a case study on how multi-agent modeling can be used to develop thought experiments in order to push theory forward. We develop a model of the evolution of gamete dimorphism (anisogamy), for which there are several competing theories in the evolutionary biology literature. We share the outcomes of our model and discuss how the model findings compare with, and contribute to previous work in the literature. The model clarifies mechanisms that can result in the evolution of anisogamy and offers a much simpler structure that is easier to understand, test, modify and extend.

1 Introduction

The most commonly used approach to model behavior of biological systems involves equational modeling with a focus on describing population-level changes based on population level descriptor variables [9]. Unfortunately, this modeling approach is limited when it comes to adding new variables or incorporating new assumptions because entirely new equations might be needed to capture even small changes [24]. In contrast, multi-agent-based modeling is a powerful approach to model complex natural and social phenomena in terms of simple micro-level agent rules that result in the emergence of macro-level patterns and regularities [21]. In this paper, we draw on Wilensky and Papert's Restructuration Theory and argue that multi-agent-based modeling can be used to develop thought experiments on complex scientific questions for novices to learn scientific domain knowledge easily, as well as domain experts to verify, modify, and even extend these models [24].

We present a multi-agent-based model about the evolution of gamete dimorphism (anisogamy) to make a case for our argument. Anisogamy is the phenomenon of males producing large numbers of small sperm cells and females producing small numbers of large egg cells for reproduction [4]. We believe this topic is a

© Springer International Publishing AG 2017
G. Sukthankar and J. A. Rodriguez-Aguilar (Eds.): AAMAS 2017 Best Papers,
LNAI 10642, pp. 51–65, 2017.
https://doi.org/10.1007/978-3-319-71682-4_4

good fit for developing a multi-agent-based thought experiment for two primary reasons: (1) there is no universally accepted theory or model in the literature [4,6,16], (2) the bulk of research in this area has been done through equation-based modeling (e.g., [5,12–14]). We begin by reviewing Restructuration Theory in detail. Then, we describe anisogamy and review the literature related to the evolution of anisogamy, as our multi-agent-based thought experiment incorporates and builds on the ideas from the existing evolutionary biology literature. We describe our model's assumptions and agent rules in detail and then present our findings. We demonstrate that our model achieves similar results to those achieved in the literature while increasing access to underlying ideas.

2   Restructuration of Scientific Domain Knowledge Through Multi-agent-based Modeling

Restructuration Theory, as proposed by Wilensky and Papert, describes how disciplinary knowledge can be re-encoded using new representational technologies in a way that can have powerful implications for science, culture and learning [24]. Many such historical restructurations are presented including the restructuration of Roman numerals to Hindu-Arabic numerals. Wilensky and Papert argue that computation offers many new opportunities for powerful restructurations and that multi-agent-based modeling can be used to create many such restructurations [24].

A good example is Wilensky and Reisman's restructuration of models of predation [22]. Traditionally, predator-prey relationships are modeled through differential equations. An example of such models is the Lotka-Volterra models that offer two equations that describe the rate of change in the densities of the predator and prey populations over time [11,19]:

$$\frac{dN_1}{dt} = b_1 N_1 - k_1 N_1 N_2 \tag{1}$$

$$\frac{dN_2}{dt} = k_2 N_1 N_2 - d_2 N_2 \tag{2}$$

In these equations, N_1 is the density of the prey population, N_2 is the density of the predator population, b_1 is the birth rate of the prey, d_2 is the death rate of the predators, and k_1 and k_2 are constants. These equations specify the dependence of the density of each population to one another. When plotted, the model shows cyclical fluctuations between the two populations: increases in the prey population will result in rising predator birth rates and increases in the predator population will result in rising prey death rates. Wilensky and Reisman's attempt to *restructurate* this problem through multi-agent-based modeling focuses on considering prey and predator as agents and describing the agent rules that emerge as population level patterns:

Rule set for wolves *(at each clock-tick)*:
   1. move randomly to an adjacent patch which contains no wolves.

2. decrease energy by E_1
3. if on the same patch as a sheep, then eat the sheep and increase energy by E_2
4. if energy < 0 then die
5. with probability R_1 reproduce

Rule set for sheep *(at each clock-tick)*:
1. move randomly to an adjacent patch and decrease energy by E_3
2. if on grassy patch, then eat grass and increase energy by E_4
3. if energy < 0 then die
4. with probability R_1 reproduce

Rule set for grass *(at each clock-tick)*:
1. if green, then do nothing
2. if brown, then wait E_4 clock-ticks and turn green

Wilensky and Papert theorize that multi-agent-based restructurations of such natural phenomena offer three powerful advantages over equation-based modeling in terms of learnability: (1) rules for agents are closer to our intuitive notions of these "objects" as distinct individuals rather than aggregate populations, (2)

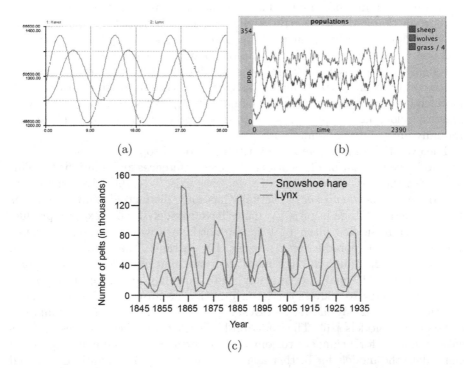

Fig. 1. Two models of predation compared to real world observations: (a) the Lokta-Volterra equational models [11, 19], (b) the Wilensky-Reisman multi-agent-based model (middle) [22], and (c) real world data from a lynx-hare population in Northern Canada [15].

equational models often require bigger changes or completely new equations even for small adjustments, (3) visualization of individual agents and their dynamics afford greater realism compared to graphs of populations [24]. These advantages make it possible for even high school students to easily learn topics that used to be hard for college graduates in related fields [22].

As Fig. 1 shows, a comparison between the real data, the equation-based model, and the multi-agent-based model shows that real world phenomena produce patterns that are more similar to the outcome of the multi-agent-based model. The outcome of the multi-agent-based model is similar to the equation-based model but with more noisy fluctuations, which the equation-based model shows less, because it is a discrete model. In this paper, we attempt a very similar restructuration of an evolutionary biology topic, which is historically studied through equational models, and re-examine it through multi-agent-based modeling.

3   Developing a Multi-agent-based Thought Experiment on the Evolution of Gamete Dimorphism

There are two main types of reproductive strategies employed by organisms: sexual reproduction and asexual reproduction [6]. The most prevalent sexual reproduction strategy is called gamete dimorphism or anisogamy. Many animal and plant species, including humans, are anisogamous: one mating type (males) provides half the chromosomes by producing small cells in large quantities (sperm) and the other mating type (females) provides half the chromosomes by producing much larger cells in much smaller numbers (egg). When two such cells, called gametes, belonging to opposite sexes fuse, a zygote is formed and this zygote gradually grows into an adult [3,14].

The evolution of anisogamy is a yet to be resolved topic in evolutionary biology and is the foundation of theories on gender differences and relations [4]. This starts with the very question of "why do sexes exist?" [4,16]. Given that asexual production (*parthenogenesis*) actually has some distinct advantages in terms of numerical advantage in progeny, many have wondered why sexual reproduction evolved in the first place [6]. It is also not known why anisogamy prevailed over other sexual reproduction strategies. For instance, there are some fungal species which reproduce through more than two mating types [10] or by producing gametes of equal size (isogamy) [14], but they are exceptions. In this paper, we attempt to address the latter question because the discussion on the evolution of anisogamy mostly revolves around the validity of the assumptions of theoretical models [16]. The equation-based methods used in these models make it harder for beginners to join the conversation and domain experts to manipulate the models for further analysis. We argue that a multi-agent-based thought experiment of anisogamy can afford domain experts the ability to easily plug new assumptions into an existing model while making it significantly easier for non-experts to learn about anisogamy [22]. In this section, we describe the process of developing one such thought experiment through reviewing the literature on anisogamy, determining model assumptions, defining agent rules and designing the user interface.

3.1 Literature Review

Evolutionary theories in general try to show how it is that a trait might be selected when there are many competing treats. In the case of reproductive strategies, there is no clear answer on why anisogamy is a more successful strategy over isogamy or multiple mating types. The most accepted theory on the evolution of gamete dimorphism is called "the Parker-Baker-Smith (PBS) model". It lays out mathematical formulations to determine the conditions for the evolution of anisogamy through a *zygotic fitness function* and a *gametic fitness function*. The PBS model makes three simple but powerful assumptions [5,14]:

1. individuals of a marine ancestor population produce a range of gametes and the fusion between pairs of gametes is at random at sea
2. each adult has only a fixed biomass available for gamete production
3. there is some sort of relationship between zygote fitness and zygote size

It is important to caution that we are far from having a model that offers a universal explanation yet. Many of these theories, including the PBS model, are actively debated [16] and there are still many questions that remain unanswered [4]. The PBS theory of evolution is generally viewed as a foundational model but not the ultimate answer [6]. Both the assumptions and the formulations of the model are challenged by other theorists [4,16]. There are also many theories that build on the PBS model and attempt to offer more explanatory value (e.g., [7]).

3.2 The NetLogo Model of Gamete Dimorphism

We develop our multi-agent-based thought experiment of anisogamy in the NetLogo agent-based modeling environment [20] as it provides powerful tools to model emergent phenomena through a beginner friendly programming environment that allows writing open, easily readable code and a rich set of visualization options [17,21]. In the model [1] [1], adults of two mating types begin with producing middle-sized gametes at approximately the same rate (isogamy). Every time an adult produces new gametes, there is a chance of a small, random mutation in the gamete size strategy. These mutations introduce a competition among multiple reproductive strategies. In this section, we describe the model's assumptions, agent rules and interface in detail.

---

[1] Source code of the NetLogo model of anisogamy is openly available through http://modelingcommons.org/browse/one_model/5007.

The Assumptions and the Agent Rules. Similar to the existing theories in the literature, our model builds on the following set of basic assumptions that we appropriated from the PBS model and its derivatives [3–5, 12–14, 16, 18]:

1. Adults have limited lifetimes.
2. Gamete production budget is fixed and the same for all adults.
3. Gametes have limited lifetimes, too, but much shorter than adults.
4. A zygote has to achieve a minimum mass to survive.
5. There are initially two isogamous mating types in the population.
6. The gamete size and the mating type traits are inherited as a bundle.
7. The chance of a zygote inheriting these traits from either gamete is equal.

Assumptions 2 and 4 directly correspond to the 2nd and 3rd assumptions of the PBS model (Sect. 3.1). We implement the 1st assumption of the PBS model by implementing a random walk algorithm in the model's code. We also implement a lifetime mechanism to simulate successive generations, although there is no mention of this in the PBS model or other equational models. Based on these assumptions, we define three agent types as *adults*, *gametes* and *zygotes* and define simple rules for each agent type.

Rule set for adults *(at each clock-tick)*:
    1. turn around randomly and move one step forward.
    2. with probability P produce gametes:
    – randomly pick the new gametes' size (m_t) through a normal distribution with *mean = my gamete size strategy* (m) and *standard deviation = σ*.
    – hatch *own mass* $(M)/m_t$ gametes of my mating-type and of the size m_t
    3. decrease the remaining lifetime by 1, die if no lifetime left.

Rule set for gametes *(at each clock-tick)*:
    1. turn around randomly and move one step forward.
    2. fuse (form a zygote) if touching a gamete of the opposite sex:
    – inherit the total mass of myself and my mating partner.
    – randomly inherit the mating type and gamete size strategy as a package
    3. decrease the remaining lifetime by 1, die if no lifetime left.

Rule set for zygotes *(at each clock-tick)*:
    1. decrease the remaining incubation time by 1. if incubation time is 0:
    – if *own mass* (M) mass is greater than the survival threshold $(M \geq \delta)$, turn into an adult.
    – if *own mass* (M) is less than the survival threshold $(M < \delta)$, die.

Interface and Parameters. NetLogo's interface affords easy manipulation of the parameters of the model, and we can observe the changes in the system visually through the model's world and plots. The *world* is a graphical window which is not a mere visualization but an actual space where the agents follow the rules and interact with each other [21], seen as the central window shown in Fig. 2. The adults are represented by circles with black dots in them. An adult's

color (blue or red) represents its mating type. The tiny arrow shaped agents
are the gametes produced by adults. They, too, are either blue or red but vary
in size depending on their parents' gamete size strategy. Lastly, the egg-shaped
agents with lighter shades of red and blue are the zygotes formed by the fusion
of two gametes.

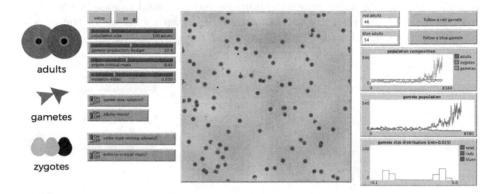

Fig. 2. The interface of the NetLogo Anisogamy model (Color figure online)

The first two plots on the right allow us to see the change in the overall
population and the number of gametes of each mating type over time. The
histogram on the bottom right shows the distribution of the gamete sizes at
the observed clock-tick. The two integer outputs on the top right (blue adults
and red adults) allow us to observe if the mating type balance is disrupted or
not. The controls on the left allow us to change the parameters of the model
so that we can test implicit and explicit assumptions. Each of these controls
corresponds to bigger questions that we want to ask through this model. For
example, one of the questions we want to ask is *"what, if any, thresholds of
zygote critical mass effect the potential evolution of anisogamy"*, so we implement
a *ZYGOTE-CRITICAL-MASS* slider that determines the threshold of mass that a
zygote needs to achieve to survive. Similarly, we want to investigate whether the
assumption of differentiation in mating types is viable, so we place the *SAME-
TYPE-MATING-ALLOWED?* switch.

4   Findings and Discussion

A comparison of our multi-agent-based model and equation-based models of
anisogamy highlights the advantages of multi-agent-based thought experiments.
In this section, we first share the outcomes of our model with the default
parameter-set, which corresponds to our basic set of assumptions (see Sect. 3.2).
In this condition, we run the model with approximately 100 adults in a confined
space. The average lifetime is 500 clock-ticks for adults and 50 clock-ticks for
gametes. Because the model's space is 256 square unit-lengths and computing

power is limited, we implement a carrying capacity mechanism. Whenever the model's adult population exceeds 100 members, some adults are randomly taken out of the population. This does not apply to gametes or zygotes. All adults are of 1 unit-length, mass of 1 unit-mass, and they move around randomly with the speed of 1 unit-length per clock-tick. Adults can use half of their mass for producing gametes. Initially, all adults have the same reproductive strategy of producing two middle sized gametes. Gametes move around randomly with the same speed, too, and they are only allowed to fuse with gametes of the opposite mating type. Lastly, the critical threshold for a zygote to survive is 0.45 unit-mass.

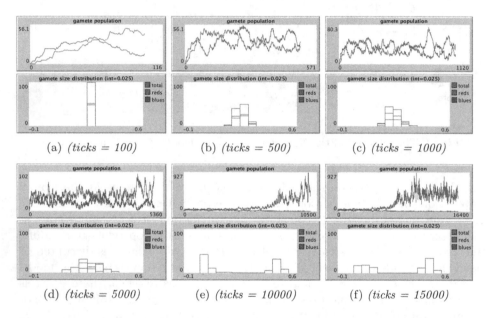

Fig. 3. The emergence of gamete dimorphism over time in the multi-agent model (Color figure online)

Figure 3 shows the outcome of a typical run with the default parameters. Each subfigure consists of two plots: a plot showing the change in the number of red versus blue gametes over time and a plot showing the distribution of gamete sizes at the presented clock-tick. Gametes of the two mating-types are represented with red and blue colored lines and bars in the graphs. As our model assumes that the reproduction budget is fixed for all the adults, a large gamete number means smaller gamete size, and vice versa. Figure 3a is a snapshot of the model after 100 clock-ticks and the subsequent subfigures are after 500, 1000, 5000, 10000 and 15000 clock-ticks. In our model, 15000 clock-ticks correspond to approximately 300 generations. This might be extremely small for such an evolutionary process in real life but in the small world of our thought experiment, it is enough to observe meaningful and consistent results.

As the first four subfigures show, the model starts with oscillations between two similar strategies. In this specific run, a stochastic disruptive event happens at about 7000 clock-ticks (Fig. 3e) resulting in one mating type getting committed to producing big gametes and the other to producing small gametes. In other words, anisogamy evolves and is sustained. Figure 4 presents the results of 300 runs with this default parameter-set over 20000 clock-ticks . Each data point presents the average number of red or blue gametes in the last 5000 clock-ticks, which provides more reliable data because the number of gametes in the model oscillates continuously. We clearly observe evolution of two distinct gamete size strategies at the end of each simulation run (Fig. 4). Statistical analysis of this data shows that there was a significant difference between the number of large gametes ($m = 1.735, sd = 0.184$) and the number of small gametes ($m = 437.675, sd = 19.505$); $t(299) = -384.221, p < 0.0005$. These findings provide a theoretical explanation of not only why but also how anisogamy might have evolved, as well as supporting previous theory on the instability of isogamy in the long run [18].

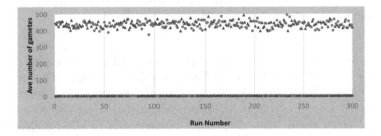

Fig. 4. Testing the model with default parameters (n = 300, ticks = 20000).

The affordances of multi-agent-based thought experiments become even more noticeable when it comes to testing assumptions of a model to answer *"what if?"* questions. In the following sections, we test an explicit and an implicit assumption of the PBS model, as well as another non-PBS assumption that is common in the literature. We not only show the ease of doing this through our model but also demonstrate how powerful the outcomes of such assumption tests can be.

4.1   Zygote Survival as a Function of Zygote Mass

We begin testing assumptions with one of the main assumptions of the PBS model concerning the relationship between viability of a zygote to its size [3, 5, 14]. We call this the ZYGOTE-CRITICAL-MASS assumption, which can be turned on and off easily with a switch on the models interface (see Fig. 2). With the default parameter-set of the model, we observe the emergence of anisogamy after 10000 ticks. We keep all the other parameters the same, but allow zygotes to survive regardless of their mass and run the model again. As seen in Fig. 5b, the gamete sizes and gamete population for both sexes fluctuate over time with the overall direction of reduction in the size. Anisogamy does not evolve when each zygote survives regardless of its mass.

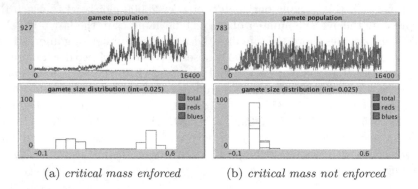

(a) *critical mass enforced*   (b) *critical mass not enforced*

Fig. 5. The comparison of the model outcomes with ENFORCE-CRITICAL-MASS? switch turned on and off. (Color figure online)

We also conducted an experiment running the model starting with 0.0 as the value of the ZYGOTE-CRITICAL-MASS variable and then incrementing it by 0.01 until 0.5 over 20000 ticks. For each value, we ran the model 3 times, so we ended up with a total of 150 experiments. Figure 6a shows the results of this experiment. Once again, each data point corresponds to the running average of the number of gametes in the last 5000 ticks of each run. The most important outcome of this test is the fact that anisogamy did not evolve and isogamy was sustained when the value of the ZYGOTE-CRITICAL-MASS parameter was below 0.1, which is consistent with the assumptions of the PBS model [5,14]. Surprisingly, we also noticed some runs which did not result in anisogamy between the range of 0.3 and 0.45. We hypothesized that anisogamy would still evolve in this parameter space in a longer experiment. Accordingly, we conducted the same experiment but this time over 200000 clock-ticks and the results confirmed our hypothesis (Fig. 6). Once again, these findings align with the PBS model's assumption that *for anisogamy to evolve, some sort of a relationship between the zygote size and zygote survival is necessary* [6,14].

4.2   Mating Types

Another affordance of multi-agent-based thought experiments is the possibility of testing implicit assumptions. For instance, the existence of two mating types is a common assumption in many models of anisogamy, but it is rarely discussed explicitly (e.g., [5]). Our model assumes two mating types, too, but it is actually possible to test this assumption indirectly by allowing gametes of the same mating type to fuse. When we run the model with this alternative

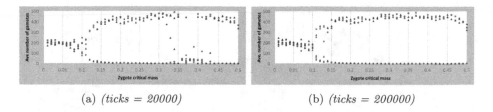

(a) *(ticks = 20000)*   (b) *(ticks = 200000)*

Fig. 6. Testing the model with a range of ZYGOTE-CRITICAL-MASS values between 0 and 0.5 (n = 150).

assumption, we observe that anisogamy does not evolve. Instead, there are two possible outcomes. In most of the runs, genetic drift [8] happens and one mating type prevails over the other (Figs. 7a and c). However, in some rare occasions, we observe almost no quantitative change in the population composition (Fig. 7b) because, by random chance, it takes more time for genetic drift to emerge in some runs (as in Sect. 4.1). These results provide support for the implicit assumption that mating types are required for anisogamy to evolve. On the other hand, our model currently does not allow testing the possibility of more than two mating types. This could be an interesting follow up on our test, and it is possible to do it with a few changes in the model's code.

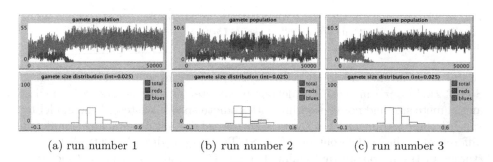

(a) run number 1   (b) run number 2   (c) run number 3

Fig. 7. The outcome of the model when fusion between two gametes of the same mating type is allowed *(ticks = 50000)* (Color figure online)

4.3   Adult and Gamete Motility

Another debated topic in models of anisogamy is the role of gamete and/or adult motility in the marine environment [4,16]. Some of the models assume that the speed of a gamete is inversely related to its mass according to Stokes Law [7], while others challenge the validity of this assumption [16]. As the actual physics of locomotion in water is somewhat complex, our point is to test whether a relationship of this sort is needed for the evolution of anisogamy.

Our model allows us to (1) make all the gametes move with the same speed or *with a variable speed that is inversely related to a gamete's size*, and (2) make

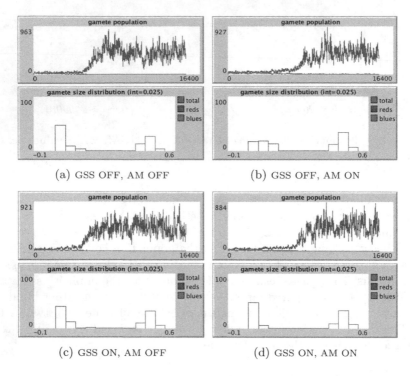

Fig. 8. The outcomes of the model when gamete-speed size relation (GSS) and adult motility (AM) assumptions are tested *(ticks = 15000)* (Color figure online)

adults move around randomly with the same speed or *remain stationary* (see Sect. 3.2). In our runs with the default parameter-set, the adults were moving and gamete size had no relationship with gamete speed. We tested the model by varying these parameters but to our surprise, we did not observe any significant differences in the model's outcome (Fig. 8). This finding directly contradicts some studies in the literature that claim that gamete motility is a critical factor in the evolution of anisogamy (e.g., [7,13]).

4.4   A Qualitative Comparison Between the Two Models of Anisogamy

In this section, we present a "relational alignment" [2,23] between our multi-agent-based model and the equational PBS model developed by Bulmer and Parker by qualitatively comparing the relationships between critical parameters of these two models and the evolution of anisogamy as a continuously stable strategy (ESS [12]). These critical parameters are gamete size (m), zygote size (S), and parameters that determine viability of gametes (α) and zygotes (β). Figure 9 shows two plots from Bulmer and Parker's mathematical formulation of the PBS model. Figure 9a is concerned with the conditions that result with anisogamy as ESS and Fig. 9b is concerned with a critical threshold for zygote survival in an anisogamous ESS [5].

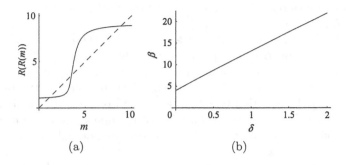

Fig. 9. Plots from Bulmer and Parker's equational PBS model of anisogamy: (a) anisogamy as ESS for given m and β values and (b) the critical value of β above which anisogamy evolves as a function of δ [5].

Bulmer and Parker use the PBS model to explore the parameter space for the parameters β and δ to find a parameter range over which anisogamy would evolve as an evolutionary stable strategy (Fig. 9b). β is a parameter that determines the shape of the response strategy function and δ is a parameter related to the gamete critical mass. In our multi-agent model of evolution of anisogamy, we demonstrate that anisogamy evolves as an ESS as reliably over the default parameter range (Fig. 4). We have also demonstrated that our multi-agent-modeling approach to evolution of anisogamy using NetLogo as a modeling environment allows as such comparison where we have investigated the parameter range for zygote-critical-mass (Fig. 6). Hence, these two models are qualitatively similar, or relationally aligned, in terms of inputs (conditions) and outputs (evolution of anisogamy as an ESS).

5   Conclusions

We argued that multi-agent-based models can be used to express scientific domain knowledge in the form of thought experiments. As a case study, we developed a multi-agent-based thought experiment on the evolution of anisogamy, which is the phenomenon of male species producing numerous small sperm cells and female species producing only a handful of large egg cells for reproductive purposes. We noted that anisogamy is a topic in evolutionary biology with direct implications on the evolution of animal and plant species, but it is yet to be resolved. We reviewed the evolutionary biology literature and developed a model in the NetLogo agent-based modeling environment building on a set of assumptions that we adopted from previously research.

Our model provided similar results to the equation-based models of anisogamy but allowed us to easily test explicit and implicit assumptions suggested by previously offered theories. For example, we were able to confirm that the existence of two mating types is a necessary prerequisite for anisogamy to emerge, and we showed why anisogamy does not evolve when any two gametes can fuse with each other [5,14]. On the other hand, we found no evidence of a possible relationship between adult or gamete speeds with the evolution of anisogamy [7,13].

Our study demonstrates that multi-agent-based thought experiments can allow scientists and theorists to explore a wide range of subtle and difficult *"what if"* questions. One can think of a new question and almost immediately manipulate the model to answer it. Even a strong mathematician may not be comfortable changing the equation-based models of anisogamy, but making changes in our multi-agent-based model of anisogamy is almost *mind-to-fingers*. More importantly, our model provides such opportunities not only to scientists but also to informed citizens and younger students without having to master all the formal mathematics. We argue that such multi-agent-based restructurations would make scientific domain knowledge more accessible for a wider population and speed up the progress in currently unresolved topics like the evolution of anisogamy.

5.1 Limitations

It is important to note that the outcomes of our model are by no means definitive as it is the case for all the other theoretical and equational models in the literature [4,6]. Because our goal was to primarily demonstrate the advantages of multi-agent-based thought experiments, we left out some theoretical considerations in this paper such as the possibility of more than two mating types existing in the population or a more comprehensive comparison between our model and the PBS model [5,14]. In future studies, we hope to focus, in greater depth, on the theoretical implications of our model for the field of evolutionary biology. We also hope to conduct research which explores the use of this multi-agent-based thought experiment and similar approaches in educational settings.

Acknowledgments. This work was made possible through generous support from the National Science Foundation (grants CNS-1138461 and CNS 1441041) and the Spencer Foundation (Award #201600069). Any opinions, findings, or recommendations expressed in this material are those of the author(s) and do not necessarily reflect the views of the funding organizations.

References

1. Aslan, U., Dabholkar, S., Wilensky, U.: NetLogo Anisogamy model. http://ccl.northwestern.edu/netlogo/models/Anisogamy. Center for Connected Learning and Computer-Based Modeling, Northwestern University, Evanston, IL (2016)
2. Axtell, R., Axelrod, R., Epstein, J.M., Cohen, M.D.: Aligning simulation models: a case study and results. Comput. Math. Organ. Theor. **1**, 123–141 (1996)
3. Bell, G.: The evolution of anisogamy. J. Theor. Biol. **73**, 247–270 (1978)
4. Blute, M.: The evolution of anisogamy: more questions than answers. Biol. Theor. **7**, 3–9 (2013)

5. Bulmer, M.G., Parker, G.A.: The evolution of anisogamy: a game-theoretic approach. Proc. R. Soc. Lond. B Biol. Sci. **269**, 2381–2388 (2002)
6. Cox, P.A.: The evolutionary mystery of gamete dimorphism. In: The Evolution of Anisogamy: A Fundamental Phenomenon Underlying Sexual Selection, pp. 1–16 (2011)
7. Cox, P.A., Sethian, J.A.: Gamete motion, search, and the evolution of anisogamy, oogamy, and chemotaxis. Am. Nat. **125**(1), 74–101 (1985)
8. Dennett, D.C.: Darwin's Dangerous Idea: Evolution and the Meanings of Life. Simon and Schuster, New York (1996)
9. Hastings, A.: Population Biology: Concepts and Models. Springer Science & Business Media, New York (2013)
10. Kues, U., Casselton, L.A.: The origin of multiple mating types in mushrooms. J. Cell Sci. **104**(2), 227–230 (1993)
11. Lotka, A.J.. Elements of physical biology. Dover, New York (1925)
12. Maynard-Smith, J.: The Evolution of Sex. Cambridge University Press, London (1978)
13. Parker, G.A.: Selection on non-random fusion of gametes during the evolution of anisogamy. J. Theor. Biol. **73**(1), 1–28 (1978)
14. Parker, G.A., Baker, R.R., Smith, V.G.F.: The origin and evolution of gamete dimorphism and the male-female phenomenon. J. Theor. Biol. **36**(3), 529–553 (1972)
15. Purves, W., Orians, G., Heller, H.: Life: The Science of Biology, 3rd edn. Sinauer Associates, Sunderland (1992)
16. Randerson, J.P., Hurst, L.D.: The uncertain evolution of the sexes. Trends Ecol. Evol. **16**(10), 571–579 (2001)
17. Tisue, S., Wilensky, U.: NetLogo: A Simple Environment for Modeling Complexity. Paper presented at the International Conference on Complex Systems. Boston (2004)
18. Togashi, T., Cox, P.A.: The Evolution of Anisogamy: A Fundamental Phenomenon Underlying Sexual Selection. Cambridge University Press, New York (2011)
19. Volterra, V.: Fluctuations in the abundance of a species considered mathematically. Nature **118**, 558–560 (1926)
20. Wilensky, U.: NetLogo (1999). http://ccl.northwestern.edu/netlogo/
21. Wilensky, U.: Modeling nature's emergent patterns with multi-agent languages. In: Proceedings of Eurologo. Linz, Austria (2001)
22. Wilensky, U., Reisman, K.: Thinking like a wolf, a sheep, or a firefly: learning biology through constructing and testing computational theories - an embodied modeling approach. Cogn. Instr. **24**(2), 171–209 (2006)
23. Wilensky, U., Rand W.: Making models match: replicating an agent-based model. J. Artif. Soc. Soc. Simul. **10**(4) (2007). http://jasss.soc.surrey.ac.uk/10/4/2.html
24. Wilensky, U., Papert, S.: Restructurations: reformulations of knowledge disciplines through new representational forms. In: Proceedings of Constructionism. Paris, France (2010)

Cooperative Multi-agent Control Using Deep Reinforcement Learning

Jayesh K. Gupta$^{(\boxtimes)}$, Maxim Egorov, and Mykel Kochenderfer

Stanford University, Stanford, USA
jkg@cs.stanford.edu, {megorov,mykel}@stanford.edu

Abstract. This work considers the problem of learning cooperative policies in complex, partially observable domains without explicit communication. We extend three classes of single-agent deep reinforcement learning algorithms based on policy gradient, temporal-difference error, and actor-critic methods to cooperative multi-agent systems. To effectively scale these algorithms beyond a trivial number of agents, we combine them with a multi-agent variant of curriculum learning. The algorithms are benchmarked on a suite of cooperative control tasks, including tasks with discrete and continuous actions, as well as tasks with dozens of cooperating agents. We report the performance of the algorithms using different neural architectures, training procedures, and reward structures. We show that policy gradient methods tend to outperform both temporal-difference and actor-critic methods and that curriculum learning is vital to scaling reinforcement learning algorithms in complex multi-agent domains.

1 Introduction

Cooperation between several interacting agents has been well studied [1–3]. While the problem of cooperation can be formulated as a decentralized partially observable Markov decision process (Dec-POMDP), exact solutions are intractable [4,5]. A number of approximation methods for solving Dec-POMDPs have been developed recently that adapt techniques ranging from reinforcement learning [6] to stochastic search [7]. However, applying these methods to real-world problems is challenging because they are typically limited to discrete action spaces and require carefully designed features.

On the other hand, recent work in single agent reinforcement learning has enabled learning in domains that were previously thought to be too challenging due to their large and complex observation spaces. This line of work combines ideas from deep learning with earlier work on function approximation [8,9], giving rise to the field of deep reinforcement learning. Deep reinforcement learning has been successfully applied to complex real-world tasks that range from playing Atari games [10] to robotic locomotion [11]. The recent success of the field leads to a natural question—how well can ideas from deep reinforcement learning be applied to cooperative multi-agent systems?

© Springer International Publishing AG 2017
G. Sukthankar and J. A. Rodriguez-Aguilar (Eds.): AAMAS 2017 Best Papers,
LNAI 10642, pp. 66–83, 2017.
https://doi.org/10.1007/978-3-319-71682-4_5

In this work, we focus on problems that can be modeled as Dec-POMDPs. We extend three classes of deep reinforcement learning algorithms: temporal-difference learning using Deep Q Networks (DQN) [10], policy gradient using Trust Region Policy Optimization (TRPO) [12], and actor-critic using Deep Deterministic Policy Gradients (DDPG) [13] and A3C [14]. We consider three training schemes for multi-agent systems based on centralized training and execution, concurrent training with decentralized execution, and parameter sharing during training with decentralized execution. We incorporate curriculum learning [15] into cooperative domains by first learning policies that require a small number of cooperating agents and then gradually increasing the number of agents that need to cooperate. The algorithms and training schemes are benchmarked on four multi-agent tasks requiring cooperative behavior. The benchmark tasks were chosen to represent a diverse variety of complex environments with discrete and continuous actions and observations.

Our empirical evaluations show that multi-agent policies trained with parameter sharing and an appropriate choice of reward function exhibit cooperative behavior without explicit communication between agents. We show that the multi-agent extension of TRPO outperforms all other algorithms on benchmark problems with continuous action spaces, while A3C has the best performance on the discrete action space benchmark. By combing curriculum learning and TRPO, we demonstrate scalability of deep reinforcement learning in large, continuous action domains with dozens of cooperating agents and hundreds of agents present in the environment. To our knowledge, this work presents the first cooperative reinforcement learning algorithm that can successfully scale in large continuous action spaces. The benchmark problems and the implementations of multi-agent algorithms can be found at https://github.com/sisl/MADRL.

2   Related Work

Multi-agent reinforcement learning has a rich literature [2, 16]. A number of algorithms involve value function based cooperative learning. Tan compared the performance of cooperative agents to independent agents in reinforcement learning settings [1]. Ono and Fukumoto identified modularity as a useful prior to simplify the application of reinforcement learning methods to multiple agents [17]. Guestrin et al. later extended this idea and factored the joint value function into a linear combination of local value functions and used message passing to find the joint optimal actions [18]. Lauer and Riedmiller tried distributing the value function into learning multiple tables but failed to scale to stochastic environments [19].

Policy search methods have found better success in partially observable environments [20]. Peshkin et al. studied gradient based distributed policy search methods [21]. Our solution approach can be considered a direct descendant of the techniques introduced in their work. However, instead of using finite state machines, our model uses deep neural networks to control the agents. This approach allows us to extend neural network controllers to tasks with continuous actions, use deep reinforcement learning optimization techniques, and consider more complex observation spaces.

Relatively little work on multi-agent reinforcement learning has focused on continuous action domains. A few notable approaches include those of Fernández and Parker who focus on discretization and Tamakoshi and Ishii who used a normalized Gaussian Network as a function approximator to learn continuous action policies [22,23]. Many of these approaches only work in fairly restricted settings and fail to scale to high-dimensional raw observations or continuous actions. Moreover, their computational complexity grows exponentially with the number of agents.

Multi-agent control has also been studied in extensive detail from the dynamical systems perspective in problems like formation control [24], coverage control [25], and consensus [26]. The limitations of the dynamical systems approach lie in its requirement for hand-engineered control laws and problem specific features. While the approach allows for development of provable characteristics about the controller, it requires extensive domain knowledge and hand engineering. Overall, deep reinforcement learning provides a more general way to solve multi-agent problems without the need for hand-crafted features and heuristics by allowing the neural network to learn those properties of the controller directly from raw observations and reward signals.

Recent research has applied deep reinforcement learning to multi-agent problems. Tampuu et al. extended the DQN framework to independently train multiple agents [27]. Specifically, they demonstrate how collaborative and competitive behavior can arise with the appropriate choice of reward structure in a two-player Pong game. More recently, Foerster et al. and Sukhbaatar et al. train multiple agents to learn a communication protocol to solve tasks with shared utility [28,29]. They demonstrate end-to-end differentiable training using novel neural architectures. However, these examples work with either relatively few agents or simple observations and do not share our focus on decentralized control problems with high-dimensional observations and continuous action spaces.

3   Background

In this work, we consider multi-agent domains that are fully cooperative and partially observable. All agents are attempting to maximize the discounted sum of joint rewards. No single agent can observe the state of the environment. Instead, each agent receives a private observation that is correlated with that state. We assume the agents cannot explicitly communicate and must learn cooperative behavior only from their observations.

Formally, the problems considered in this work can be modeled as Dec-POMDPs defined by the tuple $(\mathcal{I}, \mathcal{S}, \{\mathcal{A}_i\}, \{\mathcal{Z}_i\}, T, R, O)$, where $\mathcal{I}$ is a finite set of agents, $\mathcal{S}$ is a set of states, $\{\mathcal{A}_i\}$ is a set of actions for each agent i, $\{\mathcal{Z}_i\}$ is a set of observations for each agent i, and T, R, O are the joint transition, reward, and observation models, respectively. In this work, we consider problems where $\mathcal{S}$, $\mathcal{A}$, and $\mathcal{Z}$ can be infinite to account for continuous domains. In the reinforcement learning setting, we do not know T, R, or O, but instead have access to a generative model. It is natural to also consider a centralized model known as a multi-agent POMDP (MPOMDP), with joint action and observation models. The centralized nature of MPOMDPs makes them less effective at scaling to systems with many agents.

In the reminder of the section, we briefly describe four single-agent deep reinforcement learning algorithms, including temporal-difference, actor-critic, and policy gradient approaches. We also discuss the roles of reward shaping and curriculum learning in multi-agent settings.

3.1   Deep Q-Network

The DQN algorithm [10] is a temporal-difference method that uses a neural network to approximate the state-action value function. DQN relies on an experience replay dataset $\mathcal{D}_t = \{e_1, \ldots, e_t\}$, which stores the agent's experiences $e_t = (s_t, a_t, r_t, s_{t+1})$ to reduce correlations between observations. The experience consists of the current state s_t, the action the agent took a_t, the reward it received r_t, and the state it transitioned to s_{t+1}. The learning update at each iteration i uses a loss function based on the temporal-difference update:

$$L_i(\theta_i) = \mathbb{E}_{(s,a,r,s')\sim\mathcal{D}}\left[(r + \gamma\max_{a'} Q(s',a';\theta_i^-) - Q(s,a;\theta_i))^2\right]$$

where θ_i and θ_i^- are the parameters of the Q-networks and a target network respectively at iteration i, and the experience samples (s, a, r, s') are sampled uniformly from $\mathcal{D}$. In partially observable domains where only observations o_t are available at time t instead of the entire state s_t, the experience takes the form $e_t = (o_t, a_t, r_t, o_{t+1})$. One of the limitations of DQN is that it cannot easily handle continuous action spaces.

3.2   Deep Deterministic Policy Gradient

DDPG combines the actor-critic and DQN approaches to learn policies in domains with continuous actions. DDPG maintains a parameterized actor function $\mu(s \mid \theta^\mu)$, which deterministically maps states to actions while learning a critic $Q(s, a)$ that estimates the value of state-action pairs. The actor can be updated with the following optimization step:

$$\nabla_{\theta^\mu} J \approx \mathbb{E}_{s_t\sim\rho_\pi}[\nabla_a Q(s, a \mid \theta^Q)|_{s=s_t, a=\mu(s_t)} \nabla_{\theta_\mu}\mu(s \mid \theta^\mu)|_{s=s_t}]$$

where ρ_π are transitions generated from a stochastic behavior policy π, typically represented with a Gaussian distribution centered at $\mu(s \mid \theta^\mu)$.

3.3   Asynchronous Advantage Actor Critic

Asynchronous Advantage Actor Critic (A3C) [14] consists of global shared networks for policy $\pi(a \mid s, \theta_p)$ and value $V(s, \theta_v)$ functions. Multiple copies running independently accumulate gradients in parallel to asynchronously update this network. The policy gradients are given by:

$$\nabla_{\theta_p} \log \pi(a_t \mid s_t; \theta_p) A(s_t, a_t; \theta_v)$$

where the advantage function $A(s_t, a_t; \theta_v)$ is computed from difference between returns from n-step rollout and value function output.

The value network loss function is to minimize squared error of value function outputs from environment returns.

3.4   Trust Region Policy Optimization

TRPO [12] is a policy gradient method that allows precise control of the expected policy improvement during the optimization step. At each iteration k, TRPO aims to solve the following constrained optimization problem by optimizing the stochastic policy π_θ:

$$\underset{\theta}{\text{Maximize}} \quad \mathbb{E}_{s \sim \rho_{\theta_k}, a \sim \pi_{\theta_k}} \left[\frac{\pi_\theta(a|s)}{\pi_{\theta_k}(a|s)} A_{\theta_k}(s, a) \right]$$

$$\text{subject to} \quad \mathbb{E}_{s \sim \rho_{\theta_k}} \left[D_{KL}(\pi_{\theta_k}(\cdot|s) \| \pi_\theta(\cdot|s)) \right] \leq \Delta_{KL}$$

where $\rho_\theta = \rho_{\pi_\theta}$ are the discounted state-visitation frequencies induced by π_θ. $A_{\theta_k}(s, a)$ is the advantage function, which can be estimated by the difference between the empirical returns and the baseline. We use a linear value function baseline in our experiments. D_{KL} is the KL divergence between the two policy distributions, and Δ_{KL} is a step size parameter that controls the maximum change in policy per optimization step. The expectations in the expression can be evaluated using sample averages, and the policy can be represented by non-linear function approximators such as neural networks. The stochastic policy π_θ can be represented by a categorical distribution when the actions of the agent are discrete and by a Gaussian distribution when the actions are continuous.

3.5   Reward Structure

The concept of reward shaping [30] involves modifying rewards to accelerate learning without changing the optimal policy. When modeling a multi-agent system as a Dec-POMDP, rewards are shared jointly by all agents. In a centralized representation, the reward signal cannot be decomposed into separate components, and is equivalent to the joint reward in a Dec-POMDP. However, decentralized representations allow us an alternative local reward representation. Local rewards can restrict the reward signal to only those agents that are involved in the success or failure at a task. Bagnell and Ng have shown that such local information can help reduce the number of samples required for learning [31]. As we will note later, this decomposition can drastically improve training time. The performance of the policy is still evaluated using the global reward.

3.6   Curriculum Learning

Curriculum learning leverages the idea of learning policies for simple tasks first, and then building on that knowledge to solve more difficult tasks [15]. Formally, a curriculum $\mathcal{T}$ is an ordered set of tasks organized by increasing difficulty. In cooperative settings, the tasks in the curriculum become more difficult as the number of cooperating agents required to complete the task increases.

4   Cooperative Reinforcement Learning

This section outlines three training schemes for multi-agent reinforcement learning in cooperative settings as well as their advantages and disadvantages.

4.1   Centralized

A centralized policy maps the joint observation of all agents to a joint action, and is equivalent to a MPOMDP policy. A major drawback of this approach is that it is centralized in both training and execution, and leads to an exponential growth in the observation and actions spaces with the number of agents. We address this intractability in part by factoring the action space of centralized multi-agent systems.

We first assume that the joint action can be factored into individual components for each agent. The factored centralized controller can then be represented as a set of sub-policies that map the joint observation to an action for a single agent. In the policy gradient approach this reduces to factoring the joint action probability as $P(\boldsymbol{a}) = \prod_i P(a_i)$ where a_i are the individual actions of an agent. In practice, this means that the policy of a given agent is represented by a subset of the output nodes in the neural network. In systems with discrete actions, this reduces the size of the action space from $|\mathcal{A}|^n$ to $n|\mathcal{A}|$, where n is the number of agents and $\mathcal{A}$ is the action space for a single agent (we assume homogeneous agents for simplicity). While this is a significant reduction in the size of the action space, the exponential growth in the observation spaces ultimately makes centralized controllers impractical for complex cooperative tasks.

4.2   Concurrent

In concurrent learning, each agent learns its own individual policy. Concurrent policies map an agent's private observation to an action for that agent. Each agent's policy is independent. In the policy gradient approach, this means optimizing multiple policies simultaneously from the joint reward signal. One of the advantages of this approach is that it makes learning of heterogeneous policies easier. This can be beneficial in domains where agents may need to take on specific roles in order to coordinate and receive reward.

The major drawback of concurrent training is that it does not scale well to large numbers of agents. Because the agents do not share experience with one

---

Algorithm 1. PS-TRPO

Input: Initial policy parameters Θ_0, trust region size Δ
for $i \leftarrow 0, 1, \ldots$ **do**
    Rollout trajectories for all agents $\tau \sim \pi_{\theta_i}$
    Compute advantage values $A_{\pi_{\theta_i}}(o^m, m, a^m)$ for each agent m's trajectory element.
    Find $\pi_{\theta_{i+1}}$ maximizing Eq. (1)
        subject to $\overline{D}_{KL}(\pi_{\theta_i} \| \pi_{\theta_{i+1}}) \leq \Delta$

---

another, this approach adds additional sample complexity to the reinforcement learning task. Another drawback of the approach is that the agents are learning and adjusting their policies individually making the environment dynamics non-stationary, which can lead to instability.

4.3 Parameter Sharing

The policies of homogeneous agents may be trained more efficiently using parameter sharing. This approach allows the policy to be trained with the experiences of all agents simultaneously. However, it still allows different behavior between agents because each agent receives unique observations, which includes their respective index. In parameter sharing, the control is decentralized but the learning is not. In the remainder of the paper, all training schemes use parameter sharing unless stated otherwise.

So long as the agents can execute decentralized policies with shared parameters, single agent algorithms like DDPG, DQN, TRPO and A3C can be extended to multi-agent systems. As an example, Algorithm 1 describes a policy gradient approach that combines parameter sharing and TRPO. We refer to it as PS-TRPO. We first initialize the policy network and set the step size parameter. At each iteration of the algorithm, the policy with shared parameters is used by each agent to generate trajectories. The batch of trajectories from all the agents is used to compute the advantage value and maximize the following objective:

$$L(\theta) = \mathbb{E}_{o \sim \rho_{\theta_k}, a \sim \pi_{\theta_k}} \left[\frac{\pi_\theta(a \mid o, m)}{\pi_{\theta_k}(a \mid o, m)} A_{\theta_k}(o, m, a) \right] \tag{1}$$

where m is the agent index. The results of the optimization are used to compute the parameter update for the policy.

5 Tasks

The four multi-agent benchmark tasks are described in this section. All tasks are partially observable. For more details we refer the reader to the source code.

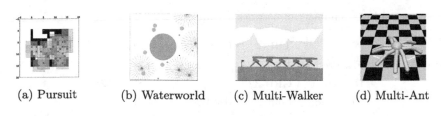

(a) Pursuit   (b) Waterworld   (c) Multi-Walker   (d) Multi-Ant

Fig. 1. Examples of the four cooperative domains. (Color figure online)

5.1   Discrete

Pursuit. Pursuit is a standard task for benchmarking multi-agent algorithms [32]. The pursuit-evasion domain consists of two sets of agents: evaders and pursuers. The evaders are trying to avoid pursuers, while the pursuers are trying to catch the evaders. The action and observation spaces in this problem are discrete. Each pursuer receives a range-limited observation of its surroundings, and must choose between five actions Stay, Go East, Go West, Go South, Go North. The observations contain information about the agent's surroundings, including the location of nearby pursuers, evaders, and obstacles. The example in Fig. 1a shows a 32×32 grid world with randomly generated obstacles, 20 pursuers (denoted by red stars), and 20 evaders (denoted by blue stars). The square box surrounding the pursuers indicates their observation range. The pursuers receive a reward of 5.0 when they surround and catch an evader, and a reward of 0.01 when they occupy the same space as an evader.

5.2   Continuous

Waterworld. Waterworld can be seen as an extension of the above mentioned pursuit problem to a continuous domain. The extension is based on the single agent waterworld domain used by [33]. In this task, agents need to cooperate to capture moving food targets while avoiding poison targets. Both the observation and action spaces are continuous, and the agents move around by applying a two-dimensional force. The agents receive a reward of 10.0 for capturing a food target, a reward of -1.0 for capturing a poison target, and an exertion penalty of $-0.01 \cdot \|a_i\|^2$.

Multi-Walker. Multi-Walker is a more difficult continuous control locomotion task based on the BipedalWalker environment from OpenAI gym [34]. The domain consists of multiple bipedal walkers that can actuate the joints in each of their legs. At the start of each simulation, a large package that stretches across all walkers is placed on top of the walkers. The walkers must learn how to move forward and to coordinate with other agents in order to keep the package balanced while navigating a complex terrain. Each agent receives a reward of 1.0 for moving the package forward 1 meter, a reward of -100.0 for falling, and a reward of -100.0 for dropping the package. An example environment with five walkers is shown in Fig. 1c.

Table 1. Summary of network architectures for each algorithm

	TRPO	DDPG/DQN	A3C
Feature Net	100-50-25	400-300	128
Recurrent	GRU-32	NA	LSTM-128
Activation	tanh	ReLU	tanh

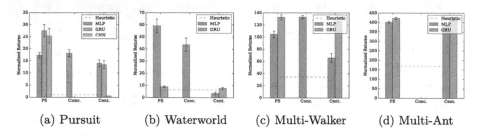

(a) Pursuit (b) Waterworld (c) Multi-Walker (d) Multi-Ant

Fig. 2. Normalized average returns for multi-agent policies trained using TRPO. Missing entries indicate the training was unsuccessful. A random policy has zero normalized average return. Error bars represent standard error. The Wilcoxon test suggests the differences are significant ($p < 0.05$) except for the difference between centralized GRU and shared parameter GRU for the waterworld domain.

Multi-Ant. The multi-ant domain is a 3D locomotion task based on the quadrupedal robot used in [35]. The goal of the robot is to move forward as quickly as possible. In this domain, each leg of the ant is treated as a separate agent that is able to sense its own position and velocity as well as those of its two neighbors. Each leg is controlled by applying torque to its two joints. An example multi-ant with ten legs is shown in Fig. 1d.

6 Experiments

This section presents empirical results that compare the performance of multi-agent extensions of TRPO, DDPG, A3C, and DQN. In continuous action domains we compare TRPO, A3C, and DDPG, while in discrete action domains we compare TRPO, A3C, and DQN. We examine both feed-forward and recurrent policies in this work. We also examine the effects of centralized, concurrent, and shared parameters training schemes as well as two reward mechanisms that are relevant to multi-agent domains. The results are compared against each other and against a heuristic hand-crafted baseline for each task. Lastly, we demonstrate the benefits of curriculum learning to scalability in cooperative domains.

The neural network architectures used in this work are summarized in Table 1. The feature net represents the number of neurons in each layer and is used as the feedforward multi-layer perceptron (MLP) policy in each algorithm. The type of the hidden cell, either GRU or LSTM, and their number is indicated for recurrent policies. The feature net serves as the observation embedding for

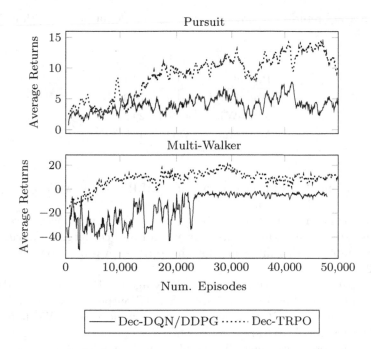

Fig. 3. Training curves comparing PS-TRPO and PS-DQN in Pursuit and PS-DDPG in Multi-Walker Domains.

recurrent policies. DQN/DDPG do not use recurrent policies, and A3C uses a single hidden layer as a feature network.

In all experiments, we use the discount factor $\gamma = 0.99$. For PS-TRPO, we set the step size to $\Delta = 0.01$, and constrain the size of each batch to a maximum of 24000 time-steps. For DDPG and DQN, we used batch sizes of 32, learning rate of 1×10^{-3} for the state-action value function and 1×10^{-4} for the policy network. For A3C, we used RMSProp [36] with an annealed learning rate starting from 5×10^{-5} with decay of 0.99.

6.1 Discrete Control Task

We first compared performances of the three training schemes on the pursuit problem using TRPO. The emergent behavior observed in TRPO policies included pursuers breaking up into teams to maximize the number of evaders that were captured. The results are summarized in Fig. 2a for a 16 × 16 grid, 8 pursuers with an observation range of 7, and 30 evaders. The figure shows that parameter sharing tends to outperform both the concurrent and centralized training schemes. Because the observation is image-like with spatial correlations present in each observation dimension, we also used a convolutional neural networks (CNN) to represent the policy in this task. The results show that with

Table 2. Average returns for parameter sharing multi-agent policies with global and local rewards

	Global	Local
Pursuit	8.1	12.1
Waterworld	−1.4	14.3
Multi-Walker	−23.3	29.9
Multi-Ant	475.2	488.1

parameter sharing, CNN policies outperform MLP policies, while GRU policies have the best overall performance.

We then compared the training behavior of global and local rewards. We found that using local rewards consistently improved convergence during training. An example of this difference for the pursuit evasion problem is shown in Table 2.

We compared the performance of PS-DQN against PS-TRPO and PS-A3C. As can be seen from Fig. 3 and Table 4, PS-A3C outperforms both PT-TRPO and PS-DQN, with PS-DQN having the worst performance. We hypothesize that PS-DQN is unable to learn a good controller due to the changing policies of other agents in the environment. This makes the dynamics of the problem non-stationary which causes experience replay to inaccurately describe the current state of the environment.

We also tested the ability of PS-TRPO to scale with very large observation spaces. The pursuit domain was set up on a 128 × 128 grid with 200 pursuers and 200 evaders with at least 16 pursuers required to capture an evader. While hundreds of agents are present in the environment, only 16 of them need to cooperate to achieve the capture task. Each observation is a four channel 21 × 21 image, making the observation space 1764 dimensional. The training curves for this task are shown in Fig. 4, and show that the MLP policy fails to learn a policy that can outperform the heuristic. However, by leveraging CNNs, we are able to outperform the heuristic in this complex domain.

Comparison to Traditional Method. Traditional reinforcement learning and Dec-POMDP approaches have difficulty solving problems with continuous action spaces and scale to problems with large numbers of agents. We also confirmed that PS-TRPO performs as well as a traditional approach for solving PS-TRPO on a small 5 × 5 grid pursuit problem. The approach we use as comparison resembles Joint Equilibrium search for policies (JESP) [37] in that it finds a policy that maximizes the joint expected reward for one agent at a time, while keeping the policies of all the other agents fixed. The process is repeated until an equilibrium is reached. In our approach, we use the fast informed bound (FIB) algorithm [38] to perform the policy optimization of a single agent.

The pursuit problem is set on a 5 × 5 grid with a square obstruction in the middle. There is a single evader and two pursuers. Both of the pursuers must

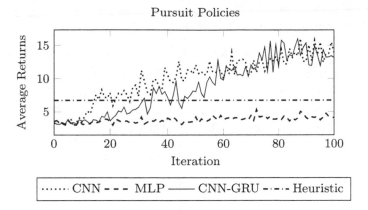

Fig. 4. Performance as a function of the number of iteration for different neural architectures in the pursuit domain with 200 agents. At least 16 agents need to occupy the same cell to capture an evader.

Table 3. Average returns on small-scale pursuit problem

	PS-TRPO	FIB
Average Returns	9.36 ± 0.52	9.29 ± 0.65

occupy the same location as the evader in order to catch it and obtain a reward. This problem has a total of 15625 states and 729 observations. The results comparing the average performance and their standard errors of PS-TRPO and FIB policies averaged over 100 simulations are shown in Table 3. The results demonstrate that PS-TRPO performs as well as the traditional approaches on the small problem, and has the ability to scale to large and continuous spaces.

6.2   Continuous Control Tasks

We next compared the performance of our algorithms on continuous control tasks. We compared the proposed training schemes with TRPO and found that parameter sharing and concurrent approaches tend to outperform centralized training for continuous tasks (Figs. 2b, c and d). GRU policies outperform MLP policies in the multi-walker and multi-ant domains. However, MLP policies perform significantly better in the waterworld domain. We believe this is caused by the difficulty of training recurrent networks compared to simpler feedforward ones with high-dimensional observations, especially when the task is relatively simple and does not require a history of observations. Visualizing the best performing policies showed consistent intelligent behavior in coordination between agents. In the waterworld domain, the pursuers learn to herd the evaders. In the multi-walker domain, the walkers learn to push the box forward without letting it fall down. In the multi-ant domain, the legs learn to avoid collision with each other.

Table 4. Average returns (over 50 runs) for policies trained with parameter sharing. DQN for discrete environment, DDPG for continuous

Task	PS-DQN/DDPG	PS-A3C	PS-TRPO
Pursuit	10.1 ± 6.3	25.5 ± 5.4	17.4 ± 4.9
Waterworld	NA	10.1 ± 5.7	49.1 ± 5.7
Multiwalker	-8.3 ± 3.2	12.4 ± 6.1	58.0 ± 4.2
Multi-ant	307.2 ± 13.8	483.4 ± 3.4	488.1 ± 1.3

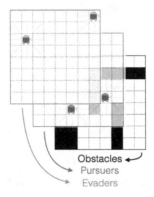

Obstacles
Pursuers
Evaders

Fig. 5. Image like representation of an observation in the pursuit evasion domain. The locations of each entity (pursuers, evaders, and obstacles) are represented as bitmaps in their respective channels.

We also compared local and global reward schemes in the continuous domain (see Table 2). Overall, local reward shaping leads to better performance, and is critical to learning intelligent behavior in the waterworld and multi-walker domains (Table 2).

Finally, we compared the performance of PS-TRPO, PS-A3C, and PS-DDPG in continuous multi-agent domains. Training curves comparing PS-DDPG and PS-TRPO are shown in Fig. 3 for the multi-walker task, while the performance of all the algorithms and tasks are compared in Table 4. The results show the PS-TRPO significantly outperforms both PS-A3C and PS-DDPG in the waterworld and multi-walker domains. The performance of PS-TRPO and PS-A3C is comparable in the multi-ant domain.

6.3   Scaling

We next studied how well the parameter sharing method scales to larger observation spaces and many agents.

Curriculum training: Figure 6 shows the degrading performance of all policies with increasing number of agents in the multi-walker domain, and the performance improvements when curriculum learning is used. The policies were all

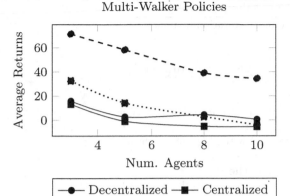

Fig. 6. Performance of multi-walker policies as a function of the number of agents during training. Each data point in the shared parameters, centralized, and concurrent curves was generated by training and evaluating a policy with a fixed number of agents. The curriculum curve was generated by evaluating a single policy with varying number of agents.

trained with TRPO. The decrease in performance is in part due to the increasing difficulty of the reinforcement learning task as the number of cooperating agents grows. As the number of agents required to complete a task increases, it becomes more difficult to reach the parts of the state space with positive rewards using naive exploration policies.

We investigated how a curriculum learning scheme can help scale the multi-walker problem in the number of agents. An intuitive curriculum for this problem is over the number of agents, and so we define a curriculum with the number of agents in the environment ranging from 2 to 10. Because the policies are decentralized even though the parameters are shared, they can be evaluated on tasks with any number of cooperating agents regardless of the number of cooperating agents present during training. Unfortunately, we found that these decentralized shared parameter policies trained on a few agents often fail to generalize to larger numbers of agents. We therefore define a Dirichlet distribution for this range of tasks with higher probability assigned to the simplest task (with 2 agents for Multi-Walker domain). We then sample an environment from this distribution over the tasks in the curriculum and optimize the policy with PS-TRPO for a few iterations. Once the expected reward for the most likely environment reaches a threshold, we change the distribution such that the next environment is most likely. We continue this curriculum until the expected reward in all environments reaches the defined threshold. Algorithm 2 describes this process. As shown earlier, the resulting policy outperforms policies trained without the curriculum. We believe this improvement in performance is due to two reasons: 1. The distribution over environments provides a regularization effect, helping avoid local

---

Algorithm 2. Curriculum Training

Input: Curriculum $\mathcal{T}$, Iteration n, Policy π_Θ, $r_{\text{threshold}}$
$\alpha_{\mathcal{T}} \leftarrow [\text{length}(\mathcal{T}), 1, 1, \ldots]$
while $r_{\text{min}} < r_{\text{threshold}}$ **do**
  {Sample task from the task distribution.}
  $w \sim \text{Dirichlet}(\alpha_{\mathcal{T}})$
  $i \sim \text{Categorical}(w)$
  {Apply optimization step for a few iterations.}
  PS-TRPO $(\mathcal{T}_i, \pi_\theta, n)$
  {e_{curr} is the task with the highest weight $\alpha_{\mathcal{T}}$.}
  $r_{e_{\text{curr}}} \leftarrow \text{Evaluate}(\pi_\theta, e_{\text{curr}})$
  if $r_{e_{\text{curr}}} > r_{\text{threshold}}$ **then**
    Circular shift $\alpha_{\mathcal{T}}$ weights to the next task
  {Find the minimum average reward across tasks.}
  $r_{\text{min}} \leftarrow \min_{\mathcal{T}} \mathbb{E}r_{\mathcal{T}}$

---

minima during optimization, and 2. It partially addresses the exploration problem by smoothly increasing the difficulty of the policy to be learned.

One potential issue with this experiment is that the curriculum scheme observed more episodes than the ones without curriculum. However, we tried training several policies with a fixed number of agents without a curriculum for an equivalent number of episodes. These policies converged before reaching the performance seen with curriculum training.

7 Conclusion

Despite the advances in decentralized control and reinforcement learning over recent years, learning cooperative policies in multi-agent systems remains a challenge. The difficulties lie in scalability to high-dimensional observation spaces and to large numbers of agents, accommodating partial observability, and handling continuous action spaces. In this work, we extended three deep reinforcement learning algorithms to the cooperative multi-agent context, and applied them to four high-dimensional, partially observable domains with many agents.

Our empirical evaluations show that PS-TRPO policies have substantially better performance than PS-DDPG and PS-A3C in continuous action collaborative multi-agent domains while PS-A3C is able to outperform PS-TRPO in the discrete domain. We suspect that DQN and DDPG perform poorly in systems with multiple learners due to the non-stationarity of the system dynamics caused by the changing policies of the agents. The non-stationary nature of the system makes experience replay samples obsolete and negatively impacts training. As evidence, we found that by disabling experience replay and instead relying on asynchronous training [14] we were able to improve on the performance of DQN and DDPG. However, we believe more hyperparameter tuning might be required to reduce the gap in overall performance in continuous domains with respect to TRPO. Finally, we presented how cooperative domains can form a natural curriculum over the number of agents required to collaborate on a task and discovered how this not only allows us to scale PS-TRPO to environments with large number of cooperating agents, but owing to the regularization effect offered, allows us to reach better local optima in general.

There are several areas for future work. To improve scalability of the proposed approach for larger numbers of cooperating agents further future work is needed. Two major challenges in multi-agent systems are accommodating reward sparsity through intelligent domain exploration and incorporating high-level task abstractions and hierarchy [39]. These are acute forms of similar challenges in the single agent learning. Recently, curiosity based information gain maximizing exploration strategy was explored by [40] . Similar ideas could be adapted to maximize information gain not only about the environment's dynamics, but the dynamics of an agent's behavior as well. Correspondingly, hierarchical value functions were integrated with deep reinforcement learning [41]. Incorporating task hierarchies in a multi-agent system would allow us to tackle learning specialization and heterogeneous behavior.

Acknowledgements. This work was supported by Army AHPCRC grant W911NF-07-2-0027. The authors would like to thank the anonymous reviewers for their helpful comments.

References

1. Tan, M.: Multi-agent reinforcement learning: independent vs. cooperative agents. In: International Conference on Machine Learning (ICML), pp. 330–337 (1993)
2. Panait, L., Luke, S.: Cooperative multi-agent learning: the state of the art. In: International Conference on Autonomous Agents and Multiagent Systems (AAMAS), vol. 11(3), pp. 387–434 (2005)
3. Bloembergen, D., Tuyls, K., Hennes, D., Kaisers, M.: Evolutionary dynamics of multi-agent learning: a survey. J. Artif. Intell. Res. **53**, 659–697 (2015)
4. Amato, C., Chowdhary, G., Geramifard, A., Ure, N.K., Kochenderfer, M.J.: Decentralized control of partially observable Markov decision processes. In: IEEE Conference on Decision and Control (CDC), Florence, Italy (2013)
5. Bernstein, D.S., Zilberstein, S., Immerman, N.: The complexity of decentralized control of Markov decision processes. In: Conference on Uncertainty in Artificial Intelligence (UAI), pp. 32–37 (2000)
6. Banerjee, B., Lyle, J., Kraemer, L., Yellamraju, R.: Sample bounded distributed reinforcement learning for decentralized POMDPs. In: AAAI Conference on Artificial Intelligence (AAAI) (2012)
7. Omidshafiei, S., Agha-mohammadi, A.-A., Amato, C., Liu, S.-Y., How, J.P., Vian, J.: Graph-based cross entropy method for solving multi-robot decentralized POMDPs. In: IEEE International Conference on Robotics and Automation (ICRA) (2016)
8. Tesauro, G.: Extending Q-learning to general adaptive multi-agent systems. In: Advances in Neural Information Processing Systems (NIPS) (2003)
9. Lin, L.-J.: Reinforcement learning for robots using neural networks, Ph.D. dissertation. Carnegie Mellon University (1992)

10. Mnih, V., Kavukcuoglu, K., Silver, D., Rusu, A.A., Veness, J., Bellemare, M.G., Graves, A., Riedmiller, M., Fidjeland, A.K., Ostrovski, G., et al.: Human-level control through deep reinforcement learning. Nature **518**(7540), 529–533 (2015)
11. Levine, S., Finn, C., Darrell, T., Abbeel, P.: End-to-end training of deep visuomotor policies. J. Mach. Learn. **17**(39), 1–40 (2016)
12. Schulman, J., Levine, S., Abbeel, P., Jordan, M., Moritz, P.: Trust region policy optimization. In: International Conference on Machine Learning (ICML) (2015)
13. Lillicrap, T.P., Hunt, J.J., Pritzel, A., Heess, N., Erez, T., Tassa, Y., Silver, D., Wierstra, D.: Continuous control with deep reinforcement learning, arXiv preprint arXiv:1509.02971 (2015)
14. Mnih, V., Badia, A.P., Mirza, M., Graves, A., Lillicrap, T.P., Harley, T., Silver, D., Kavukcuoglu, K.: Asynchronous methods for deep reinforcement learning, arXiv preprint arXiv:1602.01783 (2016)
15. Bengio, Y., Louradour, J., Collobert, R., Weston, J.: Curriculum learning. In: International Conference on Machine Learning (ICML), pp. 41–48 (2009)
16. Busoniu, L., Babuska, R., Schutter, B.D.: Multi-agent reinforcement learning: a survey. In: International Conference on Control, Automation, Robotics and Vision, vol. 527, pp. 1–6 (2006)
17. Ono, N., Fukumoto, K.: A modular approach to multi-agent reinforcement learning. In: Weiß, G. (ed.) LDAIS/LIOME -1996. LNCS, vol. 1221, pp. 25–39. Springer, Heidelberg (1997). https://doi.org/10.1007/3-540-62934-3_39
18. Guestrin, C., Lagoudakis, M., Parr, R.: Coordinated reinforcement learning. In: International Conference on Machine Learning (ICML), vol. 2, pp. 227–234 (2002)
19. Lauer, M., Riedmiller, M.: An algorithm for distributed reinforcement learning in cooperative multi-agent systems. In: International Conference on Machine Learning (ICML), pp. 535–542 (2000)
20. Singh, S.P., Jaakkola, T.S., Jordan, M.I.: Learning without state-estimation in partially observable markovian decision processes. In: International Conference on Machine Learning (ICML) (1994)
21. Peshkin, L., Kim, K.-E., Meuleau, N., Kaelbling, L.P.: Learning to cooperate via policy search. In: Conference on Uncertainty in Artificial Intelligence (UAI), pp. 489–496 (2000)
22. Fernández, F., Parker, L.E.: Learning in large cooperative multi-robot domains. Int. J. Robot. Autom. **16**(4), 217–226 (2001)
23. Tamakoshi, H., Ishii, S.: Multiagent reinforcement learning applied to a chase problem in a continuous world. Artif. Life Robot. **5**(4), 202–206 (2001)
24. Das, A.K., Fierro, R., Kumar, V., Ostrowski, J.P., Spletzer, J., Taylor, C.J.: A vision-based formation control framework. IEEE Trans. Robot. Autom. **18**(5), 813–825 (2002)
25. Cortes, J., Martinez, S., Karatas, T., Bullo, F.: Coverage control for mobile sensing networks. In: IEEE International Conference on Robotics and Automation (ICRA), vol. 2, pp. 1327–1332. IEEE (2002)
26. Olfati-Saber, R., Fax, J.A., Murray, R.M.: Consensus and cooperation in networked multi-agent systems. Proc. IEEE **95**(1), 215–233 (2007)
27. Tampuu, A., Matiisen, T., Kodelja, D., Kuzovkin, I., Korjus, K., Aru, J., Aru, J., Vicente, R.: Multiagent cooperation and competition with deep reinforcement learning, arXiv preprint arXiv:1511.08779 (2015)
28. Foerster, J.N., Assael, Y.M., de Freitas, N., Whiteson, S.: Learning to communicate with deep multi-agent reinforcement learning. In: Advances in Neural Information Processing Systems (NIPS) (2016)

29. Sukhbaatar, S., Szlam, A., Fergus, R.: Learning multiagent communication with backpropagation. In: Advances in Neural Information Processing Systems (NIPS) (2016)
30. Ng, A.Y., Harada, D., Russell, S.: Policy invariance under reward transformations: theory and application to reward shaping. In: International Conference on Machine Learning (ICML), vol. 99, pp. 278–287 (1999)
31. Bagnell, D., Ng, A.Y.: On local rewards and scaling distributed reinforcement learning. In: Advances in Neural Information Processing Systems, pp. 91–98 (2005)
32. Vidal, R., Shakernia, O., Kim, H.J., Shim, D.H., Sastry, S.: Probabilistic pursuit-evasion games: theory, implementation, and experimental evaluation. IEEE Trans. Robot. Autom. **18**(5), 662–669 (2002)
33. Ho, J., Gupta, J.K., Ermon, S.: Model-free imitation learning with policy optimization. In: International Conference on Machine Learning (ICML) (2016)
34. Brockman, G., Cheung, V., Pettersson, L., Schneider, J., Schulman, J., Tang, J., Zaremba, W.: Openai gym (2016)
35. Schulman, J., Moritz, P., Levine, S., Jordan, M., Abbeel, P.: High-dimensional continuous control using generalized advantage estimation. arXiv preprint arXiv:1506.02438 (2015)
36. Tieleman, T., Hinton, G.: Lecture 6.5-RmsProp: Divide the gradient by a running average of its recent magnitude. COURSERA: Neural Netw. Mach. Learn. **4**, 26–31 (2012)
37. Nair, R., Tambe, M., Yokoo, M., Pynadath, D., Marsella, S.: Taming decentralized POMDPs: towards efficient policy computation for multiagent settings. In: International Joint Conference on Artificial Intelligence (IJCAI) (2003)
38. Hauskrecht, M.: Incremental methods for computing bounds in partially observable Markov decision processes. In: AAAI Conference on Artificial Intelligence (AAAI) (1997)
39. Parr, R., Russell, S.: Reinforcement learning with hierarchies of machines. In: Advances in Neural Information Processing Systems (NIPS), pp. 1043–1049 (1998)
40. Houthooft, R., Chen, X., Duan, Y., Schulman, J., De Turck, F., Abbeel, P.: Variational information maximizing exploration. arXiv preprint arXiv:1605.09674 (2016)
41. Kulkarni, T.D., Narasimhan, K.R., Saeedi, A., Tenenbaum, J.B.: Hierarchical deep reinforcement learning: integrating temporal abstraction and intrinsic motivation. arXiv preprint arXiv:1604.06057 (2016)

Stereotype Reputation with Limited Observability

Phillip Taylor[1]([✉]), Nathan Griffiths[1], Lina Barakat[2], and Simon Miles[2]

[1] Department of Computer Science, The University of Warwick,
Coventry CV4 7AL, UK
Phillip.Taylor@warwick.ac.uk
[2] Department of Informatics, Kings College London, London WC2R 2LS, UK

Abstract. Assessing trust and reputation is essential in multi-agent systems where agents must decide who to interact with. Assessment typically relies on the direct experience of a trustor with a trustee agent, or on information from witnesses. Where direct or witness information is unavailable, such as when agent turnover is high, stereotypes learned from common traits and behaviour can provide this information. Such traits may be only partially or subjectively observed, with witnesses not observing traits of some trustees or interpreting their observations differently. Existing stereotype-based techniques are unable to account for such partial observability and subjectivity. In this paper we propose a method for extracting information from witness observations that enables stereotypes to be applied in partially and subjectively observable dynamic environments. Specifically, we present a mechanism for learning translations between observations made by trustor and witness agents with subjective interpretations of traits. We show through simulations that such translation is necessary for reliable reputation assessments in dynamic environments with partial and subjective observability.

1   Introduction

In multi-agent systems (MAS) agents must decide whether or not to interact with others, and can use trust and reputation to inform this decision [6,20,23]. Trust is the degree of belief, from the perspective of a trustor agent, that a trustee agent will act as they say they will in a given context [1,2,10]. A trustor with a high level of trust in a trustee is confident of a successful interaction with a good outcome. Likewise, a low level of trust in a trustee implies that the trustor agent expects a bad outcome. Whereas trust is assessed using experiences of the trustor, reputation is based on the opinions of several agents in a network.

In domains where agents join and leave with high frequencies, it can be difficult to reliably assess trust and reputation due to limited relevant experience. A trustor agent who recently joined a MAS, for instance, will have limited experience with trustees and be unable to reliably assess trust. In this case, opinions of witness agents can be used to produce a reputation assessment [9].

© Springer International Publishing AG 2017
G. Sukthankar and J. A. Rodriguez-Aguilar (Eds.): AAMAS 2017 Best Papers,
LNAI 10642, pp. 84–102, 2017.
https://doi.org/10.1007/978-3-319-71682-4_6

When a trustee agent is new to a MAS, however, no agent will have direct experience with them, preventing reliable assessments of trust and reputation.

In many domains, trustee agents exhibit traits that provide insight into their behaviour during, but can be observed prior to entering into, an interaction [2,12,16]. Such traits are referred to as stereotypes, and can be used to bootstrap trust and reputation assessments when experience is limited. If a trustor has observed a stereotype it can be used to assess stereotype-trust in a trustee, otherwise stereotype-reputation can be assessed using witnesses. A witness may be unable themselves to observe the trustee traits, however, and must assess those observed and reported by the trustor. When these trait observations are subjective and agents have different interpretations or observe different traits, communication of observations and assessing stereotype-reputation is problematic. In this paper we propose the Partially Observable and Subjective Stereotype Trust and Reputation (POSSTR) system, which enables agents in partially observable environments to translate observations from different subjective perspectives, and enables witnesses to provide reliable stereotype-reputation assessments. POSSTR does not replace existing reputation systems, but rather it should be used alongside them to provide a way to deal with partial observability and subjectivity. We make the following contributions:

- We propose a mechanism for learning a translation between traits observed by a trustor and a witness, and
- Using simulations, we show that our translation mechanism improves trust and reputation assessments in environments with partially and subjectively observable traits.

The remainder of this paper is structured as follows. Related work is discussed in Sect. 2. Section 3 describes our use case and outlines the problem with partial observability and subjective observations of traits. The POSSTR system, which overcomes challenges in such partially observable and subjective domains, is proposed in Sect. 4. The simulation environment used for evaluating POSSTR is outlined, and results from our investigation are discussed in Sect. 5. Finally, Sect. 6 concludes the paper.

2   Related Work

In many domains, trustor agents use trust and reputation to select interaction partners from sets of trustees [6]. Trust can be assessed using direct experience gathered by a trustor interacting with trustee agents. Where direct experience is lacking, reputation assessments are gathered from witness agents [1,6,23]. In highly dynamic environments, where agents leave or depart regularly, relevant experience with trustees is often insufficient to produce reliable assessments. In these cases, stereotypes can be used to bootstrap trust and reputation [2,12,16].

Trustees often exhibit traits that are observable to trustors prior to an interaction. When these traits are related to the behaviour of trustees during interactions, the trustor can form stereotypes that can be used as a surrogate for other more relevant experience in assessing trust and reputation [2,12,16]. If several trustee agents exhibit the same trait and are similarly reputable, for instance, a new trustee also exhibiting the trait may be assumed to have similar reputation. To build a stereotype-trust model, a trustor must interact with several trustees and analyse the correlations between their observable traits and reputations. If the trustor is unable to assess stereotype-trust because they lack relevant experience of the observed traits, stereotype-reputation assessments can be requested from witnesses [2,12].

To assess reputation of a trustee, the trustor combines the following:

- *Direct-trust* based on direct experience the trustor has with the trustee;
- *Witness-reputation* based on witness reports summarising their experiences with the trustee;
- *Stereotype-trust* based on common trustee traits observed by the trustor; and
- *Stereotype-reputation* based on experience and common trustee traits observed by witnesses.

Direct-trust requires the trustor to have previously interacted with the trustee being evaluated. The same is true when witnesses compute opinions about a specific trustee, to be sent to the trustor. In combination, direct-trust and witness-reputation, make up the Beta Reputation System (BRS), as proposed by Jøsang et al. [9]. Other reputation systems that combine direct-trust and witness-reputation include FIRE [7], TRAVOS [18], BLADE [15], and HABIT [17]. TRAVOS extends BRS to cope with dishonest witnesses by discounting information provided by unreliable sources, and BLADE and HABIT both use Bayesian networks to transform opinions from witnesses that are unreliable in a consistent way.

As well as direct-trust and witness-reputation, FIRE [7] also combines two other sources of information, namely certified and role-based trust. Certified trust is based on testimonials gathered by the trustee and given to the trustor, and as a result is often optimistic of their performance in an interaction. Role-based trust can be viewed as a kind of stereotype, but the roles are defined statically by trustors and as a result it is limited compared to the observation based approach used in this paper. Stereotype-trust enables assessments of trustees with whom the trustor has not previously interacted, by assuming trustees with similar observable traits behave similarly. As with witness-reputation, stereotype-reputation is gathered from witnesses who provide their opinions.

Liu et al. [12] proposed that characteristics of trustee agents, correlated with their trustworthiness, be used to separate them into groups defined by their common characteristics. When evaluating a new trustee, its observable characteristics are compared to those that define each group and the mean trustworthiness of their members is used as the stereotype and overall trust score. When a trustor is unable to determine stereotype-trust, because they lack experience with the particular characteristics, stereotype-reputation is gathered from witnesses. Similarly, Teacy et al. [17] suggest building a separate HABIT model for

overlapping groups of agents defined by stereotypes, but they do not describe
how such groups should be formed.

The bootstrapping model proposed by Burnett et al. [2] combines all four
sources of trust and reputation. Instead of the clustering approach employed by
Liu et al. [12], the trustor learns a regression model that maps observed traits to
trustworthiness. Observed characteristics of trustees are then input into the model
with the output used as a base reputation value in a probabilistic trust model.
In this way, the base trust value has less of an impact on the overall reputation
score as more direct evidence is gathered about the trustee. STAGE, proposed by
Şensoy et al. [16], combines direct-trust, stereotype-trust, and witness-reputation
in a similar way to Burnett et al. [2]. In STAGE, reports provided by witnesses for
both witness- and stereotype-reputation are discounted based on their perceived
reliability. As well as using stereotype-trust to bootstrap assessments of trustees,
STAGE also learns stereotypes for witnesses to bootstrap this reliability assessment of opinions. To avoid the need for opinions, Fang et al. [3], build a stereotype-
trust model that enables observations to generalise to others when experience for
a particular stereotype is limited.

In these existing reputation models, witness-reputation requires that the
trustee is known to the witness. This means that the trustor must be willing
to identify the trustee to the witness, and the witness must have interacted with
them previously. Likewise, stereotype-reputation as proposed by Burnett et al. [2]
requires:

- The trustee is identified and the witness can observe its traits (i.e. trustees
  are *fully observable*), and
- All agents observe trustee traits in the same way (i.e. trustee traits are *objective*).

In real-world environments, however, trustees may be *partially observable* and
such observations may often be *subjective*. If the trustees are only *partially
observable* and the witness is unable to observe the traits, the trustor must
disclose their observations of traits for the witness to provide their opinion. For
example, if a new trustee is unknown to a witness, the trustor must describe
their observations when requesting a stereotype-reputation assessment. If trait
observations are also *subjective*, those observed by a trustor may be meaningless
to a witness. In this paper, we propose the POSSTR system to overcome this
issue by translating traits observed by the trustor.

3   Problem Setting

To formalise these issues of partial observability and subjectivity, we define the
full set of traits in an environment that agents can exhibit or observe as Θ. For
example, taxi services can exhibit numerous traits, including 'airport transfer'
and 'suitcase storage'. Each individual trustee agent, te, exhibits a subset of
these traits, $\theta^{te} \subseteq \Theta$, and each trustor agent, tr has an observation function,

$\mathcal{O}_{tr} : \mathcal{P}(\Theta) \rightarrow \mathcal{P}(\Theta)$. When presented with a trustee, this observation function determines how the traits of a trustee are interpreted, $\theta_{tr}^{te} = \mathcal{O}_{tr}(\theta^{te})$.

In a fully observable setting, it is valid for all agents to observe the traits of all trustees themselves. When assessing stereotype-reputation with full observability, witness agents can apply their observation function, $\theta_w^{te} = \mathcal{O}_w(te)$, and correctly interpret any associated stereotype. With partial observability the traits of some trustees may be unavailable, such as when there is a cost to making observations or if the trustees are in different locations. In such partially observable environments, traits may only be accessible when considering whether to interact with a trustee, i.e. when assessing direct-trust or stereotype-trust. An agent that has neither visited a city nor considered using a taxi there, for example, cannot use their observation function when acting as a witness for stereotype-reputation. In such cases the traits observed by the trustor must be assessed by witnesses instead.

If traits are observed objectively by agents, then observations made by a trustor are the same as those that a witness would make, i.e. $\mathcal{O}_{tr}(te) = \mathcal{O}_w(te)$. With objective observations, therefore, there is no issue with partial observability and a witness can directly assess observations made by the trustor. With subjectivity, however, agents may have no interest in a particular trait or interpret traits differently. A customer considering a taxi service for airport transfer who is carrying hand luggage only, for example, may not notice if the taxi service is able to accommodate suitcases or not. An observation of suitcase storage may then be meaningless to this customer, resulting in a poor a stereotype-reputation assessment. In another situation, two customers may have different interpretations of suitable storage for suitcases. Such subjective observations can lead to misunderstandings of stereotype-reputation assessments, and so a translation between the two subjective observations is required.

To overcome these potential misunderstandings, we propose that the trustor or witness learns to translate observations made by the trustor agent to what the witness would have observed. After the translation is made, the witness can assess the stereotype in a meaningful way and respond with their opinion. To learn such a translation function, either the trustor or witness must provide their observations of several trustees to the other. These observations do not have to be linked to a reputation assessment for the trustee, but can have been observed during other reputation assessments. Traits of trustee agents in both sets of observations can then be analysed for correlations and a translation learned. If the trustor observes 'suitcase storage' for several taxi services for which the witness has observed 'airport transfer', for example, a translation between the two observations can be learned. If the trustor observes 'suitcase storage' for an entirely new trustee, this can be translated into the witness's stereotype for 'airport transfer' when assessing stereotype-reputation.

4 The POSSTR Model

In assessing trust and reputation it is typical to aggregate ratings of previous interactions. An interaction between tr and te is recorded in the tuple

$\langle tr, te, \theta_{tr}^{te}, r_{tr}^{te} \rangle$, where θ_{tr}^{te} are the traits of te that were observed by tr prior to the interaction, and r_{tr}^{te} is the rating given by tr. Without loss of generality, we assume that ratings are binary, with 1 indicating success and 0 indicating otherwise. A real-valued rating can be converted to binary by choosing a threshold, above which the interaction is deemed successful and otherwise it is unsuccessful. The aim of the reputation assessment is then to determine the likelihood of a future interaction with a trustee being successful.

4.1   Direct-Trust

In evaluating the direct-trust of a trustee, te, a trustor, tr, aggregates their relevant interaction records, $\mathbf{I}_{tr}^{te}$, with te. There are many possible aggregations, but as in existing work on stereotypes [2,16], and BRS [9], we use one based on Subjective Logic (SL) [8]. SL is a belief calculus that can represent opinions as degrees of belief, b, disbelief, d, and uncertainty, u, in BDU triples, (b, d, u), where $b, d, u \in [0, 1]$, and $b + d + u = 1$. In SL, a completely uncertain opinion is represented as $(0,0,1)$, and total belief is represented as $(1,0,0)$. As evidence is accrued and the opinion changes, the degrees of belief, disbelief, and uncertainty change also.

In BRS [2,9], the trustor computes a BDU triple by counting the number of successful interactions they have had with the trustee, $p_{tr}^{te} = |\mathbf{I}_{tr}^{te} : r_{tr}^{te} = 1|$, and the number of unsuccessful interactions, $n_{tr}^{te} = |\mathbf{I}_{tr}^{te} : r_{tr}^{te} = 0|$. A mapping from interaction records and ratings to the belief, disbelief, and uncertainty is provided by,

$$b_{tr}^{te} = \frac{p_{tr}^{te}}{p_{tr}^{te} + n_{tr}^{te} + 2}, \qquad d_{tr}^{te} = \frac{n_{tr}^{te}}{p_{tr}^{te} + n_{tr}^{te} + 2}, \qquad u_{tr}^{te} = \frac{2}{p_{tr}^{te} + n_{tr}^{te} + 2}. \qquad (1)$$

If there are two ratings of 1 and one rating of 0, for example, the resulting BDU triple is $(0.4, 0.2, 0.4)$. This mapping ensures that uncertainty decreases monotonically as the evidence is accumulated. Other mappings from ratings to SL are possible, such as that proposed by Wang and Singh [21,22] where uncertainty is affected by disagreement in ratings as well as the amount of evidence.

The likelihood that a future interaction with te will be successful, is then calculated as,

$$P(\hat{r}_{tr}^{te} = 1) = b_{tr}^{te} + a_{tr}^{te} \times u_{tr}^{te}, \qquad (2)$$

where $\hat{r}_{tr}^{te}$ is the future rating being predicted and a_{tr}^{te} is the Bayesian prior. The prior in BRS [9] is $a_{tr}^{te} = 0.5$, which represents that an interaction with an unknown agent for which there is no information is equally likely to be successful or unsuccessful. A prior of greater than 0.5 means that uncertain opinions lean more to belief in success, whereas priors less than 0.5 make $P(\hat{r}_{tr}^{te} = 1)$ closer to 0. As evidence is gathered, the uncertainty reduces toward 0 and the prior has less of an effect on the likelihood of success. Stereotypes, as discussed in Sects. 4.3 and 4.4, can be used to inform this prior based on observations of trustee traits.

4.2 Witness-Reputation

When the trustor has insufficient ratings of a trustee, witnesses, $w \in W$, are asked to provide theirs. The witness ratings are then combined with those of the trustor using SL as described above,

$$p^{te} = p_{tr}^{te} + \sum_{w \in W} p_w^{te}, \quad n^{te} = n_{tr}^{te} + \sum_{w \in W} n_w^{te}, \tag{3}$$

where p_w^{te} and n_w^{te} are respectively the number of positive and negative interactions reported by witness, w, about te. Witness-reputation is then computed as,

$$P(\hat{r}_{tr}^{te} = 1) = b^{te} + a_{tr}^{te} \times u^{te}, \tag{4}$$

where the Bayesian prior is again $a_{tr}^{te} = 0.5$, and

$$b^{te} = \frac{p^{te}}{p^{te} + n^{te} + 2}, \quad u^{te} = \frac{2}{p^{te} + n^{te} + 2}. \tag{5}$$

4.3 Stereotype-Trust

Stereotypes can be used to inform the Bayesian prior in environments where trustees that exhibit similar observable traits have performed similarly in interactions. For instance, the ratings given to interactions with known agents can be used as the prior for an unknown agent with similar traits. A stereotype model,

$$f_{tr} : \mathcal{P}(\mathbf{\Theta}) \to \mathbb{R}, \tag{6}$$

is learned by tr, which maps traits of a trustee agent observed by tr to a stereotype-trust value,

$$a_{tr}^{te} = f_{tr}(\theta_{tr}^{te}), \tag{7}$$

that is used as the Bayesian prior in Eqs. 2 and 4 when computing direct-trust or witness-reputation.

The stereotype model is learned by generating a training sample for each agent the trustor has previously interacted with. In each of these samples, the te traits observed by tr are the input features, θ_{tr}^{te}. The target, or class value, is the direct-trust that tr has in te, as outlined in Sect. 4.1, with a Bayesian prior of 0.5. The training data is therefore a set of samples that express observed trustee traits and their direct-trust values. A regression model is then learned to map traits observed by tr to the trust in agents that express those traits, which can be used as the Bayesian prior in Eq. 2. As before, if the trustor has high uncertainty about a trustee and the stereotype model outputs a prior close to 0, the direct-trust will be low. As the trustor gains experience with trustee, the prior will have less effect on the trust value.

As in Burnett et al. [2] and Şensoy et al. [16], we learn the mapping from features of a trustee to the likelihood of a successful interaction using the M5 model tree algorithm [13]. The M5 model tree recursively splits training samples using the values of the features that best discriminate the class labels. Whereas in typical decision trees the leaves are target values, the leaves of the M5 tree are piecewise linear regression models that output the target value. The regression models are learned using samples that were not divided in learning the tree and therefore use features not specified by the ancestors of the leaf. If all features are specified, the linear regression model defaults to outputting the mean target value of the samples in the relevant split. The splitting process stops at the level where the leaf model would have the highest accuracy on the training data. If there are many traits observed by a trustor then it may be necessary to perform feature selection to reduce their number [4].

4.4   Stereotype-Reputation

When the trustor is not confident in their stereotype-trust assessment, witnesses can be asked for their stereotype based assessment of the trustee. As with the trustor, each witness, $w \in W$, has their own stereotype model,

$$f_w : \mathcal{P}(\mathbf{\Theta}) \to \mathbb{R}, \tag{8}$$

learned using their own experience of trustee agents. The witness in some cases may have observed the trustee previously, in which case they are able both to provide a witness-reputation assessment as well as use the features they observed, θ_w^{te}, in their stereotype model. In other cases the witness may have not observed the trustee previously and must rely on the stereotype features observed by the trustor, who may have observed different features in different ways. This necessitates a translation function between the two observation capabilities,

$$f_{tr \to w} : \mathcal{P}(\mathbf{\Theta}) \to \mathcal{P}(\mathbf{\Theta}). \tag{9}$$

This function converts observed features of a trustee from the subjective perspective of the trustor, tr, to that of the witness, w. It is a multi-target learning problem with an input of stereotype features the trustor, tr, has observed, θ_{tr}^{te}, and an output vector of features that the witness, w, would observe, $\hat{\theta}_w^{te}$.

To learn the translation function, training data is generated from common observations that both the witness and the trustor have made. When requesting a stereotype assessment from a witness, either the trustor provides their observations of other trustee agents to the witness or vice versa. These observations, consist of the observed traits along with the trustee identifier. As an example, consider that the trustor has observed the traits of three trustees, $\{\theta_{tr}^{te_1}, \theta_{tr}^{te_2}, \theta_{tr}^{te_3}\}$, and a witness has observed those of two, $\{\theta_w^{te_1}, \theta_w^{te_2}\}$. Training data can then be generated by matching up the common observations, as $\{\theta_{tr}^{te_1} : \theta_w^{te_1}, \theta_{tr}^{te_2} : \theta_w^{te_2}\}$, where ':' separates the inputs and outputs. These observations may have been made without having interacted with the trustees, such as a potential customer observing traits of taxis during an assessment but without using their service. These common observations samples form the training data that can be input into a multi-target learning algorithm [14].

Multi-target learning algorithms learn mappings from input features to multiple targets. One simple yet powerful approach is the binary relevance method [19], where a separate model is built for each target. In this paper, a model is learned that maps traits observed by the trustor to each trait that would be observed by the witness. The traits observed by the trustor are then input into each of the learned models and their outputs are combined to be the traits the witness would have observed. As the base learning algorithm for each of the output traits we use Naïve Bayes, although any classification algorithm may be used in its place [4, 14].

If a witness has not observed the trustee, the trustor's observations are input into the learned translation,

$$\hat{\theta}_w^{te} = f_{tr \to w}(\theta_{tr}^{te}), \tag{10}$$

to estimate the traits that they would have observed. This output is then used in the witness stereotype model,

$$a_w^{te} = f_w(\theta_w^{te}|\hat{\theta}_w^{te}) = \begin{cases} f_w(\theta_w^{te}) & \text{if witness observed trustee,} \\ f_w(\hat{\theta}_w^{te}) & \text{if trustor provided observations,} \end{cases} \tag{11}$$

which outputs the prior from the witness perspective to be returned to the trustor. A new Bayesian prior is then computed as the mean stereotype assessment of the trustor and witnesses,

$$a^{te} = \frac{1}{|W|+1}\left(a_{tr}^{te} + \sum_{w \in W} a_w^{te}\right). \tag{12}$$

Finally, the overall reputation score is computed as,

$$P(\hat{r}_{tr}^{te} = 1) = b^{te} + a^{te} \times u^{te}. \tag{13}$$

4.5   Subjective Opinions

In many domains, witnesses cannot be assumed to rate interactions objectively or report ratings benevolently. This is the same for witness-reputation as it is for stereotype-reputation, where witnesses may be dishonest or otherwise have different opinions about a trustee or its traits. While this issue is out of the scope of this paper, there are two broad approaches to dealing with this problem. First, information provided by unreliable witnesses can be discounted, or weighted lower than more reliable information [16, 17]. In this method, opinions of a witness are compared to those of the trustor for the same trustees or traits. If there is a significant difference in opinions then the witness is deemed unreliable and their reports are discounted before being combined with others. Zhang et al. [24] evaluate the reliability of witnesses by comparing their reports to trustor ratings as well as those of other witnesses. Second, if witnesses are unreliable in a consistent way, their opinions can be reinterpreted to be from the perspective of the trustor [11,15,17]. It is worth noting that these translations are different to the observation translations proposed in Sect. 4.4, as they aim to translate a single variable (ratings) with potentially different ranges, whereas our translation is more general and aims to translate multiple observed traits. As with discounting, opinions of the witnesses and trustor are compared to learn a mapping from one to the other, but investigating either approach to subjective ratings alongside partially observable trustees and subjective stereotypes is out of the scope of this paper.

5   Evaluation and Results

To evaluate POSSTR we use a simulated marketplace based on that used by Burnett et al. [2] and Şensoy et al. [16]. The simulation consists of trustor and trustee agents that interact over 250 rounds. Each trustee agent is randomly assigned one of five profiles at the beginning of the simulation, defining a mean, standard deviation (STD), and observable traits, θ^{te}, as outlined in Table 1. The mean and STD define the Gaussian distribution from which interaction outcomes are drawn. As in Burnett et al. [2] and Şensoy et al. [16], an interaction with an outcome greater than a success threshold of 0.5 is deemed successful and given a rating of 1 by the trustor, otherwise it is rated as 0. The observable traits distinguish each of the profiles, to be used in stereotype assessments of trustees. Each element in these feature vectors can be interpreted as the trustee exhibiting a trait or not, e.g. the first trait may represent 'airport transfer'.

Table 1. Objective trustee profiles. The observations of an example observation vector, $\mathcal{O}_{tr}$, are also shown.

Profile	Description	Mean	STD	θ^{te}	$\mathcal{O}_{tr} = 001122$
1	Usually good	0.9	0.05	100001	100010
2	Often good	0.6	0.15	010100	010011
3	Often poor	0.4	0.15	001100	000011
4	Usually poor	0.3	0.05	011010	010001
5	Random	0.5	1.00	011001	010010

Each trustor and trustee agent leaves the simulation with a probability of 0.05 in each round, to be replaced by another. New trustees are assigned a profile selected uniformly at random from those in Table 1. The number of agents in the simulation is static, therefore, and in all of our simulations there were 100 trustee agents and 20 trustor agents. In each round, each trustor agent is given a random 10 available trustees from which they select the one with highest reputation as an interaction partner. Similarly, in each reputation assessment, each trustor requests witness-reputation and stereotype-reputation from 10 random witnesses.

Table 2. Reputation assessment strategies investigated listing their information sources and definitions.

Strategy	Description	Definition
Random	No information	NA
T	Direct-trust	Eq 2, where $a_{tr}^{te} = 0.5$
TR	Direct-trust + witness-reputation	Eq 4, where $a_{tr}^{te} = 0.5$
T+ST	Direct-trust + stereotype-trust	Eq 2, where $a_{tr}^{te} = f_{tr}(\theta_{tr}^{te})$
TR+ST	Direct-trust + witness-reputation + stereotype-trust	Eq 4, where $a_{tr}^{te} = f_{tr}(\theta_{tr}^{te})$
TR+STR	Direct-trust + stereotype-trust + witness-reputation + stereotype-reputation	Eq 13, where $a_w^{te} = f_w(\theta_w^{te}\|\theta_{tr}^{te})$, $\forall w \in W$, and $a_{tr}^{te} = f_{tr}(\theta_{tr}^{te})$
POSSTR	Direct-trust + stereotype-trust + witness-reputation + stereotype-reputation (with translation)	Eq 13, where $a_w^{te} = f_w(\theta_w^{te}\|\hat{\theta}_{tr}^{te})$, $\forall w \in W$, and $a_{tr}^{te} = f_{tr}(\theta_{tr}^{te})$

Trustee traits are observed subjectively through trustor observation functions, $\mathcal{O}_{tr}(\theta^{te})$, defined by an observation vector, $\mathcal{O}_{tr}$, assigned to each new trustor. The observation vector is the same length as the number of traits in the network and each value corresponds to an observable trait. A value of 0 means that the trait is observed with the correct value if it is expressed by a trustee, and 1 means that the trait is never observed (or always observed as 0). A value of 2 in the vector means that the trustor always changes the value of the trait, i.e. a trustee trait of value 0 is observed as a 1 and vice versa. An example observation vector, along with the traits observed by such a trustor, is shown in the final column of Table 1. Observation vectors are sampled from a distribution defined by subjectivity parameters, s and o, which determine the likelihoods of 0, 1, or 2. A value is 1 with a probability of o, and given that its value is not 1 it has a value of 2 with probability of s. A higher o means that more traits are ignored, and a higher s increases the likelihood that a trait is interpreted incorrectly.

In our experiments we compare each of the strategies outlined in Table 2. For example, the T+ST strategy combines direct-trust and stereotype-trust information in the assessment of trustee agents, as defined by Eq. 2. Similarly, TR+STR uses all four information sources in each reputation assessment, regardless of any confidence that may be derived from the number of experiences. At the end of each round, the mean overall utility gained by all agents is computed and recorded as the simulation utility. All results presented in this paper are averaged over 50 iterations of our simulation. In all settings and for all strategies the standard deviation of simulation utilities was less than 5% of the mean, and the standard error was less than 1% of the mean. Also, all significance results discussed are from an ANOVA followed by an all-pairs t-test, with multiple comparisons normalised using the Bonferroni correction.

5.1   Full Observability and Objective Traits

Table 3 shows the mean utilities after 250 rounds over the 50 iterations, with fully observable trustees and objectively observable traits ($s = o = 0$). The differences between each pair of strategies, excluding TR+STR and POSSTR, was significant with $p < 0.01$. The strategy that gained the lowest utility in all cases was Random, followed by using direct-trust only (T). Using witness opinions alongside direct-trust (TR), trustor agents were able to choose better interaction partners. This extra information gathered from witnesses is clearly advantageous, given that trustor exploration of the trustee population was limited and agent turnover was high.

Table 3. Fully observable trustees with objective traits. Utilities significantly smaller ($p < 0.01$) than that of POSSTR are prepended with a '*'.

Strategy	Mean utility	STD utility
Random	*126.73	5.36
T	*142.31	5.17
TR	*194.80	4.78
T+ST	*186.56	4.78
TR+ST	*217.07	3.98
TR+STR	226.48	3.36
POSSTR	227.60	2.14

Trustor agents were better able to search the trustees and gain good utilities when they combined witness-reputation with either stereotype-trust or stereotype-reputation. Combining direct-trust with stereotype-trust (T+ST) was also beneficial when compared to using only direct-trust (T) although the utility gained was significantly lower than using witness-reputation (TR). The highest utilities were gained when all four kinds of trust and reputation were used (TR+STR and POSSTR). With full observability and no subjectivity the translation was not required and there was no advantage to using POSSTR over using the observed traits directly, as in TR+STR.

5.2   Partially Observable Trustees

To model partial observability we first restricted observations of trustee traits to those made during previous assessments. If a witness had not previously assessed direct-trust of a trustee, they were unable to observe the traits themselves and used those observed by the trustor. This restriction on observations

Table 4. Utilities for strategies with different levels of subjectivity. STD shown in braces after each result and results significantly smaller ($p < 0.01$) from POSSTR are prepended with '*'.

	o = 0	o = 0.25	o = 0.5	o = 0.75
Random	*126.73 (5.36)	*125.69 (4.14)	*123.67 (4.78)	*125.40 (5.31)
T	*142.31 (5.17)	*140.49 (4.76)	*141.13 (5.71)	*141.58 (4.59)
TR	*194.80 (4.78)	*195.42 (5.16)	*195.57 (5.38)	*194.22 (4.97)
T+ST	*186.56 (4.78)	*181.89 (4.56)	*177.39 (5.79)	*156.78 (6.33)
TR+ST	*217.07 (3.98)	*214.91 (3.81)	*213.26 (3.66)	*205.89 (5.07)
TR+STR	226.51 (3.27)	223.65 (2.88)	221.49 (3.57)	214.75 (4.58)
POSSTR	226.68 (3.74)	223.65 (3.98)	223.76 (3.93)	215.54 (4.52)

(a) Observed traits are not changed (s = 0)

	s = 0	s = 0.25	s = 0.5	s = 0.75	s = 1
Random	*126.73 (5.36)	*125.04 (5.79)	*126.29 (5.53)	*125.40 (5.52)	*126.28 (4.55)
T	*142.31 (5.17)	*142.00 (5.74)	*141.77 (4.34)	*141.33 (5.00)	*141.05 (4.61)
TR	*194.80 (4.78)	*195.29 (6.15)	*194.71 (5.16)	*195.34 (6.06)	*195.38 (4.85)
T+ST	*186.56 (4.78)	*185.73 (4.78)	*186.25 (5.12)	*186.78 (5.03)	*186.76 (5.47)
TR+ST	*217.07 (3.98)	*217.08 (3.58)	*216.57 (3.53)	*216.46 (3.28)	*216.49 (3.65)
TR+STR	226.51 (3.27)	*223.01 (3.68)	*220.11 (3.42)	*222.46 (4.00)	226.47 (3.57)
POSSTR	226.68 (3.74)	226.94 (3.19)	226.22 (3.04)	226.44 (2.83)	226.08 (3.14)

(b) All traits are observed (o = 0)

was also applied when generating training data for the translation function in POSSTR. Tables 4(a) and (b) show the utilities gained for strategies in this assessment-restricted partially observable setting, for different levels of subjectivity. The results in Table 4(a) are for different values of o with $s = 0$ (agents observed traits with different likelihoods) and the results in Table 4(b) are for different values of s with $o = 0$ (agents flipped values of traits with different likelihoods). As with full observability, a significant difference was observed between each pair of strategies, other than TR+STR and POSSTR, within each subjectivity and observability condition. For all levels of subjectivity, the utilities gained by strategies that do not use stereotypes, namely Random, T, and TR, were the same as with full observability. Similarly, with objective traits, i.e. $o = s = 0$, the utilities for TR+STR and POSSTR, which both use stereotype-reputation, were not significantly different ($p > 0.05$) from with full observability. This is because observations made by a trustor were the same as those that a witness would have made in this setting.

For all strategies that use stereotypes, higher values of o led to lower utilities, with performance being substantially lower when $o = 0.75$. This is likely due to there being fewer traits observed by the trustor agents, meaning there is less distinction between the trustee profiles. The value of s had no significant effect ($p > 0.05$) on the performance of strategies that did not use stereotype-reputation, including T+ST and TR+ST. When trustors had to communicate their observed traits to witnesses subjectively, i.e. when $0 < s < 1$, the TR+STR, which does employ stereotype-reputation, was negatively affected. POSSTR did not suffer any significant loss ($p > 0.05$) in utility gain over all values of s, as a result of it successfully translating observed traits before computing stereotype-reputation. When $0 < s < 1$, therefore, POSSTR significantly outperformed TR+STR ($p < 0.01$), again after pairwise t-tests with the Bonferroni correction. This means that POSSTR reliably assessed reputation with partially observable trustees and subjectively interpreted traits, while TR+STR did not.

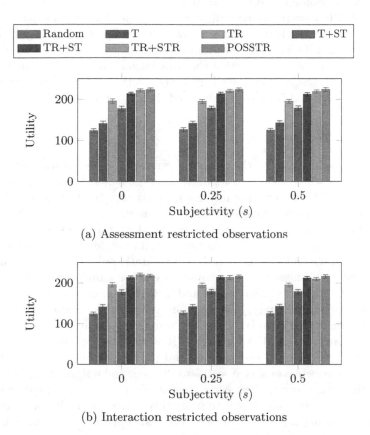

(a) Assessment restricted observations

(b) Interaction restricted observations

Fig. 1. Utilities for different levels of subjectivity with $o = 0.5$, with partially observability and observations are restricted to previous (a) assessments and (b) interactions. The error bars show the STD.

Figure 1(a) shows the utilities gained by strategies for $o = 0.5$ and $s = 0$, 0.25, and 0.5. The results are similar to those found in Table 4(b), where $o = 0$, and show that POSSTR had significantly the highest performance in all cases. The results for $s = 0.75$ and $s = 1$ are omitted from this plot for clarity, as the utilities under these conditions were mirrored by those when $s = 0.25$ and $s = 0$ respectively.

To restrict observability further we limited observations to interactions, meaning that a witness must have interacted with a trustee for their traits to be available when assessing stereotype-reputation. These results, for different values of s and $o = 0.5$, are presented in Fig. 1(b), where again the strategies that do not use witness-stereotypes were unaffected by the observability of trustee traits. The TR+STR and POSSTR strategies both gained lower utilities when traits were subjectively observed in this setting than when observability was restricted to assessments. With $o = 0.5$ and $s = 0.5$, POSSTR again significantly ($p < 0.01$) outperformed all other strategies but was outperformed by TR+STR ($p < 0.05$) when $s = 0$ and traits were objective. No significant difference between TR+ST, TR+STR, and POSSTR was found ($p > 0.05$) when $s = 0.25$. These results indicate that the translation function is outputting traits as they would be observed by the witness incorrectly, possibly due to the lack of training data gathered from traits observed during interactions.

5.3   Private Trustees

In this case, the identifier of the trustee agent being assessed was not disclosed to the witnesses when asking for reputation assessments. While this is extreme, a trustor may wish to keep their interest in particular trustee agents private for several reasons, including competition, embarrassment, or affects to reputation. For example, a trustor's interest in a particular doctor may reveal private health information or their interest in a particular subprovider may negatively affect their own reputation. This case is also representative of when trustees are regularly unknown to witnesses, such as when they are in a different locations. When unable to use witness-reputation, TR and TR+ST are equivalent to the T and T+ST strategies respectively, and therefore gained the same utilities. Utilities gained over simulation rounds for the five remaining strategies are presented in Fig. 2, which includes utilities for POSSTR with training data for the translation function limited to (a) assessments, and (b) interactions. In all simulations, subjectivity parameters of $s = o = 0.5$ were used, and a significant difference in overall utility was observed between all pairs of strategies presented ($p < 0.01$). The TR+STR model is equivalent to using direct-trust and witness-stereotypes, or T+STR, causing its performance to drop significantly over this simulation compared to those in Sects. 5.1 and 5.2.

After an initial learning phase lasting fewer than 10 rounds POSSTR (a), which learned the translation function using observations made in previous assessments, gained by far the highest utilities in each round. With less training data, POSSTR (b) gained lower utility, but still outperformed all the other strategies that either did not translate traits or did not use stereotypes. After around 50 rounds the utilities gained per round for all of the strategies stabilised. Interestingly, in this setting there was still a benefit to using witness-stereotypes (TR+STR) over just direct-trust (T), even though witnesses may have misinterpreted the observations made by the trustor. Without the translation function the best strategy was to use only direct evidence in the form of direct-trust and stereotype-trust (T+ST).

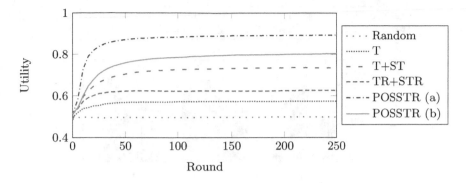

Fig. 2. Utilities for strategies with private trustee identifiers, where $o = s = 0.5$.

5.4 Fewer Available Advisors

The results in Fig. 3(a) show the utilities gained when the number of witnesses available was reduced to two. In these results observations were limited to assessments, the subjectivity parameters were $o = 0.5$ and $s = 0, 0.25$, or 0.5. As a result, all strategies that use witness information gained less utility than with ten advisors. In these results using only direct-trust and witness-reputation (TR) was outperformed by the combination of direct-trust and stereotype-trust, with $p < 0.01$. The extra utility gained by POSSTR compared to TR+STR when either $s = 0.25$ or $s = 0.5$, was also significantly ($p < 0.01$) more than with ten witnesses.

5.5 Increased Dynamism

Fig. 3(b) shows results for simulations with increased dynamism, where agents departed with an increased probability of 0.2. Again, in these results observability was restricted to assessments, $o = 0.5$, and $s = 0, 0.25$, or 0.5. All strategies other than Random performed less well in this setting, and gained lower utilities than in the less dynamic scenario where agents left with a probability of 0.05. Also in this highly dynamic setting POSSTR gained much more utility than the other strategies with subjectivities of $s = 0.25$ or 0.75 ($p < 0.01$).

5.6 Summary

In summary, we found POSSTR gained significantly more utility than all other strategies, including TR+STR, in environments where partial observability was combined with subjectivity. The difference was greatest when witnesses were unable to observe traits themselves due to the trustor withholding their identities, where the performance of POSSTR was affected much less than the other strategies. In simple environments, with either full observability or objective traits, the performance of POSSTR was not significantly different to that of TR+STR, and both gained more utility than the other strategies.

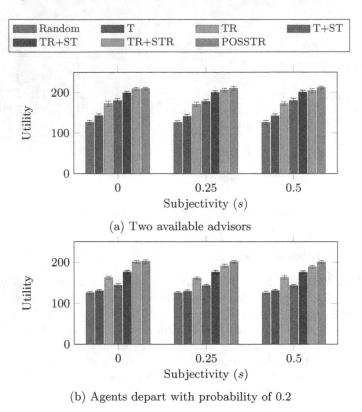

Fig. 3. Utilities for strategies with different levels of subjectivity with $o = 0.5$, assessment partial observability, for (a) fewer advisors and (b) increased dynamism. The error bars show the STD.

6   Conclusion

In this paper we have presented the POSSTR reputation system, which combines direct-trust, witness-reputation, stereotype-trust, and stereotype-reputation, and is robust to various levels of partial and subjective observability. Using simulations we have shown that a translation function is necessary when communicating observed traits to witnesses in partially observable and subjective environments. We found that POSSTR provided significantly more reliable reputation assessments compared to other strategies in such settings.

In settings without partial observability, where witnesses were able to observe all trustee traits themselves, the utilities gained by POSSTR and TR+STR, which both use direct-trust, witness-reputation, direct-stereotypes, and witness-stereotypes, were not significantly different. This was because the translation function employed in POSSTR has no effect when agents can observe the traits of all trustee agents. With no observability, where trustors concealed the identities of trustees they were assessing, using witness stereotypes without translation provided lower utilities than using only direct evidence. With translations, however, POSSTR was able to retain much of the performance observed in much less restricted settings with full observability.

Investigating subjectivity and dishonesty in interaction ratings is left as future work, but could be solved using a strategy such as TRAVOS [18] or HABIT [17]. Either of these strategies can be applied directly to witness-reputation described in this paper, but applying them to subjective-reputation may require some alterations. Another approach is to learn a mapping, akin to the translation function for observed traits, to translate reputation-assessments from one perspective to the other.

Another limitation is that concept drift, where the profile parameters or traits change over time, is not considered. To overcome such drift a learning window is often sufficient, but determining an appropriate window size is non-trivial. Another approach may be to apply techniques from the concept drift literature [5], to both detect when a change has occurred in the underlying profiles and adapt the model accordingly.

References

1. Artz, D., Gil, Y.: A survey of trust in computer science and the semantic web. Web Semant. Sci. Serv. Agents World Wide Web **5**(2), 58–71 (2007)
2. Burnett, C., Norman, T., Sycara, K.: Stereotypical trust and bias in dynamic multiagent systems. ACM Trans. Intell. Syst. Technol. **4**(2) (2013)
3. Fang, H., Zhang, J., Şensoy, M., Thalmann, N.: A generalized stereotypical trust model. In: International Conference on Trust, Security and Privacy in Computing and Communications, pp. 698–705, June 2012
4. Frank, E., Hall, M., Witten, I.: Data Mining: Practical Machine Learning Tools and Techniques, 4th edn. Morgan Kaufmann, Burlington (2016)
5. Gama, J., Žliobaitė, I., Bifet, A., Pechenizkiy, M., Bouchachia, A.: A survey on concept drift adaptation. ACM Comput. Surv. **46**(4), 44:1–44:37 (2014)
6. Hendrikx, F., Bubendorfer, K., Chard, R.: Reputation systems: a survey and taxonomy. J. Parallel Distrib. Comput. **75**, 184–197 (2015)
7. Huynh, T., Jennings, N., Shadbolt, N.: An integrated trust and reputation model for open multi-agent systems. Auton. Agents Multi-Agent Syst. **13**(2), 119–154 (2006)
8. Jøsang, A.: A logic for uncertain probabilities. Int. J. Uncertain. Fuzziness Knowl. Based Syst. **09**(03), 279–311 (2001)
9. Jøsang, A., Ismail, R.: The Beta reputation system. In: Electronic Commerce Conference, pp. 41–55 (2002)
10. Jøsang, A., Ismail, R., Boyd, C.: A survey of trust and reputation systems for online service provision. Decis. Support Syst. **43**(2), 618–644 (2007)
11. Koster, A., Schorlemmer, M., Sabater-Mir, J.: Engineering trust alignment: theory, method and experimentation. Int. J. Hum. Comput. Stud. **70**(6), 450–473 (2012)
12. Liu, X., Datta, A., Rzadca, K.: Trust beyond reputation: a computational trust model based on stereotypes. Electron. Commer. Res. Appl. **12**(1), 24–39 (2013)
13. Quinlan, R.: Learning with continuous classes. In: Australian Joint Conference on Artificial Intelligence, pp. 343–348. World Scientific, Singapore (1992)

14. Read, J., Reutemann, P., Pfahringer, B., Holmes, G.: MEKA: a multi-label/multi-target extension to WEKA. J. Mach. Learn. Res. **17**(21), 1–5 (2016)
15. Regan, K., Poupart, P., Cohen, R.: Bayesian reputation modeling in e-marketplaces sensitive to subjectivity, deception and change. In: National Conference on Artificial Intelligence, vol. 2, pp. 1206–1212, July 2006
16. Şensoy, M., Yilmaz, B., Norman, T.: STAGE: stereotypical trust assessment through graph extraction. Computat. Intell. **32**(1), 72–101 (2016)
17. Teacy, L., Luck, M., Rogers, A., Jennings, N.: An efficient and versatile approach to trust and reputation using hierarchical Bayesian modelling. Artif. Intell. **193**, 149–185 (2012)
18. Teacy, L., Patel, J., Jennings, N., Luck, M.: Coping with inaccurate reputation sources: experimental analysis of a probabilistic trust model. In: International Joint Conference on Autonomous Agents and Multiagent Systems, pp. 997–1004 (2005)
19. Tsoumakas, G., Katakis, I.: Multi-label classification: an overview. Int. J. Data Warehous. Min. **3**(3), 64–74 (2007)
20. Wahab, O., Bentahar, J., Otrok, H., Mourad, A.: A survey on trust and reputation models for web services: single, composite, and communities. Decis. Support Syst. **74**, 121–134 (2015)
21. Wang, Y., Singh, M.: Formal trust model for multiagent systems. In: International Joint Conference on Artifical Intelligence, pp. 1551–1556. Morgan Kaufmann, San Francisco, January 2007
22. Wang, Y., Singh, M.: Evidence-based trust: a mathematical model geared for multiagent systems. ACM Trans. Auton. Adapt. Syst. **5**(4), 14:1–14:28 (2010)
23. Yu, H., Shen, Z., Leung, C., Miao, C., Lesser, V.: A survey of multi-agent trust management systems. IEEE Access **1**, 35–50 (2013)
24. Zhang, J., Cohen, R.: Evaluating the trustworthiness of advice about seller agents in e-marketplaces: a personalized approach. Electron. Commer. Res. Appl. **7**(3), 330–340 (2008)

Working Together: Committee Selection and the Supermodular Degree

Rani Izsak[✉]

Department of Computer Science and Applied Mathematics,
Weizmann Institute of Science, Rehovot, Israel
rani.izsak@gmail.com

Abstract. We introduce a voting rule for committee selection that captures positive correlation (synergy) between candidates. We argue that positive correlation can naturally happen in common scenarios that are related to committee selection. For example, in the movies selection problem, where prospective travelers are requested to choose the movies that will be available on their flight, it is reasonable to assume that they will tend to prefer voting for a movie in a series, only if they can watch also the former movies in that series. In elections to the parliament, it can be that two candidates are working extremely well together, so voters will benefit from being represented by both of them together.

In our model, the preferences of the candidates are represented by set functions, and we would like to maximize the total satisfaction of the voters. We show that although computing the best solution is $\mathcal{NP}$-hard, there exists an approximation algorithm with approximation guarantees that deteriorate gracefully with the amount of synergy between the candidates. This amount of synergy is measured by a natural extension of the supermodular degree [Feige and Izsak, ITCS 2013] that we introduce – the joint supermodular degree. To the best of our knowledge, our results represent the first voting rule that capture synergy between specific candidates.

1 Introduction

Consider the following scenario (see, e.g., [9, 21]). An airline wishes to increase the satisfaction of the travelers by letting them choose the set of movies that will be available on their flight. It is decided to store on the airplane some fixed number k of movies. The airline surveys the preferences of the prospective passengers of the flight, and aims to make the best decision given their preferences. Two questions arise. First, how should the preferences of the prospective travelers be modeled? Second, given the preferences of the travelers, how should the set of movies be chosen? This problem of choosing some fixed number of candidates to the satisfaction of the voters is a fundamental problem. Generally speaking,

© Springer International Publishing AG 2017
G. Sukthankar and J. A. Rodriguez-Aguilar (Eds.): AAMAS 2017 Best Papers,
LNAI 10642, pp. 103–115, 2017.
https://doi.org/10.1007/978-3-319-71682-4_7

in the k-COMMITTEE SELECTION problem, we have a set V of n voters and a set C of m candidates, and we would like to select k candidates out of the m, such that the voters will be most satisfied. The answers to the two questions above vary in the literature. For example, by the Chamberlin-Courant rule we have a value for each of the candidates, by each of the voters, and the satisfaction of a voter is measured by the highest value she has for any elected candidate. The overall satisfaction is either the sum of the values of the voters or the value of the least satisfied voter (utilitarian [5] or egalitarian [2] variant, respectively). Other possibilities are to aggregate for every voter her value for every elected candidate or to give higher weight for candidates ranked higher by her (e.g. Borda rule). In a recent work, Skowron, Faliszewski and Lang [21] introduce an elegant model that captures the latter examples as well as others. They model the preferences of each voter by an intrinsic value for each of the candidates. Then, they calculate the value of a possible set of k candidates by a voter, by ordering her k intrinsic values for the k candidates, and multiplying them by some weight that corresponds to their rank in the order. This vector of weights is called "OWA operator" (Ordered Weighted Average). Skowron, Faliszewski and Lang [21] study their model for different restrictions on the OWA vector. Among their results, they show a $(1 - 1/e)$-approximation algorithm for the case of non-increasing weights OWA vectors, by showing it is captured by submodular set functions.[1]

However, none of the models above capture positive correlation (i.e. synergy) between specific candidates (see Sect. 2.2 for further discussion). Positive correlation can happen in various cases: from two candidates to the parliament that are working great together (see Woolley et al. [22] for a research about collective intelligence), to a series of movies that people tend to prefer watching the latter parts only after watching the former parts. In this paper we suggest a voting rule that captures positive correlation between specific candidates. Specifically, our answers to the two questions above are:

- The preferences of each of the candidates are modeled by a non-decreasing monotone set function from subsets of candidates to non-negative real numbers.
- A set of k candidates that maximizes the sum of values of the voters is elected.

We formally present our model in Sect. 4. In order to measure the amount of synergy between different candidates, we extend the supermodular degree [10], by introducing the joint supermodular degree (Sect. 4.1).

We also study applications for the model. In Sect. 5, we justify the naturalness of the joint supermodular degree from an applicative view point. In Sect. 6, we demonstrate how preference elicitation can be practically done.

---

[1] A submodular set function is a function $f : 2^M \to \mathbb{R}^+$, such that for every $S' \subseteq S \subseteq M$, and every $j \in M$, $f(j \mid S') \geq f(j \mid S)$, where $f(j \mid S) = f(\{j\} \cup S) - f(S)$ is the marginal value of j with respect to S. That is, the marginal values are monotone non-increasing.

Finally, in Sect. 7, we study the computability of our voting rule. On the bright side, we show that although computing the optimum is, generally, $\mathcal{NP}$-hard, one can approximate the optimum with a guarantee that depends on the amount of synergy between different candidates, as measured by the joint supermodular degree. On the flip side, we show that the same results cannot be achieved for the supermodular degree.

2 Preliminaries

The definitions below are taken from the works [10, 11]. Let C be a set of items (e.g. candidates in election, movies to watch on an airplane) and let $f : 2^C \to \mathbb{R}^+$ be a set function (e.g. of preferences of one of the voters). The following definition is standard.

Definition 1. *Let $c \in C$. The* **marginal set function** $f_c : 2^{C \setminus \{c\}} \to \mathbb{R}^+$ *is a function mapping each subset $S \subseteq C \setminus \{c\}$ to the marginal value of c given S:*

$$f_c(S) \stackrel{def}{=} f(S \cup \{c\}) - f(S) .$$

We denote the marginal value $f_c(S)$ by $f(c \mid S)$. For $S' = \{c_1, \ldots, c_{|S'|}\} \subseteq C$ and $S \subseteq C \setminus S'$ we also use either of the notations $f(c_1, \ldots, c_{|S'|} \mid S)$ or $f(S' \mid S)$ to indicate $f(S \cup S') - f(S)$.

The following definitions were introduced by Feige and Izsak [10].

Definition 2. *Let $c \in C$. The* **supermodular dependency set** *of c by f is the set of all items $c' \in C$ such that there exists $S \subseteq C \setminus \{c, c'\}$ such that $f(c \mid S \cup \{c'\}) > f(c \mid S)$. We denote the supermodular dependency set of c by $\mathcal{D}_f^+(c)$. We sometimes omit f, when it is clear from the context.*

Definition 3. *The* **supermodular degree** *of f is defined as $\mathcal{D}_f^+ \stackrel{def}{=} \max_{c \in C} |\mathcal{D}_f^+(c)|$.*

2.1 Representation of Set Functions

Let $f : 2^C \to \mathbb{R}^+$ be a set function. Then, f associates values to $2^{|C|}$ possible subsets. If we want our algorithms to run in time polynomial in $|C|$, they, of course, cannot read an input that is exponential in C. Therefore, it is crucial to consider the representation of set functions. One common way to represent set functions is by queries. Another is by an explicit representation. In this section, we mention both.

Queries. The arguably simplest queries are the following.

Definition 4. *Value queries for f are defined as follows:*
Input: A subset $S \subseteq C$.
Output: $f(S)$.

That is, if we assume our algorithm has access to value queries for a given set function, we merely assume it can ask for the value of a subset by the function. Another type of queries that we use (see [10]) is the following.

Definition 5. *Supermodular queries for f are defined as follows:*
Input: An item (i.e. a candidate) $c \subseteq C$.
Output: $\mathcal{D}_f^+(c)$.

That is, given a candidate we can ask with whom she has a positive correlation as defined by the supermodular dependencies. In the context of movies, we can ask for a movie that is part of a series, what are the other movies in that series. See Sect. 5 for further discussion.

An Explicit Representation. Another way to represent set functions is by an explicit representation. For example, any set function can be represented in a unique way by a hypergraph with weighted edges (see [1,6,8]). In this representation, a vertex is introduced for each of the items in the ground set of f. The weights in the sub-hypergraph induced by a set of vertices sum up exactly to the value of the subset with the respective items, by f. To see how weights can be allocated, consider the following iterative process. To hyperedges of size 1, we allocate weights that are the values of the respective singleton subsets. Note that this allocation of weights to hyperedges is unique. Then, for hyperedges of size 2, we allocate weights that are the difference between the value of the respective subset and the sum of the weights of their two singleton subsets. Note that this allocation is unique, as well. Also note that after iteration ℓ, the values by the hypergraph representation are correct for subsets of size up to ℓ. We proceed iteratively till we arrive to the unique edge of size $|C|$, and then we have a representation of the set function for any size of subset.

A succinct representation. We say that a representation of a set function is succinct if its size is polynomially bounded by the size of the ground set of the function. Note that in the hypergraph representation, we can list only the edges of value different from 0. So, sometimes this representation can be succinct. In particular, for additive set functions we clearly allocate non-zero values only for the hyperedges of size 1.

2.2   Related Work

We list here some of the voting rules from the literature, mostly based on Masthoff [16], and also on the works [7,9,15–17,21].

- Plurality: When electing a single candidate, plurality means selecting the candidate who is ranked first among the candidates, for the highest number of voters. When "ranked first" can mean that by the voting rule, preferences are ranks of candidates, or alternatively, that there are values for the candidates by the different voters that are used in order to get the candidates' ranks. In order to use this rule for choosing k candidates, one can just repeat it k times, while removing the winner at each iteration.

- Utilitarian: Each voter has a value for each of the candidates, and these values are summed up. The k candidates with the largest sums win.
- Borda [4]: This voting rule assumes the preferences of the candidates are modeled as a list of ranks, and it converts this list to values, with higher values for higher ranks: $m - 1, m - 2, \ldots, 0$ for ranks $1, \ldots, m$, respectively (m is the number of candidates). These values are summed up and highest scores win, similarly to the utilitarian rule above.
- Copeland: The score of a candidate is the number of pairwise elections she wins (by plurality) minus the number of pairwise elections she loses (ties do not count). Values are again, summed up, and higher scores win.
- Maximin: The score of a candidate c with respect to a candidate c' is the number of voters that prefer c over c' (we denote it by $\text{score}_c(c')$). The score of a candidate c is the minimum score of c with respect to a candidate (i.e. $\text{argmin}_{c'}\ \text{score}_c(c')$). For example, if for a candidate c, there exists a candidate that is preferred by all of the voters, then c will get a value of 0. If for a candidate c, *all* the voters prefer it over *all* the candidates, then (and only then) she will get the maximal score of n (i.e. the number of voters).
- Approval voting: Each voter either approves or disapproves every candidate. The k candidates with largest number of approvals win.

Positional scoring. Positional scoring is a bunch of voting rules, where the preferences of the voters are just an ordering of the candidates and the rule is defined by a vector of size m of values corresponding to positions by the voters. The total value of a candidate is the sum of these values of the voters. Note that plurality is a positional scoring rule with the vector $(1, 0, \ldots, 0)$ and Borda is a positional scoring rule with the vector $(m - 1, m - 2, \ldots, 0)$. There is also a rule called "Veto" where the vector is $(1, \ldots, 1, 0)$, so a voter actually chooses one candidate she prefers *not* to include in the selected committee.

Weighted aggregation of preferences of a voter. Skowron et al. [21] introduced the following family of voting rules for choosing k out of m candidates. The preferences of the voters are intrinsic values for the different candidates, and additionally, there is a vector of size k that is called OWA (ordered weighted average). When calculating the value for a set of k candidates by the preferences of a single voter, we do the following. We order the k candidates by their values according to the voter, in an increasing order of values, and then we sum up the values multiplied by the OWA vector (inner product). That is, every value is multiplied by a weight appearing in the OWA vector that corresponds to the rank of the candidate by the voter. To calculate the overall value of a subset of k candidates, we sum up the values of this set of candidates by the voters (utilitarian model). Skowron, Faliszewski and Lang [21] show that when the OWA vector is non-increasing (that is higher ranked candidates by a voter are multiplied by higher (or equal) weights), then the preferences of the voters can be represented by a submodular set function, and therefore a $(1 - 1/e)$-approximation guarantee can be achieved in polynomial time, by using the classical algorithm of Fisher, Nemhauser and Wolsey [14]. When the OWA vector is not non-increasing, some positive correlation between the candidates can happen, but not between specific candidates. For example, in the min OWA vector $(0, \ldots, 0, 1)$, only the worst candidate in the selected committee counts. This means, roughly speaking, that all the candidates should be adequate by a voter in order to have an adequate score by her. In terms of set functions, it means as follows. The marginal value of a candidate is 0 with respect to any committee that contains a worse (or equal) candidate. The marginal value of a candidate with respect to a committee that contains only better candidates is the difference between the intrinsic values of the new candidate and of the worst candidate in the committee. For example, adding a candidate with an intrinsic value of 1 to a committee, when the worse candidate in it has an intrinsic value of 10 means a marginal value of (-9). On the other hand, if there is also a candidate with an intrinsic value of 2 in the committee, then the marginal value of the new one will be (-1). That is, the marginal value of the new candidate increased because of the inclusion of the candidate with a value of 2. However, it is clear that this does not model synergy between these two candidates. Moreover, positive correlation between *specific* candidates cannot be modeled using OWA vectors, as described above, since they cannot relate to specific candidates differently. This means that in scenarios like the movies example described earlier, a positive correlation within a series of movies cannot be modeled. In this sense, our model adds new possibilities with respect to the model of Skowron, Faliszewski and Lang [21].

Another relevant model was studied by Fishburn and Pekec [13]. Fishburn and Pekec [13] studied an approval voting model, where each of the voters can approve a few candidates, and a committee is approved by a voter if it contains a sufficient number of candidates that are approved by the voter.

3   Our Contribution

This paper introduces a new model for voting rules, based on set functions, together with the required conceptual framework. Our model can be used to model both synergy between candidates (i.e. compliments) and substitutes (e.g., two candidates that each of them is worth 1 and both of them together are worth 1, as well). Since general set functions might be highly complex, we introduce the joint supermodular degree, which we see as a natural extension of the supermodular degree [10]. We demonstrate applications for our model in Sect. 5. In particular, we suggest practical preference elicitation that is tailored for the joint supermodular degree in Sect. 6.

Finally, in Sect. 7, we show how the joint supermodular degree enables one to easily use existing algorithms for function maximization that are tailored for the supermodular degree to achieve approximations for our voting rule. Since there exist such algorithms both for offline and online settings, one can use either and immediately get approximation guarantees for our voting rule in the corresponding setting. Moreover, future algorithms for the supermodular degree can also be easily used by our framework, to get computational results for committee selection. Conceptually speaking, the result of the approximation algorithms can also be seen as the voting rule itself (see Skowron, Faliszewski and Lang [21]). We complement our algorithmic result with a proof of computational hardness.

To the best of our knowledge, our results represent the first voting rules that capture synergy between specific candidates.

4   The Model

We formally define our model. Let $V = \{v_1, \ldots, v_n\}$ be a set of n voters, let C be a set of m candidates and let k be an integer. Let $f_1, \ldots, f_n : 2^C \rightarrow \mathbb{R}^+$ be preference (set) functions, associated with the voters $v_1, \ldots, v_n$, respectively. We assume that the preferences functions are normalized (i.e., $\forall_i f_i(\varnothing) = 0$) and non-decreasing monotone (i.e., $\forall_{i, S' \subseteq S \subseteq M} f_i(S') \leq f_i(S)$). Our aim is to choose a set $C_{max} \subseteq C$ of size k that maximizes the satisfaction of the voters by their personal preferences:

$$C_{max} = \operatorname*{argmax}_{S \subseteq C || S| = k} \sum_{i=1}^{n} f_i(S).$$

We refer to this problem as (the) k-COMMITTEE SELECTION problem and to the selected subset as the **selected committee**. Note that this problem can be seen as a voting rule. Alternatively, an approximation algorithm to this problem can be seen as the voting rule (see also Skowron, Faliszewski and Lang [21]).

4.1   The Joint Supermodular Degree

We introduce the following natural extensions of the definitions of Feige and Izsak [10] to a collection of set functions.

Definition 6. *Let* $f_1, \ldots, f_t$ *be set functions for some* $t \in \mathbb{N}$ *and let* $c \in C$. *The joint supermodular dependency set of* c *by* $f_1, \ldots, f_t$ *is* $\bigcup_{i=1}^{t} \mathcal{D}_{f_i}^+(c)$.

Definition 7. *The joint supermodular degree of* $f_1, \ldots, f_t$ *is the maximum cardinality among the cardinalities of joint dependency sets of items of* C *by* $f_1, \ldots, f_t$.

The main property of the joint supermodular degree that we use is that the sum function of functions with joint supermodular degree of at most d has supermodular degree of at most d.

We think this definition is natural for voting rules, since it means that positive correlation between the candidates can be modeled, when it is inherent to the candidates themselves, and not to the perspective of the voters about them.

For example, if a candidate is working well together with 2 other candidates, then each of the voters has the possibility to give these 3 candidates or any subset of them a score that is higher than the sum of their individual scores. However, if a candidate does not work well with some other candidate, then none of the voters has the possibility to give them together a score that is higher than the sum of their individual scores. That is, the set of other candidates that the candidate has synergy with depends on her. The decision of whether to take this into account depends on each of the voters. So, the supermodular dependency set of a candidate c, by any of the preference functions of the voters, will contain only other candidates that have synergy (i.e. are working well together) with c.

We discuss applications of our model with respect to the joint supermodular degree in Sect. 5. In particular, we suggest preference elicitation in Sect. 6.

5   Applications

We discuss in this section applications of our model, together with the joint supermodular degree. Specifically, we demonstrate its merits for two real world examples (see [9]).

- Parliamentary elections: In voting to the parliament, it is possible that candidates complement each other, and work better together. It was actually shown by Woolley et al. [22] that there is a measure for the collective intelligence of a group of people that is different from the intelligence quantities of different people in the group. So, it seems reasonable to allow the voters to give extra value for choosing *together* a pair of candidates that are known to work well together on, e.g., suggesting complex laws in the parliament. Note that the fact that two candidates are working well together is related to the candidates and not to the voters, and indeed, the joint supermodular degree of the voters will reflect the synergies between the candidates.
- Movie selection: Consider the problem of choosing k movies to be available on an airplane (passengers can watch on their flight movies from the selected set). It seems reasonable that people would prefer to watch latter parts of a series only after the former. Moreover, it might be unreasonable to consider a series of movies as one movie, if, e.g., physical storage is a limitation. Then, it is plausible to give the prospective passengers the possibility to give higher values for movies in the series, given that all the former are selected, as well. Additionally, movie selection can admit submodular behaviour (i.e. substitutes). For example, since the time of the flight is bounded, the number of movies one can watch out of the k selected movies is bounded, as well. This means that, if for example, $k = 100$ and the time of the flight allows one passenger to watch up to 5 movies, then any movie out of the k that is not among the 5 best for that passenger is redundant for her. So her value will not increase given that we add to the selected set other great movies. On the other hand, we do want to allow k to be large enough to allow different passengers to enjoy different movies. The latter behaviour is submodular. Synergy between selected movies is supermodular. Our model enables one to express such preferences. Furthermore, submodularity does not hurt the approximation guarantees, since it does not increase the joint supermodular degree of the preference functions (see Sect. 7).

6   Preference Elicitation

Consider the movies selection example. When a prospective passenger is asked to express her preferences about possible movies, it seems unreasonable to require her to specify her values for all the exponentially many possibilities. We briefly demonstrate a simple user interface to elicit users' preferences in that case, while enabling them to benefit from the possibility of expressing positive correlations.

The user interface will be as follows. Each of the prospective passengers will be able to give a value for each of the possible movies (these are the values of the singleton subsets). In addition, the prospective passengers will be able to add for each of the movies other values – the marginal values of a movie, with respect to a subset of its joint supermodular dependency set (i.e., other movies in the same series). In order to select such a subset of the movies, a list of the movies in the joint supermodular dependency set will be presented, and a passenger will be able to select the relevant movies (e.g. by checking them by a 'V'). In order to enforce the preference functions of the prospective passengers to be well defined (i.e. a single value for each of the subsets), we will let the prospective passengers check by a 'V' only the movies that were former to a movie in a series.

Note that the supermodular dependency is symmetric (see [10] for a proof). So, in a series of movies, also the former movies are dependent on the latter movies. As an example, one can think of two movies, where each of them is worth 1, but the second one is worth 10 with respect to the first. Then, both movies together are worth 11, and the marginal contribution of each of them with respect to the other is 10, instead of 1 (as it is with respect to the empty set).

Generally speaking, this example interface can be extended in any way that enforces the preference functions to be well defined (e.g. by ordering the items and letting the prospective passengers to check a dependency by 'V' only if it is before the current item in that ordering).

To see the power of combining supermodular dependencies with submodular behaviour, note that we can also ask each passenger how many movies she would like to watch in her flight (with a maximum that depends on the duration of the flight), and then calculate as her preference, the best subset of that number of movies, from any input subset of movies.

Note that it is easy to emulate both value and supermodular queries using such a representation, and then to use the algorithms of Feldman and Izsak [11], as described in Sect. 7.

7 Computational Results

7.1 General Results Using the Joint Supermodular Degree

The following theorem shows that there exists an approximation algorithm with approximation guarantee that is linear in the amount of synergy between the candidates, as measured by the joint supermodular degree of the preference functions of the voters. For submodular set functions, the result described by the theorem coincides with the optimal result for submodular set functions of Fisher, Nemhauser and Wolsey [14] that is used by Skowron, Faliszewski and Lang [21].

Theorem 1. *When the joint supermodular degree of the preferences functions of the voters is d, the k-committee selection problem admits an approximation algorithm with guarantee $(1 - e^{-1/(d+1)}) \geq 1/(d+2)$. The algorithm gets access to the preference functions by value queries and supermodular queries, and its running time is $\texttt{Poly}(n, m, 2^d)$.*

Note that the above result captures the example of movies selection from the introduction (see Sect. 5 for further discussion). Note also that the proof of the above result applies to the case of committee selection subject to a *general matroid constraint* (cardinality constraint is a special case of a matroid constraint), but with an approximation guarantee of $1/(d+2)$, by using the respective algorithm of Feldman and Izsak [11].

Moreover, one can use the algorithms of Feldman and Izsak [12] in order to get an online (secretary like) version of Theorem 1, when the candidates arrive one by one in an online fashion, and we need to decide on the spot, irrevocably, whether to elect a candidate or not, based on the preferences of the voters (for exact details of the model, see [12]). As an example, consider hiring a team to a project, where each of the candidates meets with a few interviewers. Then, an optimal team of candidates should be hired, according to the preferences of the interviewers.

By using the algorithm of Feldman and Izsak [12] for a cardinality constraint, one gets an approximation guarantee polynomial in the joint supermodular degree. Any approximation guarantee that depends only on the joint supermodular degree gives a constant approximation guarantee, if the candidates admit synergy only with a constant number of other candidates (e.g. if there is a positive correlation only within series of movies, and all the series suggested are of length up to 3). See also Oren and Lucier [18] for a different secretary like model.

Additionally, we show a hardness result for the case of non-bounded joint supermodular degree, even when the supermodular degree of all the set functions is bounded by 1. For this, we use a reduction from the k-dense subgraph problem (see e.g. Bhaskara et al. [3]).

Definition 8. *The k-dense subgraph problem is the following. We are given as input a graph $G = (V, E)$ and an integer $k \in \mathbb{N}$, and our aim is to select k vertices such that the number of edges in their induced subgraph is maximized.*

This problem is NP-hard and it is highly believed it is hard to approximate it within any constant guarantee. Actually, no efficient algorithm is currently known that approximates it within a guarantee better than n^c, for some constant c (see e.g. [3,19,20]).

Theorem 2. *The k-committee selection problem is at least as hard as the k-dense subgraph problem, even if the supermodular degree of the set functions is 1, and even if an explicit representation of the preference functions is given. This means, in particular, that it is NP-hard[2] and SSE-hard (see [19] and also [20]).*

---

[2] NP-hardness is actually true also for submodular set functions, i.e. supermodular degree of 0.

Proof (Proof of Theorem 1). Let V be the set of n voters, let C be the set of m candidates, let k be the requested number of elected candidates and let $f_1, \ldots, f_n : 2^C \to \mathbb{R}^+$ be the preference functions of the voters. We prove that since the joint supermodular degree of $f_1, \ldots, f_n$ is upper bounded by d, then the supermodular degree of their summation function $f_\Sigma(S) \overset{\text{def}}{=} \sum_{i=1}^n f_i(S)$ is upper bounded by d, as well. Note that this would not be necessarily true if only the supermodular degree of $f_1, \ldots, f_n$ was bounded by d (or even by 1). Actually, Theorem 2 serves as a counter example to the latter for $d = 1$.

To prove the bound on the supermodular degree of the summation function f_Σ, we show that every supermodular dependency by f_Σ induces the same supermodular dependency by one of the f_is in the sum. Let $c, c' \in C$ and $S \subseteq C$ be such that $f_\Sigma(c \mid S \cup \{c'\}) > f_\Sigma(c \mid S)$. Then, by the definition of f_Σ, $\sum_{i=1}^n f_i(c \mid S \cup \{c'\}) > \sum_{i=1}^n f_i(c \mid S)$. So, $\exists_{1 \le i \le n}$ s.t. $f_i(c \mid S \cup \{c'\}) > f_i(c \mid S)$, as claimed.

Now, we can just use the algorithm of [11] for monotone function maximization subject to uniform matroid constraint (i.e. cardinality constraint) on the function f_Σ with a constraint k. Note that the latter algorithm gives an optimal approximation guarantee for submodular set functions, and generally its guarantee deteriorates linearly with the supermodular degree. Moreover, its running time is as required by the Theorem. This concludes the proof of Theorem 1.

Proof (Proof of Theorem 2). The proof is somewhat similar to the proof of $\mathcal{SSE}$-hardness for maximizing set function subject to cardinality constraint, given by [11]. Given an algorithm for solving the k-committee selection problem within approximation guarantee α, we show how to solve any input instance of the k-dense subgraph problem within approximation guarantee α. Let $G = (S, E)$ be an instance of the k-dense graph problem. Then, our set of candidates C will be S (the set of vertices of G). We also introduce a voter v_e for every edge $e = \{v_{e1}, v_{e2}\} \in E$ and let $V = \bigcup_{e \in E} \{v_e\}$. For every voter v_e, her preference set function is:

$$f_e = \begin{cases} 1 & \text{if } v_{e_1} \text{ and } v_{e_2} \text{ are both elected.} \\ 0 & \text{otherwise} \end{cases}$$

That is, in this instance of the k-committee selection problem, our aim is to find a subset of k candidates (where the set of candidates corresponds exactly to the set S of vertices of G), such that the number of pairs of candidates, that correspond to the preference functions of the voters, is maximized (where these pairs of candidates are exactly the edges E of G). This is exactly the k-dense subgraph problem. That is, given a solution to this instance of k-committee selection problem, we just output the subset of vertices of S that corresponds to the candidates in C that were selected, as a solution to the input instance of the k-dense subgraph problem. This gives us a feasible solution with the same value, and thus with the same approximation guarantee α. This concludes the proof of Theorem 2.

7.2   Important Special Cases

Subsequently to this work, we designed tailored algorithms for instances that are the result of the preference elicitation presented in Sect. 6. In particular, we designed an exact algorithm for instances of the k-COMMITTEE SELECTION problem with disjoint subsets of dependent items of some fixed size d (this corresponds to a joint supermodular degree of $d - 1$).

We also designed a $(1 - 1/e)$-approximation algorithm for the more general case, where each player can also express a budget for the maximum number of items (candidates) that can be useful for her (e.g. the maximum number of movies that she might watch during her flight).

Note that both of the latter algorithms have (constant) approximation guarantees that do not deteriorate with the joint supermodular degree. Only the running time is dependent on the parameter d.

8   Conclusions

We suggest a new voting rule for committee selection that enables the voters to express positive correlation between the candidates. We also introduce the joint supermodular degree that enables us to use existing computational results for the supermodular degree, and get efficient approximation algorithms for our voting rule. We see our work as a proof of concept, and hope that it will lead to further study of committee selection with positive correlation between the candidates.

Acknowledgments. Work supported in part by the Israel Science Foundation (grant No. 1388/16). I would like to thank Uri Feige for many useful discussions and for his contributions to this paper. I would like to thank Nimrod Talmon for useful discussions and for directing me to the paper of Skowron, Faliszewski and Lang [21]. I would also like to thank Moshe Babaioff, Shahar Dobzinski and Moshe Tennenholtz for useful discussions.

References

1. Abraham, I., Babaioff, M., Dughmi, S., Roughgarden, T.: Combinatorial auctions with restricted complements. In: EC, pp. 3–16, New York, NY, USA. ACM (2012)
2. Betzler, N., Slinko, A., Uhlmann, J.: On the computation of fully proportional representation. Artif. Intell. Res. **47**, 475–519 (2013)
3. Bhaskara, A., Charikar, M., Chlamtac, E., Feige, U., Vijayaraghavan, A.: Detecting high log-densities: an $O(n^{1/4})$ approximation for densest k-subgraph. In: Proceedings of the 42nd ACM Symposium on Theory of Computing, STOC 2010, Cambridge, Massachusetts, USA, 5–8 June 2010, pp. 201–210 (2010)
4. Borda, J.: Memmoire sur les elections au scrutine. Histoire de l'Academie Royale des Sciences (1781)
5. Chamberlin, B., Courant, P.: Representative deliberations and representative decisions: Proportional representation and the borda rule. Am. Polit. Sci. Rev. **77**, 718–733 (1983)

6. Chevaleyre, Y., Endriss, U., Estivie, S., Maudet, N.: Multiagent resource allocation in k-additive domains: preference representation and complexity. Ann. Oper. Res. **163**, 49–62 (2008)
7. Conitzer, V., Sandholm, T., Lang, J.: When are elections with few candidates hard to manipulate? J. ACM **54**(3) (2007)
8. Conitzer, V., Sandholm, T., Santi, P.: Combinatorial auctions with k-wise dependent valuations. In: AAAI, pp. 248–254 (2005)
9. Elkind, E., Faliszewski, P., Skowron, P., Slinko, A.: Properties of multiwinner voting rules. In: AAMAS, pp. 53–60 (2014)
10. Feige, U., Izsak, R.: Welfare maximization and the supermodular degree. In: ITCS, pp. 247–256 (2013)
11. Feldman, M., Izsak, R.: Constrained monotone function maximization and the supermodular degree. In: APPROX-RANDOM, pp. 160–175 (2014)
12. Feldman, M., Izsak, R.: Building a good team: secretary problems and the supermodular degree. In: SODA (2017)
13. Fishburn, P., Pekec, A.: Approval voting for committees: Threshold approaches (2017). Working paper
14. Fisher, M.L., Nemhauser, G.L., Wolsey, L.A.: An analysis of approximations for maximizing submodular set functions - II. In: Polyhedral Combinatorics, volume 8 of Mathematical Programming Study, pp. 73–87. North-Holland Publishing Company (1978)
15. Goldsmith, J., Lang, J., Mattei, N., Perny, P.: Voting with rank dependent scoring rules. In: AAAI, pp. 698–704 (2014)
16. Masthoff, J.: Group modeling: Selecting a sequence of television items to suit a group of viewers. User Model. User-Adapt. Interact. **14**(1), 37–85 (2004)
17. Masthoff, J.: Group recommender systems: combining individual models. In: Recommender Systems Handbook, pp. 677–702 (2011)
18. Oren, J., Lucier, B.: Online (budgeted) social choice. In: Proceedings of the Twenty-Eighth AAAI Conference on Artificial Intelligence, 27–31 July 2014, Québec City, Québec, Canada, pp. 1456–1462 (2014)
19. Raghavendra, P., Steurer, D.: Graph expansion and the unique games conjecture. In: STOC, pp. 755–764 (2010)
20. Raghavendra, P., Steurer, D., Tulsiani, M.: Reductions between expansion problems. In: IEEE Conference on Computational Complexity, pp. 64–73 (2012)
21. Skowron, P., Faliszewski, P., Lang, J.: Finding a collective set of items: from proportional multirepresentation to group recommendation. In: AAAI, pp. 2131–2137 (2015)
22. Woolley, A.W., Chabris, C.F., Pentland, A., Hashmi, N., Malone, T.W.: Evidence for a collective intelligence factor in the performance of human groups. Science **330**, 265–294 (2010)

On the Deployment of Factor Graph Elements to Operate Max-Sum in Dynamic Ambient Environments

Pierre Rust[1,2(✉)], Gauthier Picard[2], and Fano Ramparany[1]

[1] Orange Labs, Cesson-Sévigné, France
{pierre.rust,fano.ramparany}@orange.com
[2] MINES Saint-Etienne, Laboratoire Hubert Curien UMR CNRS 5516,
Saint-Étienne, France
picard@emse.fr

Abstract. Using belief-propagation based algorithms like Max-Sum to solve distributed constraint optimization problems (DCOPs) requires deploying the factor graph elements on which the distributed solution operates. In some utility-based multi-agent settings, this deployment is straightforward. However, when the problem gains in complexity by adding other interaction constraints (like n-ary costs or dependencies), the question of deploying these shared factors arises. Here, we address this problem in the particular case of smart environment configuration (SECP), where several devices (e.g. smart light bulbs) have to coordinate as to reach an optimal configuration (e.g. find the most energy preserving configuration), under some n-ary constraints (e.g. physical models and user preferences). This factor graph deployment problem (FGDP) can be mapped to an optimization problem, then solvable in a centralized manner. But, when dealing with the dynamics of the environment (e.g. new sensed data which activates some rules, adding new devices, etc.) we cannot afford restarting the system or relying on a centralized solver. Thus, the system has to achieve on-line and local deployment adaptations. In this paper, we present some solutions and experiment them on a simulated smart home environment.

1 Introduction

A common problem when using distributed belief-propagation techniques as Max-Sum [7] is to decide where to host computations related to variable and factor message assessments. Indeed, Max-Sum operates on a factor graph which represents the problem to solve, by sending messages from variables to factors, and *vice versa*. Assessing messages to send requires computations to be hosted by some agents. In some settings, this mapping is straightforward; this is the case in purely utility-based problems, where each agent owns a variable and a utility factor connected to some other variables owned by other agents [6]. However,

G. Sukthankar and J. A. Rodriguez-Aguilar (Eds.): AAMAS 2017 Best Papers,
LNAI 10642, pp. 116–137, 2017.
https://doi.org/10.1007/978-3-319-71682-4_8

in more complex settings such as interaction-based problems, where some factors and variables are shared by several agents, the question of deploying these elements arises.

An example of such settings is the smart environment configuration (SECP) we addressed in [11]. In SECP, several devices have to self-configure as to satisfy user requirements and to minimize energy consumption. SECP is modeled as a DCOP composed of n-ary factors corresponding to user rules (e.g. "setting the light level at 60 when someone is in the room"), physical models (e.g. "the effect of the light bulbs and the shutter on the light level in the room") and shared variables corresponding to physical properties (e.g. "the light level in the room"). Due to the dynamicity of the environment (e.g. devices appearing/disappearing, user adding/changing rules, sensed data updates, etc.), and the constrained communication and computation capabilities of our devices, deploying the factor graph is a key issue that cannot be solely solved off-line in a centralized manner. On-line repair approaches based on local techniques are needed.

The paper is structured along our contributions, as follows. Section 2 briefly exposes the SECP framework and its DCOP formulation. Section 3 introduces the factor graph deployment problem (FGDP) and presents an ILP to solve FGDP optimally in a centralized manner. The next sections discuss different cases of dynamics impacting the deployment: infrastructure changes (Sect. 4), problem and sensed environment changes (Sect. 5). Some preliminary experiments and related analysis are provided in Sect. 6. Finally, Sect. 7 concludes the paper.

2 Smart Environment Configuration

In this section, we expose the smart environment configuration problem we address in this paper, and some useful notations, as defined in [11].

Scenario. We consider the following Ambient Intelligence scenario. Our system is made of several smart devices (light bulbs, roller shutters, a TV set, etc.) and sensors (luminosity, presence, etc.). Each device is defined by (i) a unique identifier, (ii) its location (e.g. living room), (iii) a list of capabilities (e.g. emitting light or playing videos), (iv) a list of actions, (v) a consumption law that associates an energy cost to each action. The user can use an application on a dedicated device (a powerful computing device with an user interface e.g. home computer, tablet, etc.) to configure simple behaviors (or *scenes*), using the value of the sensors or the state of actuators as triggers for implementing smart home actions. For example, one could configure the system such that a luminosity level of 60 is reached in the living room whenever somebody is in this room. Once this behavior is configured, the dedicated device can be removed (shutdown or disconnected) from the system, which then autonomously decides the best way to achieve this target. Devices may be added or removed and are automatically integrated into the system. We want our system to choose the most energy-saving configuration for a given scene.

Problem Definition and Notations. In [11], this configuration problem can be seen as an optimization problem with values to assign to actuators (e.g. a light bulb is assigned a power) and user's target values (e.g. the light level in living room is 60 lumens), while maximizing the adequacy to user-defined scenes and minimizing the overall energy consumption.

Problem 1 (SECP). Given a set of actuators $\mathfrak{A}$ (and their related costs $c_i \in \mathfrak{C}$), a set of sensors $\mathfrak{S}$, a set of scene rules $\mathfrak{R}$ (and their related utility functions in $u_k \in \mathfrak{U}$), and a set of physical dependency models Φ, the *Smart Environment Configuration Problem* (or *SECP*) $\langle \mathfrak{A}, \mathfrak{C}, \mathfrak{S}, \mathfrak{R}, \mathfrak{U}, \Phi \rangle$ amounts to finding the configuration of actuators that maximizes the utility of the user-defined rules, whilst minimizing the global energy consumption and fulfilling the physical dependencies.

Let $\mathfrak{A}$ the set of available actuators. We note $\nu(\mathfrak{A})$ the set of variables that represent the states of actuators $i \in \mathfrak{A}$ (e.g. the power assigned to a bulb). We use $\mathbf{x}_i$ to refer to a possible state of $x_i \in \nu(\mathfrak{A})$, that is $\mathbf{x}_i \in \mathcal{D}_{x_i}$ (domain of x_i). Activating an actuator i incurs a cost, noted $c_i : \mathcal{D}_{x_i} \to \mathbb{R}$, derived from the consumption law of each device. We note $\mathfrak{C} = \{c_i | i \in \mathfrak{A}\}$.

Let $\mathfrak{S}$ be the set of available sensors, and $\nu(\mathfrak{S})$ the set of variables encapsulating their states. We note $\mathbf{s}_\ell \in \mathcal{D}_{s_\ell}$ the current state of sensor $\ell \in \mathfrak{S}$. Sensor values are not controllable by the system: they are *read-only* values.

Let $\mathfrak{R}$ the set of user-defined scene rules. Each scene k is specified as a condition-action rule expressed using the set devices. The condition part is specified as a conjunction of boolean expressions using state of actuators or sensors. The action part defines *target* values for either (i) some direct actions on actuators or (ii) indirect actions on abstract concepts (e.g. light level in living room) – both called *scene action variables.*

These scene action variables are therefore either (i) some $x_i \in \nu(\mathfrak{A})$ or (ii) other values constrained by values assigned to some actuators. We note $y_j \in \nu(\Phi)$ the state of such an *indirect* scene action j (e.g. the current level of light in a room), and $\mathbf{y}_j$ a possible state of y_j, that is $\mathbf{y}_j \in \mathcal{D}_{y_j}$. We note $\mathbf{x}_i^k$ (resp. $\mathbf{y}_j^k$) the target value defined by the user for the scene action variable x_i (resp. y_j) in the rule k. Obviously, $\mathbf{x}_i^k \in \mathcal{D}_{x_i}$ and $\mathbf{y}_j^k \in \mathcal{D}_{y_j}$ for all i, j and k. Note that a scene action variable can be used in several rules, but that a rule can only specify a unique target value for the scene action variable.

A scene rule can be either *active* or *inactive* depending on the state of devices appearing in the condition part of the rule. Each active scene has also a utility to be implemented, noted $u_k : \prod_{s \in \sigma(u_k)} \mathcal{D}_s \to \mathbb{R}$, with $\sigma(u_k) \subseteq \nu(\mathfrak{A}) \cup \nu(\Phi)$ being the scope of the rule (the subset of variables used in the rule). The more the states of the scene action variables (from $\nu(\mathfrak{A})$ and $\nu(\Phi)$) are close to the user's target values for this scene, the higher the utility. Moreover, if the condition to activate the rule (from $\nu(\mathfrak{A})$ and $\nu(\mathfrak{S})$) are not met, the utility should be neutral, i.e. equals to 0. We can therefore consider u_k's to be functions of the distance between the states of the scene action variables x_i's (resp. y_j's) and the target values $\mathbf{x}_i^k$ (resp. $\mathbf{y}_j^k$). We note $\mathfrak{U} = \{u_k | k \in \mathfrak{R}\}$.

Each scene action variable y_j depends physically on the values of several actuators. We note the model of this dependency $\phi_j : \prod_{\varsigma \in \sigma(\phi_j)} \mathcal{D}_\varsigma \to \mathcal{D}_{y_j}$, where $\sigma(\phi_j) \subseteq \nu(\mathfrak{A})$ is the scope of the model, i.e. the set of variables influencing y_j. Let $\Phi = \{\phi_j\}$ be the set of all physical models between actuators and user-defined values. In a more general form, a physical dependency model links a set of devices –with a given capability (e.g. emitting light, like a bulb or a TV set), in a given location (e.g. living room)– to a physical value (e.g. light level) that can be measured by some sensor (e.g. light sensor).

Formulation of SECP as a DCOP. SECP can be formulated as a DCOP $\langle \mathcal{A}, \mathcal{X}, \mathcal{D}, \mathcal{C}, \mu \rangle$ where: $\mathcal{A}$ is a set of smart devices; $\mathcal{X} = \nu(\mathfrak{A}) \cup \nu(\Phi)$; $\mathcal{D} = \{\mathcal{D}_{x_i} | x_i \in \nu(\mathfrak{A})\} \cup \{\mathcal{D}_{y_j} | y_j \in \nu(\Phi)\}$; $\mathcal{C} = \mathfrak{U} \cup \mathfrak{C} \cup \Phi$; μ is a function that maps variables and constraints to smart devices; with the following objective, where $\omega_u, \omega_c > 0$ are weights used to *normalize* the range of u_k's and c_i's:

$$\underset{\substack{x_i \in \nu(\mathfrak{A}) \\ y_j \in \nu(\Phi)}}{\text{maximize}} \; \omega_u \sum_{k \in \mathfrak{R}} u_k - \omega_c \sum_{i \in \mathfrak{A}} c_i + \sum_{\varphi_j \in \Phi} \varphi_j \qquad (1)$$

Here, we note Φ the corresponding set of φ_j's.

$$\varphi_j(x_j^1, \ldots, x_j^{|\sigma(\phi_j)|}, y_j) = \begin{cases} 0 & \text{if } \phi_j(x_j^1, \ldots, x_j^{|\sigma(\phi_j)|}) = y_j \\ -\infty & \text{otherwise} \end{cases} \qquad (2)$$

SECP Factor Graph. Such a DCOP can be represented as a bipartite factor graph, noted $\mathcal{G} = \langle V_x, V_f, E \rangle$, which is a generalization of classical constraint graphs [7]. For SECP, variable nodes are taken from $V_x = \nu(\mathfrak{A}) \cup \nu(\Phi)$, connected through factors in $V_f = \mathfrak{U} \cup \mathfrak{C} \cup \Phi$ by applying the following rules: each $x_i \in \nu(\mathfrak{A})$ is a variable node, each $x_i \in \nu(\mathfrak{A})$ is connected to a unary factor c_i specifying its cost, each $y_j \in \nu(\Phi)$ is a variable node, each y_j and all $x_i \in \nu(\mathfrak{A})$ in the scope of a physical dependency model ϕ_j are connected to a factor φ_j, each scene rule $k \in \mathfrak{R}$ is represented by a utility factor u_k connected to all the $x_i \in \sigma(u_k)$ and $y_j \in \sigma(u_k)$.

Example 1 (Factor graph). Figure 1a represents a factor graph where x_1, x_2, x_3 are the state of light bulbs; c_1, c_2, c_3 are their activation costs; u_1 is the factor representing the scene rule and defining the utility depending on a target value $\mathbf{y}_1^1$ for variable y_1; y_1 represents the theoretical light level in lumen; φ_1 is the physical dependency model between the light level and the state of actuators; s_1 and s_2 are read-only variable nodes, corresponding to sensor measurements, represented as dotted diamonds.

Variables and factors imply some computations on the hosting agents, depending on the size of the domains of the variables, the arity of the factors, and more generally on the complexity of the factors. Moreover, each link between two elements which are not hosted on the same agent implies some

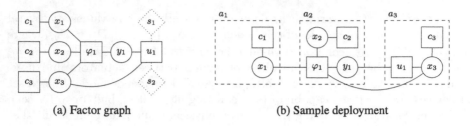

(a) Factor graph   (b) Sample deployment

Fig. 1. Factor graph (1a) and a possible deployment (1b) on 3 nodes (a_1, a_2 and a_3)

communication cost, depending on the size of the variable. Therefore, defining the mapping function μ which assigns each factor graph element to an agent is a key issue, called here the *deployment problem*.

Our devices are assumed to be resource constrained and the communication link between them is implemented with a low power network with limited throughput (typically 250 kps). Sensing-only devices run as *sleepy nodes*, meaning that they only turn their communication interface on when they want to emit a new value. These nodes cannot be reached most of the time and are not good candidates to host the computations needed for the variable and factor computations. On the other hand, actuators, and especially light sources, are usually connected to the main power line and always reachable. As a result our FG is only hosted on actuator devices, as illustrated in Fig. 1b. In the reminder, we use the terms "agents" and "nodes" interchangeably to denote these devices.

3   Optimal Deployment of Factor Graph Elements

As discussed in [11], the problem of deploying the factor graph elements on a set of nodes is equivalent to graph partitioning, which typically falls under the category of NP-hard problems [2,5]. Typically, such problem can be modeled as a mathematical optimization problem. To scale up, we propose here an integer linear program for general purpose, inspired by graph partitioning techniques from [3,5], and introduce some constraints which are specific to SECP.

Problem 2 (FGDP). Given a factor graph $FG = \langle V_x, V_f, E \rangle$ and a set of agents $\mathcal{A}$ the *Factor Graph Element Deployment Problem (FGDP)* amounts to assign each element of FG to an agent, while minimizing overall communications between agents.

First, we introduce some notations. As previously stated, μ is a function that maps elements of the FG (variables or factors) to nodes. We note $\mu_x^{-1}(a_k)$ (resp. $\mu_f^{-1}(a_k)$) the set of variables (resp. factors) hosted by agent a_k.

In the SECP model it is usual that each actuator node has a computation capability. We consider variables $x_i \in \nu(\mathfrak{A})$ and constraints $f_j \in \mathfrak{C}$ related to each actuator to be *owned* by their actuator's node, meaning they will always be deployed on this specific node. We note $\rho(e) \in \mathcal{A} \cup \{\varnothing\}$ the owner of element

$e \in V_x \cup V_f$, with $\phi(e) = \varnothing$ iff e does not belong to actuator node in $\mathcal{A}$ (i.e. $e \notin \mathfrak{C} \cup \nu(\mathfrak{A})$). We note $\rho_x^{-1}(a_k)$ (resp. $\rho_f^{-1}(a_k)$) the set of variables (resp. factors) owned by agent a_k. Remark that an owned element is always hosted on its owner, i.e. if $\rho(e) = a_k$ then $\mu(e) = a_k$.

We note $\mathbf{com}(x_i, f_j)$ the communication load induced by the interaction between x_i and f_j. For instance, $\mathbf{com}(x_i, f_j)$ is the size of the messages exchanged between the variable and the factor:

$$\forall x_i \in V_x, f_j \in V_f, \quad \mathbf{com}(x_i, f_j) = \begin{cases} a \cdot |\mathcal{D}_{x_i}| + b, & \text{if } (x_i, f_j) \in E \\ 0, & \text{otherwise} \end{cases} \tag{3}$$

where a is the number of bytes to represent a value from the domain of variable x_i and b is the size of the message header. Let $\mathbf{mem}(e), e \in V_x \cup V_f$ be the memory footprint for the computation of factor graph element e. For instance, this is the size in bytes of the hypercube representing the costs in a factor. We also note $\mathbf{cap}(a_k)$ the memory capacity in bytes of node $a_k \in \mathcal{A}$.

Let's introduce the variables that map factor graph elements to agents, i.e. x_i^k (resp. f_j^k) denotes whether variable x_i (resp. factor f_j) is deployed in node a_k:

$$\forall x_i \in V_x, \quad x_i^k = \begin{cases} 1, & \text{if } \mu(x_i) = a_k \\ 0, & \text{otherwise} \end{cases} \tag{4}$$

$$\forall f_j \in V_f, \quad f_j^k = \begin{cases} 1, & \text{if } \mu(f_j) = a_k \\ 0, & \text{otherwise} \end{cases} \tag{5}$$

Moreover, for linearization purpose we introduce another set of variables (the α_{ijk}'s) which link variables to factors:

$$\forall x_i \in V_x, f_j \in V_f, a_k \in \mathcal{A}, \quad \alpha_{ijk} = x_i^k \cdot f_j^k \tag{6}$$

Now, we are ready to model the factor graph element deployment problem (FGDP) as a linear program:

$$\operatorname*{minimize}_{x_i^k, f_j^k} \quad \sum_{(x_i, f_j) \in E} \sum_{a_k \in \mathcal{A}} \mathbf{com}(x_i, f_j) \cdot (1 - \alpha_{ijk}) \tag{7}$$

subject to

$$\forall x_i \in V_x, \quad \sum_{a_k \in \mathcal{A}} x_i^k = 1 \tag{8}$$

$$\forall f_j \in V_f, \quad \sum_{a_k \in \mathcal{A}} f_j^k = 1 \tag{9}$$

$$\forall a_k \in \mathcal{A}, \quad \sum_{x_i \in V_x} x_i^k + \sum_{f_j \in V_f} f_j^k \geq 1 \tag{10}$$

$$\forall (x_i, f_j) \in E, \quad \alpha_{ijk} \leq x_i^k \tag{11}$$

$$\forall (x_i, f_j) \in E, \quad \alpha_{ijk} \leq f_j^k \tag{12}$$

$$\forall (x_i, f_j) \in E, \quad \alpha_{ijk} \geq x_i^k + f_j^k - 1 \tag{13}$$

Objective (7) minimizes communications between factor graph elements which are not deployed on the same node. Constraints (8) and (9) force each factor graph element to be deployed on exactly one node. Constraint (10) enforces the use of all the available nodes. Finally, inspired by the linearization proposed in [1,3], constraints (11) to (13) link x_i^k's and f_j^k's to α_{ijk} in a linear way.

Problem 3 (ILP-FGDP). We term ILP-FGDP the 0/1 integer linear program consisting of objective (7) and constraints (8) to (13) which encodes the FGDP Problem 2.

While Problem 3 is an ILP, and thus it is NP-hard, it can be solved in reasonable time in a centralized manner with branch-and-cut algorithm [10], especially when the coefficient matrix is sparse (which is our case here).

To take into account SECP specificities, we add the following constraints that will reduce the search space by ensuring that each owned element is hosted by its owner:

$$\forall a_k \in \mathcal{A}, \forall x_i \in \rho_x^{-1}(a_k), \quad x_i^k = 1 \tag{14}$$

$$\forall a_k \in \mathcal{A}, \forall f_j \in \rho_f^{-1}(a_k), \quad f_j^k = 1 \tag{15}$$

Finally, as SECP deals with devices with limited memory, we add a constraint to avoid memory capacity overflow:

$$\forall a_k \in \mathcal{A}, \quad \sum_{x_i \in V_x} \mathbf{mem}(x_i) \cdot x_i^k + \sum_{f_j \in V_f} \mathbf{mem}(f_j) \cdot f_j^k \leq \mathbf{cap}(a_k) \tag{16}$$

Problem 4 (ILP-SECP-FGDP). The integer linear program consisting of objective (7) and constraints (8) to (16) is an encoding of Problem 2 for SECP problems.

Problem 4 adds more constraints but also reduces the number of variables and thus strongly prunes the search space. Thus, as proposed in [11], each time the user modify the factor graph by adding/removing/updating a rule, the optimal deployment is computed. This can also be done in a fully distributed way by using a distributed simplex, as in [4]. However, computing the solution of this ILP in a computationally limited node is not realistic. Therefore, we will discuss in the next section some techniques to repair deployments following some changes.

3.1 Heuristic-Based Deployment for SECP

A simple heuristic is used in [11] for this deployment problem. However, no formal definition, nor evaluation, of the quality of this heuristic is given. We include here a brief description of the heuristic,with the notation previously introduced.

As previously, owned actuator variables $x_i \in \nu(\mathfrak{A})$ (resp. constraints $f_j \in \mathfrak{C}$) are naturally hosted their owner agent $\rho(x_i)$ (resp. $\rho(f_j)$). Variables and factors that do not belong to an actuator node must be distributed on existing agents. Physical models are distributed first, then rules.

Each pair $\langle y_j, \varphi_j \rangle$ representing a physical model is hosted on the agent a_k for which the set $\{x_i | \mu(x_i) = a_k$ and $x_i \in \sigma(\varphi_j)\}$ has the highest cardinality, which is simply the agent hosting, at this point, the highest number of variables from the scope of φ_j.

Once all physical models have been distributed, rules factors u_l are hosted, using the same principle, on the agent a_k selected such that $\{x_i | \mu(x_i) = a_k$ and $x_i \in \sigma(u_l)\}$ has the highest cardinality, i.e. on the agent already hosting the highest number of variables from their scope.

Notice that this heuristic does not explicitly take into account the memory constraints of the devices when deploying the FG and attempts to minimize communication simply by grouping related elements, without any formal definition of the communication load induced by the interaction between two elements of the factor graph.

3.2   Dynamic SECP

SECP has not been defined as a dynamic problem, but our implementation aims at optimizing an SECP instance each time a change occurs in the problem definition (e.g. the value of a sensor changes, which triggers a rule). While not strictly anytime, Max-Sum maintains a continuously updated estimate of the best assignment, which is very convenient for this setting ; each time a change occurs the involved variables and factors will send new messages. However, in the ambient dynamic and open environment we consider here, some issues due to dynamicity may arise, especially concerning the deployment of the factor graph on which Max-Sum operates.

4   Dynamics in the Infrastructure

We want to cope with changes in the infrastructure – i.e. the set of available agents/devices. Indeed, some questions arise: (i) how to manage factors and variables hosted by a device which disappeared? (ii) how to re-deploy factors and variables when a new device appears?

As stated in Sect. 2, the only powerful device in the system is the user-interface device, which is only available be the user interacts directly with the system to configure it. As a consequence one major constraint is that when such appearance and disappearance occur, the devices have to self-adapt without help of a central computer.

4.1   Notion of Neighborhood

Solving the whole ILP-SECP-FGDP problem for each device appearance and disappearance cannot be performed on one of the constrained devices the SECP is made of. Instead we consider adapting the deployment of the factor graph locally, by only considering a reduced set of agents (termed neighborhood) and a portion of the factor graph (set of elements hosted by the neighbors). Still, the solution to this problem may not be the global optimum w.r.t. ILP-SECP-FGDP, but potentially requires far less computation than solving the ILP-SECP-FGDP over the whole FG.

Let's define the notion of neighborhood as follows:

Definition 1. *Given the current assignment μ, the* neighborhood *of an agent a_k is defined as follows: $\mathcal{A}[a_k] = \{a_\ell \mid \exists(x_i, f_j) \in E, \mu(x_i) = a_k, \mu(f_j) = a_\ell\} \cup \{a_\ell \mid \exists(x_i, f_j) \in E, \mu(f_j) = a_k, \mu(x_i) = a_\ell\} \cup \{a_k\}$, if the agent a_k hosts at least one FG element, and $\mathcal{A}[a_k] = \mathcal{A}$ otherwise.*

Similarly we define the set of edges connected to the neighborhood as $E[a_k] = \{(x_i, f_j) \mid \mu(x_i), \mu(f_j) \in \mathcal{A}[a_k]\}$ and the set of neighborhood variables (resp. factors) as $V_x[a_k] = \{x_i \mid (x_i, f_j) \in E[a_k]\}$ (resp. $V_f[a_k] = \{f_j \mid (x_i, f_j) \in E[a_k]\}$).

Example 2 (Neighborhood). Figure 2 represents a SECP with four light bulbs (with associated variables x_1 to x_4 and cost factors c_1 to c_4), two physical models ($\langle \phi_1, y_1 \rangle$ and $\langle \phi_2, y_2 \rangle$) and two rules ($u_1$ and u_2). The resulting factor graph is deployed on four devices a_1 to a_4. The neighborhood $\mathcal{A}[a_2]$ is composed of the agents a_1, a_2 and a_4 and is represented in red in Fig. 2. Associated sets $E[a_2]$, $V_x[a_2]$ and $V_f[a_2]$ are also represented in red.

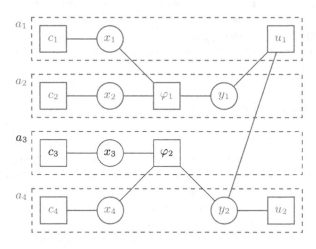

Fig. 2. The neighborhood for agent a_2, noted $\mathcal{A}[a_2]$, is represented in red (Color figure online)

4.2   Adaptation to Device Arrival

Here the initial deployment should be revised as to benefit from new computation and memory capacities. However, we cannot afford solving again Problem 4 on our constrained devices. Thus we propose here to restrict the revision of the factor graph to the neighborhood of the newcomer. There are two situations to consider:

(i) The device is an *actuator* owning, and thus already hosting, variables and factors (e.g. adding a light bulb with its decision variable and cost factor). The new device, depending on its capabilities (e.g. emitting light in room #1) has to connect to corresponding physical factors (e.g. light model for room #1) and/or rule factors (e.g. a rule switching on all light bulb in room #1)[1]. If we simply add the newcomer to the system, the resulting factor graph deployment will not be optimal in the sense of Problem 4 and could generally be improved by migrating some elements as to reduce communication costs.

(ii) The device is a *non-actuator* which only provides computation and memory, without already hosting variables nor factors. In order to benefit from these capabilities, existing elements must be relocated to the newcomer. In this case, the re deployment process amounts to selecting the elements to migrate as to optimize communication costs.

We analyse here two mechanisms, that can be used for both cases.

Restricted ILP-SECP-FGDP. The idea here is to encode the deployment revision problem as a cut version of Problem 4, restricted to the neighborhood of the newcomer. For each device in this neighborhood, the problem consists in choosing the elements to host, with respect to communication and memory capacities.

Problem 5 (ILP-SECP-FGDP$[a_k]^+$). ILP-SECP-FGDP$[a_k]^+$ consists in ILP-SECP-FGDP (4) restricted to the set of agents $\mathcal{A}[a_k]$ and to factor graph $\langle V_x[a_k], V_f[a_k], E[a_k] \rangle$.

This problem can be solved either by one agent (if the size of the problem is not too large) or by the agents composing the neighborhood. In both case it only requires local and limited knowledge on the global DCOP, which makes it ideal for large and complex systems. Prior to solving this problem, agents have to share their elements and costs with other agents involved in the revision. The worst case, when a newcomer is connected to all other agents, is equivalent to solve ILP-SECP-FGDP on the whole FG, which may not be reasonable in terms of response time and communication load. In this case one could devise a method to select another subset of agents as a neighborhood to solve ILP-SECP-FGDP$[a_k]^+$ on.

In the distributed solving case, several distributed optimization techniques could meet the requirements like the distributed simplex method designed for multi-agent assignments [4], keeping exactly the same encoding as ILP-SECP-FGDP, or dual decomposition methods like the efficient AD^3 method [9], that requires ILP-SECP-FGDP to be encoded using tractable high order potentials [12], and then implement a distributed decoding of the LP relaxation to assign integer values to decision variables. However, while providing good optimality,

---

[1] This discovery phase is not discussed in this paper.

both distributed simplex and AD^3 may require several rounds (thus message exchanges) to reach good quality solutions. For instance, from a conjecture in [4], the average time complexity of this technique is linear in the diameter of the graph ($\mathcal{O}(diam(FG))$), with polynomial communication load. In SECP case, the diameter of the FG is not bounded but mainly depends on the number of rules and models, and their interdependencies. In the case of real smart home settings, models and rules will mostly influence local areas (rooms, floor, etc.) and interdependencies, thus diameters, will be limited.

Newcomer Decision Problem. As to avoid high communication load induced by the previous techniques, we can consider a more newcomer-centric approach: the newcomer calls for proposals to move some computations. Based on the costs of the proposed computations and its own memory capacity, the newcomer has to choose a set of factor graph elements to host. Let's formulate this newcomer decision problem.

Problem 6 (SECP-NDP). Given a newcoming agent and a set of proposed computations to migrate coming from its neighborhood, the *Newcomer Decision Problem* (SECP-NDP) amounts to choose computations amongst proposed computations, so that communication load is minimized and memory constraints are fulfilled.

Each neighbor $a_\ell \in \mathcal{A}[a_k]$ sends its proposal in message $\langle V^{\ell \to k}, E^{\ell \to k}, \mathbf{com} \rangle$, where: $V^{\ell \to k} \subset V_x \cup V_f$ is the set of elements (factors and variables) it proposes; $E^{\ell \to k} = \{(e_i, e_j) \mid (e_i, e_j) \in E, e_i \in V^{\ell \to k} \text{ or } e_j \in V^{\ell \to k}\}$ is the set of edges connected to elements in $V^{\ell \to k}$; and $\mathbf{com}$ is the communication cost function (potentially restricted to elements in $E^{\ell \to k}$). We note $V^k = \bigcup_\ell V^{\ell \to k}$ and $E^k = \bigcup_\ell E^{\ell \to k}$. We assume the communication cost $\mathbf{com}(e_i, e_j)$ can be assessed only using information sent by proposers. Let e_i^k be a binary variable stating whether the newcomer a_k chooses to host computation e_i. The cost of selecting a set of computations can be formulated as follows:

$$\sum_{(e_i, e_j) \in E^k} \mathbf{com}(e_i, e_j)(e_i^k + e_j^k - 2 \cdot e_i^k \cdot e_j^k) \tag{17}$$

$$- \sum_{(e_i, e_j) \in E^k} \mathbf{com}(e_i, e_j) \cdot e_i^k \cdot e_j^k \tag{18}$$

which is composed of the sum of the communication costs for the set of edges which are cut in the new distribution (17), i.e. those for which e_i^k XOR e_j^k holds true; minored by the communication costs for the set of edges whose both ends are now hosted on the same agent (18), i.e. those for which x_i^k AND e_j^k holds true. This sum can be simplified and used as the optimization objective for the newcomer a_k, as follows:

$$\underset{e_i^k, e_j^k}{\textbf{minimize}} \sum_{(e_i, e_j) \in E^k} \textbf{com}(e_i, e_j)(e_i^k + e_j^k - 3 \cdot e_i^k \cdot e_j^k) \tag{19}$$

$$\textbf{subject to} \sum_{e_i \in V^k} \textbf{mem}(e_i) \cdot e_i^k \leq \textbf{cap}(a_k) \tag{20}$$

Problem 7 (IQP-SECP-NDP). We term IQP-SECP-NDP the 0/1 integer quadratic program consisting of quadratic objective (19) and linear constraints (20) which encodes SECP-NDP.

This problem falls into the quadratic knapsack problem (QKP) framework. Indeed, Eq. (19) can be reformulated as follows:

$$\underset{e_i^k, e_j^k}{\textbf{minimize}} \sum_{e_i \in V^k} e_i^k \cdot \textbf{p}(e_i) + \sum_{\substack{e_i \in V^k \\ e_j \in V^k}} e_i \cdot e_j \cdot \textbf{P}(e_i, e_j) \tag{21}$$

with

$$\textbf{p}(e_i) = \sum_{e_j \in V^{k+}} \textbf{com}(e_i, e_j), \quad \forall e_i \in V^k \tag{22}$$

$$\textbf{P}(e_i, e_j) = \begin{cases} -3 \cdot \textbf{com}(e_i, e_j), & \text{if } (e_i, e_j) \in E^k \\ 0, & \text{otherwise} \end{cases} \tag{23}$$

and $V^{k+} = \{e_i \mid (e_i, e_j) \in E^k \text{ or } (e_j, e_i) \in E^k\}$ is the set of elements connected to at least one edge in E^k, even the ones that are not movable (thus, not necessarily proposed for migration).

QKP can be linearized [1] and then solved using a centralized branch-and-cut method or a distributed optimization method, as discussed earlier. Alternatively, QKP is solvable by dynamic programming, but without optimality guarantees. Only requiring $\mathcal{O}(\textbf{cap}(a_k).|V^k|)$ space, and $\mathcal{O}(\textbf{cap}(a_k).|V^k|^2)$ time, such a dynamic programming approach seems realistic in our case [8].

Besides, instead of using its whole memory capacity $\textbf{cap}(a_k)$, device a_k may also set a limit capacity below its maximum one (e.g. the average memory used by its neighbors) as not to host more computation than others, in general. As to respect constraint (10) in Problem 4, we can add the following constraints to a_k's decision to enforce hosting at least one element:

$$|\rho^{-1}(a_k)| + \sum_{e_i \in V^k} e_i^k \geq 1 \tag{24}$$

$$\forall a_\ell \in \mathcal{A}[a_k] \setminus a_k, \ |\mu^{-1}(a_\ell)| - \sum_{e_i \in V^{\ell \to k}} e_i^k \geq 1 \tag{25}$$

Constraint (24) ensures that the newcomer hosts at least one computation. Constraint (25) avoids migrating all elements from one of the proposing agents and requires the newcomer to know the number of elements hosted by a_ℓ.

From the proposer side, the decision of choosing which elements to propose is also an issue, that may impact the newcomer's decision. In this paper, we only consider proposing all the "movable" elements, i.e. those that are hosted but not owned by the agent (physical factors and variables, and rules utility factors).

4.3 Device Removal

Another common problem in ambient environments is that some devices may fail or get unreachable for some reason. In this case, for belief-propagation algorithms to operate properly, we need to fix the deployment: factor and variables *owned* by the departed agent simply disappear from the factor graph while elements *hosted* on it must be relocated to an available agent. We can identify two cases: *safe* removal (the device leaves the system voluntarily and can migrate elements before leaving), and *unsafe* removal (when the device fails without migrating its hosted elements).

Safe removal is equivalent to the device arrival, but the element allocation is made over the neighborhood of the removed agents (excluded). Here, solving IQP-SECP-NDP, is not relevant, because no agent is at the center of the decision. We will not elaborate on this case.

Unsafe removal is more complex, and implies some technicalities. Here, we assume that devices are aware of disappearance of any device from their neighborhood (using *keepalive* signals). In such case, one could solve ILP-SECP-FGDP for the entire new set of devices, which we cannot afford within a single device. Therefore, we opt here for a more local and heuristic approach. The idea is to solve ILP-SECP-FGDP restricted to agents that were directly connected to the dead device.

We note $V_x[a_k]^- = V_x[a_k] \smallsetminus \rho_x^{-1}(a_k)$, $V_f[a_k]^- = V_f[a_k] \smallsetminus \rho_f^{-1}(a_k)$ and $E[a_k]^- = E[a_k] \cap (V_x[a_k]^- \times V_f[a_k]^-)$ the sets of elements and edges involved in the redistribution. We note $\mathbf{cap}^-(a_k) = \mathbf{cap}(a_k) - \sum_{e_i \in \rho^{-1}(a_k)} \mathbf{mem}(e_i)$ the memory capacity of any a_k, obtained by subtracting from $\mathbf{cap}(a_k)$ the memory footprint of computations hosted on a_k and not involved in the redistribution.

Problem 8 (ILP-SECP-FGDP$[a_k]^-$). ILP-SECP-FGDP$[a_k]^-$ consists in ILP-SECP-FGDP restricted to the set of agents $\mathcal{A}[a_k] \smallsetminus \{a_k\}$, the graph $\langle V_x[a_k]^-, V_f[a_k]^-, E[a_k]^- \rangle$, and where $\mathbf{cap}$ is replaced by $\mathbf{cap}^-$.

To solve ILP-SECP-FGDP$[a_k]^-$ some requirements are to be fulfilled: (a) devices have to know how to compute elements which share an edge with an element they host and the corresponding communication costs; (b) devices have to know to which devices they send messages; (c) devices need to have enough memory to host new elements.

Requirement (a) imply giving such information during initial deployment and revision phases. Requirement (b) depends upon the discovery mechanism when an agent hosts new elements, for example when new devices are added. Finally, requirement (c) may not be reached if the neighboring agents of a disappearing one don't have enough memory all together. In this case, the neighborhood can be extended by neighbors of neighbors until memory is sufficient. Once these requirements met, agents can solve ILP-SECP-FGDP$[a_k]^-$ on the limited set of elements and the neighboring devices, as it is the case for a newcoming device (see Sect. 4.2).

5 Discussion on Other Dynamics

Up to now, we have discussed dynamics involved by adding or removing devices. However, even for a same set of devices, other dynamics may occur, since our socio-technical system is both connected to users and a sensed environment.

5.1 Changing SECP Elements

One reason to alter the deployed FG, is for a user to add or remove a rule to the system. However, in this case, since the user is interacting with the system through a generally powerful dedicated device (e.g. home computer, tablet) we can rely on this device to perform a full deployment of the FG, as in Sect. 3. The second reason to alter the FG is to add/remove actuator variables and costs, which only occurs when adding/removing devices, as in Sects. 4.2 and 4.3. The third reason to alter the FG, is to add new physical models. This is the most difficult part. Indeed, here we assume these physical models are provided (e.g. a calibration phase). Adding a new physical model only makes sense when the user specifies a new rule with a new physical model (related to a new sensor) he has obtained. For instance, a user installs a sound level sensor and add a new rule which exploits the sound level somehow. Such a situation, once again, only occurs when the user interacts through his dedicated device with the system. Thus, deploying the FG can be done in a centralized way, as in Sect. 3.

5.2 Changing Factors Following an Environmental Change

Our system is deployed in a dynamic environment, where newly sensed data may imply that some rules activates or not. Up to now, we discussed the deployment of the whole FG, but practically, all rules are not necessarily active all the time. Rules, and therefore some factors and variables are only active when some sensed state is reached. Such activation/deactivation may greatly impact the performance of the system, by adding/removing computations and loops in the factor graph. Our deployment model does not take this into account. In fact, it considers the worst case when all the elements are active.

6 Experiments

As to evaluate the performances of the proposed repair techniques, we simulate a smart home where devices perform Max-Sum as to find a good configuration considering user preferences and energy consumption. Note that the approaches discussed here for the FGDP aims as deploying the FG on which Max-Sum operates to solve the SECP proposed in [11].

6.1 Simulated Smart Home Scenario

In our simulations, two types of events may occur: device arrival (in) and unsafe device removal (out). In case of device arrival, we use either ILP-SECP-FGDP$[a_k]^+$, which is solved using a classical ILP solver within one node (using GLPK in our simulator), or IQP-SECP-NDP, which is solved using a dedicated dynamic program (embedded in our simulator, in Python)[2], both defined in Sect. 4.2. In case of device removal, as discussed in Sect. 4.3, ILP-SECP-FGDP$[a_k]^-$ is solved using GLPK. Whatever the type of event, the best ILP-SECP-FGDP solution (computed with GLPK), and the solution provided by centralized deployment heuristic from [11] are computed to benchmark aforementioned methods.

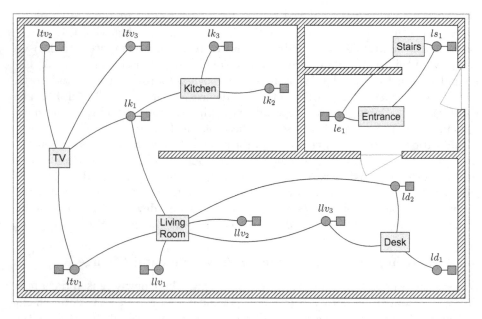

Fig. 3. Map of a simulated smart home, and corresponding initial physical models, actuators and costs.

In a first series of experiments, we simulate the first floor of a smart home, as represented in Fig. 3, which is initially composed of 13 actuators (light bulbs and their respective costs), 6 physical models (one for each space), and 5 user rules (not represented in Fig. 3, for clarity). For communication costs, $a = 5$, $b = 100$, and the default agent memory capacity (**cap**) is set to 200 memory units (one unit represents the space to store one value, e.g. 32 bits). Figure 4 traces performances of repair solutions on a scripted scenario where devices are

---

[2] Such discrepancies in terms of solution method implementation are the reason not to plot computation times.

added and removed at runtime. Each of these 40 events is followed by a repair phase using the proposed methods.

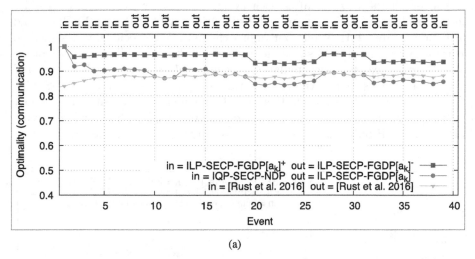

(a)

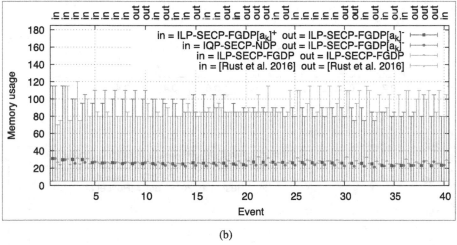

(b)

Fig. 4. Optimality (4a), and memory usage (4b) of the deployment during the simulation (standard deviation, min and max).

Figure 4a shows the optimality of the repaired deployments, computed as the ratio between the repaired cost and the best cost (real ILP-SECP-FGDP optimum). The centralized deployment heuristic used in [11], labeled "Rust et al. 2016" on the figure, is also plotted for comparison.

Clearly, with both approaches, out events tends to degrade the optimality of the deployment, while still maintaining it at a very competitive level, compared to a full deployment of the whole factor graph. Interestingly, in events improve optimality, meaning that in real systems where on average out are approximately balanced by in, the deployment should keep a very good quality level.

Figure 4b presents the standard deviation (and min and max) of memory usage over all the devices, after each event. For comparison, it also includes the values obtained with and optimal distribution (with ILP-SECP-FGDP) and the heuristic from [11]. While our approaches are not specifically designed to ensure a fair memory load share among devices, both distributed methods do not lead to an excessive accumulation of computations on a single device and perform at least as well as the two centralized approaches. Solving ILP-SECP-FGDP$[a_k]^+$ is a better choice in this regard, which can be explained by the fact that it allows relocation of computation on the full neighborhood, while solving IQP-SECP-NDP only allows migration of computations to the newcomer.

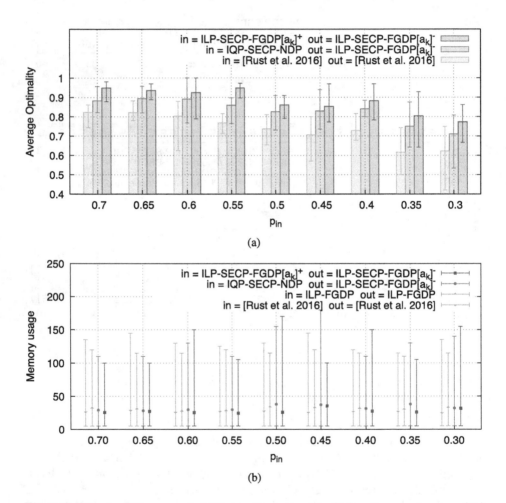

Fig. 5. Influence of the p_{in} probability on the optimality (5a) and memory usage (5b)

In a second series of experiments, we simulate the whole house with 23 actuators, 9 physical models and 9 rules. Here we evaluate the robustness of each repair techniques with more and more device removal. Figure 5 shows the average performances over 10 simulations after 20 events, in terms of optimality (computed as previously) with a varying event type probability. At each event generation, its type is determined using p_{in}, i.e. the probability for en event to be in. The higher p_{in}, the easier the adaptation is, since more devices are probably added. ILP-SECP-FGDP$[a_k]^+$ combined with ILP-SECP-FGDP$[a_k]^-$ presents very good resilience, since it offers more than 80% optimality with $p_{in} \geq 0.35$ (approx. 2 removals for 1 arrival). IQP-SECP-NDP combined with ILP-SECP-FGDP$[a_k]^-$ is always 5 to 15% lower. It is remarkable that these local repair techniques yield better distributions, from a communication point of view, than the heuristic from [11], even though it is a centralized approach and has access to information about the whole factor graph.

Figure 5b represents the standard deviation (and min and max) of memory usage over all devices, at the last event of the scenario for each value of p_{in}. No notable difference can be observed among the various techniques, both local and centralized.

Finally, ILP-SECP-FGDP$[a_k]^+$ presents better optimality, but requires much more information to be computed, whilst IQP-SECP-NDP is in average 10% worse in communication cost, and equivalent in average memory usage.

6.2   Randomly Generated SECPs

In a third series of experiments, we evaluate the influence of the number of rules in the SECP on the performance of the distribution techniques. Here we generate 10 pairs of SECP and scenarios (containing 20 events) for each combination of p_{in} and n_r where $0.3 \leq p_{in} \leq 0.7$ and $10 \leq n_r \leq 50$ is the number of rules (with a step of 10). All SECP are generated randomly with 30 lights and 7 models and map to connected factor graphs, meaning that an increase on the number of rules also results in an increase on the factor graph density.

Figure 6 shows the average performance in term of communication optimality. We can see that the good resilience of the local distribution repair approaches is not really impacted by the number of rules in the system; results are very similar to those of the second experiment for both ILP-SECP-FGDP$[a_k]^+$ combined with ILP-SECP-FGDP$[a_k]^-$ and IQP-SECP-NDP combined with ILP-SECP-FGDP$[a_k]^-$. However, we notice that the heuristic from [11] performs much better than on the second experiment and consistently returns better distribution that IQP-SECP-NDP combined with ILP-SECP-FGDP$[a_k]^-$. This can be explained by the fact that the SECP used here are generated randomly while the SECP used for previous experiments were modeling actual real smart homes. Real SECP, even when their factor graph is connected, tends to have some locally semi-independent subgraphs, which roughly maps the various rooms and zones of a house. This structure is not present is random SECP, which tends to be much more uniform. This exhibits the high impact of the topology of the factor

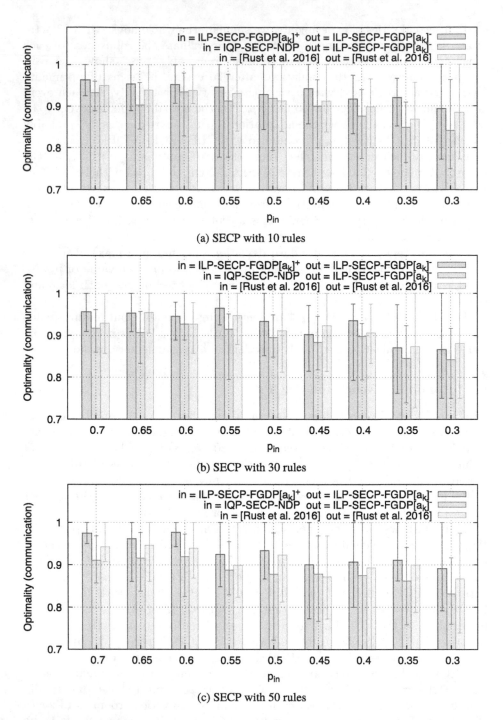

Fig. 6. Influence of the p_{in} probability on the optimality for SECP with an increasing number of rules

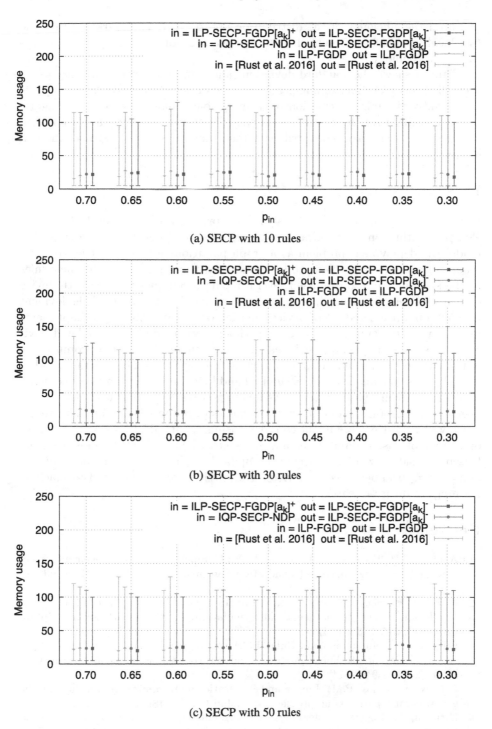

(a) SECP with 10 rules

(b) SECP with 30 rules

(c) SECP with 50 rules

Fig. 7. Influence of the p_{in} probability on the memory usage (standard deviation, min and max) for SECP with an increasing number of rules

graph on the efficiency of the distribution approach. It is remarkable that the two local approaches presented in this paper are not much impacted by this change in topology.

Figure 7 shows the standard deviation (and min and max) of memory usage over all devices. As for communication, the number of rules does not impacts significantly the fair distribution of memory load among devices. Here again, the factor graph topology seems to impact the results: all approaches tends to produce more well-balanced distribution than they did with SECP modeling real houses.

7   Conclusions

In this paper we discussed and analyzed the problem of deploying factor graph elements within an open infrastructure composed of constrained devices. We model the deployment problem as a graph partitioning problem, encoded as a binary integer linear problem, to be solved each time the user pushes new rules in the system. We also discussed several repair techniques to cope with device arrival and removal occurring at runtime, by solving the original deployment problem on a restricted set of devices and factor graph elements, or implementing a newcomer-centric approach. Experiments we made on a simulated environment show that the proposed local and heuristic techniques have competitive optimality levels in comparison to restarting the deployment from scratch. Additionally, these techniques only use limited and local knowledge and thus could be used in arbitrarily large systems. As mentioned when dealing with newcoming agents, the decision of choosing which elements to propose is also an issue we did not investigate. Here, we might consider basing agents' decisions on preferences or history of past computations and messages exchange, as to assess elements to send. Besides, we didn't discuss the update of physical models following the appearance/disappearance of devices. We let this problem, related to machine learning, to future research.

References

1. Adams, W.P., Forrester, R.J., Glover, F.W.: Comparisons and enhancement strategies for linearizing mixed 0–1 quadratic programs. Discrete Optim. **1**(2), 99–120 (2004)
2. Bichot, C.E., Siarry, P. (eds.): Graph Partitioning. Wiley, New York (2011)
3. Boulle, M.: Compact mathematical formulation for graph partitioning. Optim. Eng. **5**(3), 315–333 (2004)
4. Bürger, M., Notarstefano, G., Bullo, F., Allgöwer, F.: A distributed simplex algorithm for degenerate linear programs and multi-agent assignments. Automatica **48**(9), 2298–2304 (2014)
5. Fan, N., Pardalos, P.M.: Linear and quadratic programming approaches for the general graph partitioning problem. J. Global Optim. **48**(1), 57–71 (2010)
6. Farinelli, A., Rogers, A., Jennings, N.R.: Agent-based decentralised coordination for sensor networks using the max-sum algorithm. Auton. Agent. Multi-Agent Syst. **28**(3), 337–380 (2014)

7. Farinelli, A., Rogers, A., Petcu, A., Jennings, N.R.: Decentralised coordination of low-power embedded devices using the max-sum algorithm. In: International Conference on Autonomous Agents and Multiagent Systems (AAMAS 2008), pp. 639–646 (2008)
8. Fomeni, F.D., Letchford, A.N.: A dynamic programming heuristic for the quadratic Knapsack Problem. INFORMS J. Comput. **26**(1), 173–182 (2014)
9. Martins, A.F.T., Figueiredo, M.A.T., Aguiar, P.M.Q., Smith, N.A., Xing, E.P.: AD3: alternating directions dual decomposition for map inference in graphical models. J. Mach. Learn. Res. **16**, 495–545 (2015)
10. Mitchell, J.E.: Branch-and-cut algorithms for combinatorial optimization problems. Handbook of Applied Optimization, pp. 65–77. Oxford University Press, New York (2002)
11. Rust, P., Picard, G., Ramparany, F.: Using message-passing DCOP algorithms to solve energy-efficient smart environment configuration problems. In: International Joint Conference on Artificial Intelligence (IJCAI). AAAI Press (2016)
12. Tarlow, D., Givoni, I.E., Zemel, R.S.: HOP-MAP: efficient message passing with high order potentials. In: Teh, Y.W., Titterington, D.M. (eds.) Proceedings of the Thirteenth International Conference on Artificial Intelligence and Statistics (AISTATS-2010), vol. 9, pp. 812–819 (2010)

Optimizing Peer Teaching to Enhance Team Performance

Zheyuan Shi[1] and Fei Fang[2]([✉])

[1] Swarthmore College, Swarthmore, PA 19081, USA
zshi1@swarthmore.edu
[2] Carnegie Mellon University, Pittsburgh, PA 15213, USA
feifang@cmu.edu

Abstract. Collaboration among human agents with different expertise and capabilities is becoming increasingly pervasive and important for developing new products, providing patient-centered health care, propelling scientific advance, and solving social issues. When the roles of the agents in such collaborative teamwork are highly interdependent, the performance of the team will rely not only on each team member's individual capabilities but also on their shared understanding and mutual support. Without any understanding in other team members' area of expertise, the team members may not be able to work together efficiently due to the high cost of communication and the individual decisions made by different team members may even lead to undesirable results for the team. To improve collaboration and the overall performance of the team, the team members can teach each other and learn from each other, and such peer-teaching practice has shown to have great benefit in various domains such as interdisciplinary research collaboration and collaborative health care. However, the amount of time and effort the team members can spend on peer-teaching is often limited. In this paper, we focus on finding the best peer teaching plan to optimize the performance of the team, given the limited teaching and learning capacity. We (i) provide a formal model of the Peer Teaching problem; (ii) present hardness results for the problem in the general setting, and the subclasses of problems with additive utility functions and submodular utility functions; (iii) propose a polynomial time exact algorithm for problems with additive utility function, as well as a polynomial time approximation algorithm for problems with submodular utility functions.

Keywords: Peer teaching · Teamwork · Optimization

1 Introduction

As we welcome a new age of knowledge segmentation, teamwork is not only escalating in its importance but also shifting in its nature. Its traditional focus on distributing and sharing workload is quickly replaced by the focus on sharing

© Springer International Publishing AG 2017
G. Sukthankar and J. A. Rodriguez-Aguilar (Eds.): AAMAS 2017 Best Papers,
LNAI 10642, pp. 138–150, 2017.
https://doi.org/10.1007/978-3-319-71682-4_9

the knowledge and expertise needed for the relevant goal. Such transition is evidenced by, for example, researchers from different domains collaborating on an interdisciplinary research project, or nurse-physician interprofessional collaboration in health care. However, forming a team with diverse skill sets is far from the end of the story. Richter et al. [16] show that putting students into interdisciplinary teams and even teaching teamwork skills are not sufficient for effective interdisciplinary collaboration. It is often more desirable for agents to learn some of their teammates' knowledge, than having each agent solely responsible for her own expertise. Bridges et al. [4] find that understanding others' professions in the healthcare team is important in interprofessional collaboration and helps team member better understand her own duties. To have shared knowledge and further enhance the team performance, an efficient and effective way is to have the team members learn from each other. In health and social care, such peer-teaching is viewed as part of the interprofessional education [6,21], which enables effective collaborative practice [2]. However, the amount of time and effort the team members can spend on peer-teaching is often limited, and it is impossible to ask the team members to gain all of their teammates' knowledge. As such, determining an optimal peer teaching plan is crucial in boosting team performance.

The growing attention to teamwork in the society gives rise to the rapid development of research on team collaboration. Team formation, for instance, addresses the problem of selecting the best team member under limited resources [9,10,18]. Various models for team coordination have been proposed, especially when team members can hardly communicate with each other [1,19]. Works have also been done in the performance measures for teams [13], human-agent teams [22], and communication models [5]. However, few works explicitly consider leveraging the team's diversity and enhancing team performance by having team members teach each other.

Although existing literature on team collaboration does not emphasize peer teaching, the process of, possibly informal, learning from teammates does happen in many teamwork scenarios. Team members often help each other to enhance knowledge in certain topics, to build certain skills, to improve certain ability, or to develop certain capabilities.[1] However, there lacks a formal model to study and optimize this process. Our first contribution is the formalization of the peer teaching problem. We characterize a group based on its members and relevant expertise. By quantifying the choices and limits of teaching and learning inside the group, we model the peer teaching problem as a constrained optimization problem.

After formalizing the model, it is natural and important to find the best plan for peer teaching, and we focus on this problem in the rest of the paper. We show that the peer teaching problem in its most general form is hard. However, we analyze two key settings with additive and submodular utility functions and propose two algorithms to find the optimal peer teaching plan. In the first case,

---

[1] In the rest of the paper, we collectively refer to these abstract concepts as skills which can be taught and learned.

we present an exact polynomial time algorithm, and in the second case, we present a polynomial time approximation algorithm.

2   Related Work

Peer teaching relates to yet differentiates itself from several topics in teamwork which have been studied. Liemhetcharat and Veloso [11] introduce teams with learning agents, where agents have access to external training resources rather than learning from their teammates. Jumadinova et al. [7] treat the peer teaching process as part of a decision problem. However, their work does not explicitly consider the teaching and learning constraints, which are essential to the structure of the peer teaching problem. Compared to the study of cross-training, which refers to agents being trained the expertise of their teammates and is shown to improve the team's performance [14], peer teaching emphasizes the notion of agents autonomously learning from teammates and is thus bounded by various capacity constraints. Several pieces of work on team formation consider the diversity of skills and the synergy among team members [10,12,18]. Peer teaching problem differs from these works in treating a team as given and studies the teaching plan to optimize team performance. Other works focus on the scenario where team members from diverse communities can hardly coordinate prior to collaboration and only loosely coordinate during the collaboration [1,19]. The peer teaching problem applies to this setting and provides the learning and teaching dynamics which the above-mentioned works do not consider.

Much work has been done on multiagent MDP to study the coordination among individual agents on a team [3]. One specific line of research is information sharing, which studies how agents decide when and what observations to share in a partially observable multiagent MDP framework [17,23]. Peer teaching differs from this area of research in the special nature of knowledge and puts less emphasis on the duration of the process.

We also observe the recent attention on cross-domain collaboration. While this is a place where the process of peer teaching arises frequently, works in this area [20,24] usually focus on partner recommendation, which is in nature different from our problem.

3   Peer Teaching Problem

In a peer teaching problem, we have a set of agents with the same goal but with different areas of expertise. Before they start working as a team, they can help other team members gain expertise through teaching, and such peer teaching can lead to an improvement in the team performance. However, often there is a limit on how much time and effort an agent can spend on teaching and learning. Therefore, we need to find the best feasible peer teaching plan which can lead to the highest improvement in team performance.

We model the peer teaching problem as a constrained optimization problem defined over a group profile.

Definition 1. *The group profile* $\mathcal{G} = (A, S, M, f)$ *contains the following*

- $A = \{a_1, \ldots, a_n\}$ *denotes the set of agents.*
- $S = \{s_1, \ldots, s_m\}$ *denotes the set of areas of expertise, where the area of expertise can be a skill or a type of knowledge.*
- $M \subseteq A \times S$ *denotes the initial agent-expertise mapping of the group.* $(a, s) \in M$ *means agent a has expertise in s before any peer teaching takes place. We denote by* $M_a(a_i) = \{s_j | (a_i, s_j) \in M\}$ *the set of areas of expertise that agent i has and* $M_s(s_j) = \{a_i | (a_i, s_j) \in M\}$ *the set of agents that has expertise in* s_j. *We denote by* $\bar{M} = A \times S \setminus M$ *the complement of* M.
- $f : 2^{\bar{M}} \to \mathbb{R}$ *is the utility function. A learning profile* $M' \subseteq \bar{M}$ *is a set of learning events, and a learning event* $(a, s) \in M'$ *means agent a gains some expertise in s from some other agent during peer teaching. The utility function indicates how much improvement a learning profile can bring to the team performance.*

Next, we introduce several definitions towards the definition of the collection of all feasible learning profiles. $T = \{T_1, \ldots, T_n\}$ denotes the teaching capabilities of the agents. $T_i \subseteq M_a(a_i)$ and $s_j \in T_i$ means agent a_i is capable of helping other agents to gain expertise in s_j through teaching. Differentiating T_i from $M_a(a_i)$ provides a way to quantify the level of expertise of each agent, as one might expect that the ability to teach others implies a high level of proficiency. A peer teaching plan is defined as a set of triplets of (teacher, expertise, learner), i.e., $\Theta = \{(a_{i_1}, s_j, a_{i_2}) | (a_{i_1}, s_j) \in T_{i_1}, (a_{i_2}, s_j) \in \bar{M}\}$. A peer teaching plan is feasible if it satisfies teaching and learning capacity constraints defined by c_i^t, c^t, and c_i^l. c_i^t represents the maximum number of expertise agent i can teach, c^t represents the maximum number of agents that any agent can teach simultaneously for one expertise and c_i^l represents the maximum number of expertise one can gain through peer teaching. A learning profile M' is feasible if there exists a feasible peer teaching plan $\Theta_{M'}$ which realizes all learning events in M'. Given a group profile $\mathcal{G}$ and the collection of all feasible learning profiles L, the peer teaching problem is to find a learning profile $M^* \in L$ to optimize the utility function.

Figure 1 provides an example of the peer teaching problem. For illustration purpose, it has three agents and three areas of expertise. The right side of the graph is a bipartite graph which represents each agent's teaching capability. In this example, we assume $T_i = M_a(a_i)$, i.e., an agent is capable of teaching anything that he currently has expertise in. The bipartite graph on the left shows all possible learning events.

4   Optimizing Peer Teaching

The definition of the peer teaching problem leaves much freedom for deciding the dynamics of the peer teaching process. As we show below, without further structures in the problem, the peer teaching problem is hard.

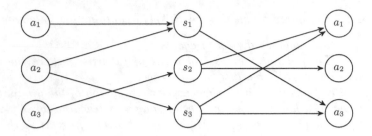

Fig. 1. The knowledge graph and the graph of all possible learning events

Theorem 1. *The peer teaching problem in its general form is NP-hard.*

Proof. We prove the hardness of the peer teaching problem by reducing from the maximum cut problem. For an arbitrary weighted undirected graph $G = (V, E, W)$ with $|V| = n$ nodes, we consider a corresponding peer teaching problem $\mathcal{G} = (A, S, M, f)$, where $|A| = |S| = n$, $T_i = M_a(a_i) = \{s_j | j \neq i\}$, and $M_s(s_j) = \{a_i | i \neq j\}$. Thus, we have $\bar{M} = \{(a_i, s_i) | i = 1, \ldots, n\}$. For any subset $M_I = \{(a_i, s_i) | i \in I\}$ of $\bar{M}$, we assign the weight of the cut $V_I = \{v_i | i \in I\}$ in graph G to be the utility value $f(M_I)$. The value $f(M_I)$ can be computed in polynomial time. In addition, we ignore teaching and learning capacity constraints by setting $c_i^t, c^t, c_i^l > 1$. Therefore, a subset of nodes V_I in G yields the maximum cut if and only if the corresponding learning profile M_I maximizes the utility function f in the peer teaching problem $\mathcal{G}$.   □

4.1 Additive Utility Function

The hardness result for the general peer teaching problem calls for more structure in the problem setup. In this subsection, we study a particular type of peer teaching problem characterized by additive utility functions, and present an exact polynomial time algorithm for finding the optimal peer teaching plan.

Consider a group profile $\mathcal{G} = (A, S, M, f)$, where f is such that for all $P, Q \subseteq \bar{M}$, $f(P \cup Q) + f(P \cap Q) = f(P) + f(Q)$. Equivalently, we assign a utility v_{ij} to each learning event $(a_i, s_j) \in \bar{M}$, and define $f(P) = \sum_{(a_i, s_j) \in P} v_{ij}$ for a learning profile $P \subseteq \bar{M}$. This model is natural, for example, when a team is assessed based on the ability of its members individually. As defined in Sect. 3, each agent a_i has teaching capacity c_i^t and learning capacity c_i^l. For this subsection, we assume all teaching happens in a one-on-one fashion, i.e., $c^t = 1$. We define $l_{ij} = 1$ if agent a_i learns skill s_j, and $t_{ij} = 1$ if agent a_i teaches skill s_j, and zero otherwise. The problem can then be formulated as the following integer linear program (ILP).

$$\text{minimize} \quad -\sum_i \sum_j v_{ij} l_{ij}$$

$$\text{subject to} \quad \sum_j l_{ij} \leq c_i^l, \quad \forall i = 1, \ldots, n \quad (X)$$

$$\sum_j t_{ij} \leq c_i^t, \quad \forall i = 1, \ldots, n \quad (Y)$$

$$\sum_i l_{ij} = \sum_i t_{ij}, \qquad \forall j = 1, \ldots, m \tag{Z}$$

$$l_{ij}, t_{ij} \in \{0, 1\}, \qquad \forall i = 1, \ldots, n, \forall j = 1, \ldots, m$$

While solving ILP is hard in general, this problem has the structure of network flow and thus solving the linear program relaxation of the ILP can directly lead to an optimal integer solution [15].

Theorem 2. *The peer teaching problem with additive utility function and $c^t = 1$ can be solved in polynomial time.*

Proof. Recall that a square, integer matrix B is unimodular if $\det(B) = \pm 1$. An integer matrix A is totally unimodular if every square, nonsingular submatrix of A is unimodular. As a sufficient condition, an integer matrix A whose only nonzero entries are ± 1 is totally unimodular if no column of A contains more than two nonzero elements and we may partition the rows of A into I_1 and I_2 such that

- if a column has two entries of the same sign, their rows are in different sets;
- if a column has two entries of different signs, their rows are in the same set.

If A is totally unimodular, and b, u, l are integer vectors, then all the vertices of the polyhedron $P = \{x \mid Ax = b, u \le x \le l\}$ are integer points.

Consider the linear program (LP) relaxation of the ILP above. The feasible polyhedron of the relaxed LP can be written as $P = \{x \mid Ax = b, 0 \le x \le 1\}$, where we collect all the constraints in X, Y, Z into the equation $Ax = b$ by adding the necessary slack variables. Observe that each column of A has at most two nonzero entries, which are 1 or -1. Furthermore, we may partition the rows of A into two sets, one containing all constraints in X, the other containing all constraints in Y and Z. Such a partition satisfies the conditions for a totally unimodular matrix as mentioned above. Therefore, it follows that solving the relaxed LP will guarantee us an integer optimal solution. Applying ellipsoid method for the relaxed LP leads to an algorithm that finds the optimal solution in polynomial time. □

Below we show the running time of the LP compared to two baseline algorithms: the brute force algorithm and the greedy algorithm. The brute force algorithm examines the utility value of all possible learning profiles. The greedy algorithm starts with the learning event with highest utility, and adds the most beneficial learning event as long as the learning profile remains feasible. The utility values v_{ij} are generated independently from a uniform distribution on integers between 0 and 1000. The teaching and learning capacities c_i^l, c_i^t are generated independently uniformly on integers between 1 and $m = |S|$. We use the *linprog* function in MATLAB R2016a and run on a PC with Intel Core i7-4700MQ processor and 4 GB RAM.

In the experiments we fix the number of agents $|A| = n$ and vary the number of skills $|S| = m$. As shown in Fig. 2a, the brute force algorithm quickly blows up, making it infeasible to test its running time beyond $n = 3$, while the running

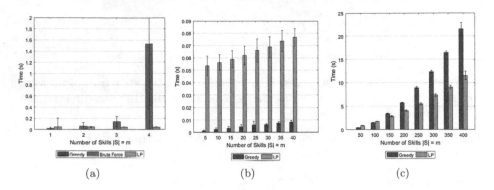

Fig. 2. The running time of the three algorithms. For (a), $|A| = n$ is fixed at 3; for (b), n is fixed at 20; for (c), n is fixed at 200. The standard deviation across five runs is also shown. For (a) and (b), the data are averages over 1000 runs; for (c), the data are averages over 30 runs.

time of the greedy algorithm is negligible compared to others. In Fig. 2b, where the problem size is relatively small, the greedy algorithm outperforms the LP in running time. However, as shown in Fig. 2c, the LP becomes the better one as the problem grows larger.

We also measure the accuracy of the greedy algorithm by the ratio between its output and the LP optimal utility, as shown in Fig. 3. In general, it gives a relatively good approximation, and its accuracy improves as the problem size grows.

4.2   Submodular Utility Function

Under many circumstances, more teaching may not benefit the team as much if the agents are already learning a lot from each other. For instance, given the limited time in a hackathon, students should not bother learning their teammates'

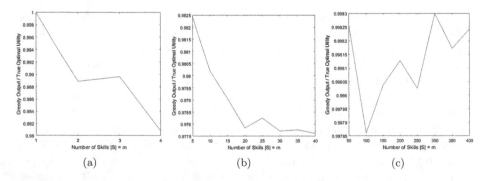

Fig. 3. The accuracy of the greedy algorithm. For (a), $|A| = n$ is fixed at 3; for (b), n is fixed at 20; for (c), n is fixed at 200. For (a) and (b), the data are averages over 1000 runs; for (c), the data are averages over 30 runs.

programming languages if for each potentially useful language there are already two or three members who can use it. We may use submodular utility functions to model this diminishing return. To proceed, recall the following definitions.

Definition 2. *A function $f : 2^X \to \mathbb{R}$ is submodular if for any $A, B \subseteq X$, $f(A) + f(B) \geq f(A \cup B) + f(A \cap B)$. X is referred to as the ground set of f.*

Definition 3. *Let $I \subseteq 2^X$. If a pair $N = (X, I)$ satisfies*

- *Downwards closure: If $P \in I, Q \subseteq P$, then $Q \in I$.*
- *Exchange property: If $P, Q \in I, |P| > |Q|$, then there exists $p \in P \backslash Q$ such that $Q \cup \{p\} \in I$.*

then N is a matroid. I is called the collection of independent sets.

In this subsection, we assume that each learning profile $P \subseteq \bar{M}$, feasible or not, is assigned a utility $f(P)$, where f is a submodular function. The solution to the peer teaching problem is a feasible learning profile $P^* \in L$ which maximizes the utility among all feasible learning profiles. To add more structures to the problem, we make the following rules (R1, R2) and assumptions (A1, A2).

- (R1) An agent can teach at most one expertise, but to multiple agents possibly. Equivalently, we set $c_i^t = 1$ for all $a_i \in A$, and set $c^t = n$.
- (R2) An agent can learn at most one expertise. Equivalently, we set $c_i^l = 1$ for all $a_i \in A$.
- (A1) An agent may have multiple expertise but is only able to teach one or two. Equivalently, we assume $|T_i| = 1$ or 2 for all $a_i \in A$.
- (A2) For each expertise, at least two agents can teach it. Equivalently, we assume for all $s_k \in S$, there exist $i \neq j$ such that $s_k \in T_i \cap T_j$.

The two assumptions might not seem very realistic, but we will relax them later. Recall that we refer to $n = |A|$ as the number of agents, and $m = |S|$ as the number of skills. Consider the knowledge graph in Fig. 1. By A1, the number of outgoing edges from agent nodes is less than or equal to $2n$. By A2, the number of incoming edges to skill nodes is greater than or equal to $2m$. Thus, we have $|A| \geq |S|$. These two assumptions allow us to exploit the structure in the feasible learning profiles. More specifically, assuming all the given conditions in this subsection, we have the following theorem.

Theorem 3. *For a given group profile $\mathcal{G} = (A, S, M, f)$, $N = (\bar{M}, L)$ is a matroid.*

Proof. Downwards closure is obvious, we prove the exchange property. Let $P, Q \in L$ be two feasible learning profiles, and $|P| > |Q|$. By R2 and $|P| > |Q|$, there exists an agent a_i who is taught in P but not in Q. Suppose in a peer teaching plan Θ_P corresponding to P, agent a_j teaches a_i the expertise s_k. Let Θ_Q be a peer teaching plan corresponding to Q. We discuss the following possible cases:

Case 1: In Θ_Q, *someone is teaching the expertise* s_k. According to R1, we can add the learning event (a_i, s_k) to Q, and will maintain feasibility.

Case 2: In Θ_Q, *no one is teaching the expertise* s_k, *but* a_j *is not teaching anything.* We can let a_j teach this expertise s_k to a_i, and add the learning event (a_i, s_k) to Q, and will maintain feasibility.

Case 3: In Θ_Q, *no one is teaching the expertise* s_k, *and* a_j *is teaching something else.* By A2, suppose a_j and a_l know expertise s_k and are both teaching something else, say s_j and s_l. Let $a_{j'}$ be another person who knows s_j, and $a_{l'}$ be another person who knows s_l. Note that $a_{j'}$ and $a_{l'}$ can be the same agent, but they must be different from a_j and a_l by A1. If $a_{j'}$ (or $a_{l'}$) is teaching something else, we repeat the same argument. As this argument propagates, we must be able to find an agent a^* who is not teaching anything, because $n \geq m$ and s_k is not being taught, there must be an agent who is not teaching. Furthermore, by A1, this agent a^* knows some expertise. Then, we can propagate back, and eventually find one of a_j and a_l to teach s_k, without impacting the group's other teaching ability. Once an agent is teaching s_k, we can add the learning event (a_i, s_k) to Q, and will maintain feasibility.   □

With this theorem, the peer teaching problem reduces to maximizing a submodular function subject to a matroid constraint. Unlike minimization, maximizing a submodular function is NP-hard, however. To find the best peer teaching strategy, we may use the algorithm, which we refer to as MAX, proposed by Lee et al. [8]. This is a polynomial time algorithm which achieves a $1/(4 + \epsilon)$-approximation, assuming a value oracle model, i.e. given a learning profile $P \subseteq \bar{M}$, the algorithm can access the utility value $f(P)$.

Algorithm 1. FEASIBLE

Input: Learning profile $P \subseteq \bar{M}$
  if any agent learns more than 1 expertise **then**
    Output: false
  end if
  Get the set of expertise S' that are being learned. Find a maximum matching R on the knowledge graph (Fig. 1) restricted to A and S'.
  if $|R| = |S'|$ **then**
    Output: true
  else
    Output: false
  end if

The main routine is MAX. LOCAL-SEARCH is a greedy algorithm which improves the current learning profile by adding, deleting, or substituting one learning event at a time. At each step, it checks whether the proposed better learning profile is feasible. Lee et al. [8] do not explicitly provide an algorithm for checking whether a set is independent. However, in our setting the feasibility of a learning profile is not trivial to verify. In FEASIBLE, we consider the bipartite

---

Algorithm 2. LOCAL-SEARCH

Input: Ground set V, value oracle access to submodular utility f

   Set $P = \{e_0\}$, where e_0 is the single learning event with highest utility

  while we can do one of the following operations **do**

     Delete: If $\exists e \in P$ such that $f(P\backslash\{e\}) \geq (1 + \epsilon/|V|^4)f(P)$, then set $P = P\backslash\{e\}$.

     Exchange: If $\exists e \notin P, e' \in P \cup \{\phi\}$ such that $f(P\backslash\{e'\} \cup \{e\}) \geq (1 + \epsilon/|V|^4)f(P)$

     and $P\backslash\{e'\} \cup \{e\}$ is feasible, then set $P = P\backslash\{e'\} \cup \{e\}$.

  end while

  Output: learning profile P

Algorithm 3. MAX

  Set $V_1 = \bar{M}$.

  Do LOCAL-SEARCH with ground set V_1, get solution P_1.

  Set $V_2 = \bar{M}\backslash P_1$.

  Do LOCAL-SEARCH with ground set V_2, get solution P_2.

  Output: RETURN the learning profile P_i whose $f(P_i)$ is greater

---

graph representing the current knowledge of the agents (Fig. 1). If all learned expertise are being matched in a maximum matching between agents and the expertise being learned in the learning profile, the learning profile is feasible because by R1 each agent can teach at most one expertise. Finally, by doing LOCAL-SEARCH twice, the algorithm MAX achieves the approximation bound of $\frac{1}{4+\epsilon}$.

While MAX is guaranteed to run in polynomial time and achieves a good approximation bound, we wish to relax the Assumptions A1 and A2. First, we consider A2, that for each expertise, there are at least two agents who can teach it.

Theorem 4. *Given A1, we may replace A2 with the assumption that $|A| = n \geq m = |S|$. If we treat n as fixed, we may remove A2, while still having a polynomial time algorithm with the same approximation bound.*

Proof. It is an uninteresting case where no agent knows some skill s_i. Suppose only one agent a_j can teach some expertise s_i, i.e., $s_i \notin T_k$ if $k \neq j$. If a_j can only teach this expertise s_i, then this particular violation of A2 does not fail $N = (\bar{M}, L)$ from being a matroid. Consider the proof of Theorem 3: if a_j is the teaching agent a_j that we picked in P, then we would not even get to Case 3. Otherwise, the pair (a_j, s_i) can be viewed as isolated, and the argument for Case 3 still holds.

If in addition to expertise s_i, agent a_j can also teach expertise s_k, i.e. $T_j = \{s_i, s_k\}$. If s_k also violates A2 such that nobody besides a_j can teach s_k, we may run the algorithm MAX twice and take the better output, where in each run a_j can only teach one of s_i and s_k. If at least one other agent a_l can teach s_k, then we can consider the expertise a_l has, and trim and rearrange their expertise in a way where the only violations of A2 are isolated agent-expertise pairs and the collection of feasible learning profiles L remain unchanged. This is achievable because we instead assume $n \geq m$.

In fact, once we replace A2 with the assumption that $n \geq m$, we may also relax this condition. Suppose we have more expertise than agents, i.e. $m > n$. Let $\{S_i^n\}$ be the collection of all subsets of S where $|S_i^n| = n$. We apply the algorithm MAX to the each of the induced group profiles $\mathcal{G}_i = (A, S_i^n, M_i, f_i)$. Since we assume each agent teaches at most one expertise, this modified algorithm achieves the same approximation bound as if we had only one problem to solve. It is worth noting, however, that naively restricting the problem may violate A1 by having some agents not able to teach any expertise in the subproblem. This can be fixed by assigning all agents who cannot teach any expertise in the subproblem to an imaginary expertise. Meanwhile, we add an imaginary agent who can also teach this imaginary expertise to maintain $m = n$. Then, we extend f_i by assigning the same utility to learning profiles which contain events involving imaginary knowledge or agent as without those events. This preserves the submodularity, and we can continue with the above-proposed procedure. $\square$

We may also consider relaxing assumption A1. Assuming A2 holds, if agent a_p can teach p expertise where $p \geq 3$, we can initially split the problem into p subproblems, and in each subproblem a_p can only teach one expertise. We may trim and rearrange the knowledge graph so that in each subproblem the matroid is maintained, and collectively all feasible learning profiles can be reached. However, we may have a combinatorial number of subproblems because to preserve the matroid other agents might require us to divide their outgoing edges in the knowledge graph as well.

5   Conclusion and Future Work

Team collaboration is gaining more attention in the society, and the research community as the segmentation of knowledge continues to grow. It is likely that a team's performance might improve if members are learning from each other and hence have a better sense of the work. In this paper, we focused on the problem of how teammates should teach and learn from their peers with limited resources to boost group performance. We formalized this process as the peer teaching problem. This problem in its most general case is hard, yet we provided good algorithmic solutions for some two specific setups of the problem, which are still general enough to model many real-world scenarios. We showed that with additive utility functions, we could solve the peer teaching problem with a linear program. In the case of submodular utility, a polynomial time approximation algorithm for maximizing submodular functions can be leveraged to find the optimal peer teaching plan.

There are many future directions to consider. One possible extension of the current peer teaching model is to explicitly quantify to what extent an agent has learned a skill instead of only considering whether or not she has learned the skill. One piece of knowledge often builds on another. Thus, studying planning with precedence graph could better characterize the peer teaching dynamics. Another direction is to consider multi-round collaboration, where at each round agents have different teaching and learning capacity and utility function. Some learning profiles might not be optimal considering the extended duration of collaboration, even if it achieves the best utility at a certain round. The learning aspect of the peer-teaching problem also needs further investigation. In this paper, we assumed agents could access the utility value through a value oracle. It is interesting to study the problem where such access comes with noise, and agents can learn the utility function across time or through available data. Such scenarios appear when, for example, researchers collaborate on interdisciplinary projects. Furthermore, in this paper we only model the member-skill relationship, while one may also consider the familiarity between team members, for example, if the members (or some of the members) have previously worked together on similar projects. In addition, knowing the optimal peer teaching strategy could offer insights into the team formation problem. When selecting group members, candidates' current expertise matter as well as the potential learning outcome they as a group could achieve. Last but not least, in real world problems, it is also useful to study how peer teaching interacts with learning from other resources.

Acknowledgments. The research presented in this paper was done when the first author was a visiting student at Massachusetts Institute of Technology, and the second author was a postdoctoral fellow at Harvard University. The research is supported by Harvard Center for Research on Computation and Society fellowship. We thank Barbara Grosz for her comments and helpful discussions.

References

1. Amir, O., Grosz, B.J., Gajos, K.Z.: Mutual influence potential networks: enabling information sharing in loosely-coupled extended-duration teamwork. In: Proceedings of the Twenty-Fifth International Joint Conference on Artificial Intelligence, IJCAI 2016, New York, NY, USA, 9–15 July 2016, pp. 796–803 (2016)
2. World Health Organization: Framework for action on interprofessional education and collaborative practice. World Health Organization, Geneva (2010)
3. Boutilier, C.: Planning, learning and coordination in multiagent decision processes. In: Proceedings of the 6th Conference on Theoretical Aspects of Rationality and Knowledge, pp. 195–210. Morgan Kaufmann Publishers Inc. (1996)
4. Bridges, D.R., Davidson, R.A., Odegard, P.S., Maki, I.V., Tomkowiak, J.: Interprofessional collaboration: three best practice models of interprofessional education. Med. Educ. Online **16**, 1–10 (2011)
5. Fussell, S.R., Kraut, R.E., Lerch, F.J., Scherlis, W.L., McNally, M.M., Cadiz, J.J.: Coordination, overload and team performance: effects of team communication strategies. In: Proceedings of the 1998 ACM Conference on Computer Supported Cooperative Work, pp. 275–284. ACM (1998)
6. Gilbert, J.H., Yan, J., Hoffman, S.J.: A who report: framework for action on interprofessional education and collaborative practice. J. Allied Health **39**(3), 196–197 (2010)
7. Jumadinova, J., Dasgupta, P., Soh, L.K.: Strategic capability-learning for improved multiagent collaboration in ad hoc environments. IEEE Trans. Syst. Man Cybern. Syst. **44**(8), 1003–1014 (2014)
8. Lee, J., Mirrokni, V.S., Nagarajan, V., Sviridenko, M.: Non-monotone submodular maximization under matroid and knapsack constraints. In: Proceedings of the Forty-First Annual ACM Symposium on Theory of Computing, pp. 323–332. ACM (2009)

9. Liemhetcharat, S., Veloso, M.: Modeling mutual capabilities in heterogeneous teams for role assignment. In: 2011 IEEE/RSJ International Conference on Intelligent Robots and Systems (IROS), pp. 3638–3644. IEEE (2011)

10. Liemhetcharat, S., Veloso, M.: Modeling and learning synergy for team formation with heterogeneous agents. In: Proceedings of the 11th International Conference on Autonomous Agents and Multiagent Systems, vol. 1, pp. 365–374. International Foundation for Autonomous Agents and Multiagent Systems (2012)

11. Liemhetcharat, S., Veloso, M.: Team formation with learning agents that improve coordination. In: Proceedings of the 2014 International Conference on Autonomous Agents and Multi-Agent Systems, pp. 1531–1532. International Foundation for Autonomous Agents and Multiagent Systems (2014)

12. Liemhetcharat, S., Veloso, M.: Weighted synergy graphs for effective team formation with heterogeneous ad hoc agents. Artif. Intell. **208**, 41–65 (2014)

13. MacBryde, J., Mendibil, K.: Designing performance measurement systems for teams: theory and practice. Manag. Decis. **41**(8), 722–733 (2003)

14. Marks, M.A., Sabella, M.J., Burke, C.S., Zaccaro, S.J.: The impact of cross-training on team effectiveness. J. Appl. Psychol. **87**(1), 3 (2002)

15. Papadimitriou, C.H., Steiglitz, K.: Combinatorial Optimization: Algorithms and Complexity. Courier Corporation, Mineola (1982)

16. Richter, D.M., Paretti, M.C., McNair, L.D.: Teaching interdisciplinary collaboration: learning barriers and classroom strategies. In: proceedings of 2009 ASEE Southeast Section Conference (2009)

17. Roth, M., Simmons, R., Veloso, M.: What to communicate? Execution-time decision in multi-agent POMDPs. In: Gini, M., Voyles, R. (eds.) Distributed Autonomous Robotic Systems, vol. 7, pp. 177–186. Springer, Tokyo (2006). https://doi.org/10.1007/4-431-35881-1_18

18. Marcolino, L.S., Jiang, A.X., Tambe, M.: Multi-agent team formation: diversity beats strength? In: Proceedings of the Twenty-Third International Joint Conference on Artificial Intelligence, IJCAI 2013, Beijing, China, pp. 279–285. AAAI Press (2013). http://dl.acm.org/citation.cfm?id=2540128.2540170. ISBN 978-1-57735-633-2

19. Stone, P., Kaminka, G.A., Kraus, S., Rosenschein, J.S., Agmon, N.: Teaching and leading an ad hoc teammate: collaboration without pre-coordination. Artif. Intell. **203**, 35–65 (2013)

20. Tang, J., Wu, S., Sun, J., Su, H.: Cross-domain collaboration recommendation. In: Proceedings of the 18th ACM SIGKDD International Conference on Knowledge Discovery and Data Mining, pp. 1285–1293. ACM (2012)

21. Barr, H., Koppel, I., Reeves, S., Hammick, M., Freeth, D.: Effective Interprofessional Education: Assumption, Argument and Evidence. Blackwell Publishing, Oxford (2005)

22. Traum, D., Rickel, J., Gratch, J., Marsella, S.: Negotiation over tasks in hybrid human-agent teams for simulation-based training. In: Proceedings of the Second International Joint Conference on Autonomous Agents and Multiagent Systems, pp. 441–448. ACM (2003)

23. Wu, F., Zilberstein, S., Chen, X.: Online planning for multi-agent systems with bounded communication. Artif. Intell. **175**(2), 487–511 (2011)

24. Wu, S., Sun, J., Tang, J.: Patent partner recommendation in enterprise social networks. In: Proceedings of the Sixth ACM International Conference on Web Search and Data Mining, pp. 43–52. ACM (2013)

A Protocol for Mixed Autonomous and Human-Operated Vehicles at Intersections

Guni Sharon[✉] and Peter Stone[✉]

Department of Computer Science, The University of Texas at Austin,
Austin, TX, USA
gunisharon@gmail.com, pstone@cs.utexas.edu

Abstract. Connected and autonomous vehicle technology has advanced rapidly in recent years. These technologies create possibilities for highly efficient, AI-based, transportation systems. One such system is the Autonomous Intersection Management (AIM), an intersection management protocol designed for the time when all vehicles are fully autonomous and connected. Experts, however, anticipate a long transition period during which human and autonomously operated vehicles will coexist. Unfortunately, AIM has been shown to provide little or no improvement over today's traffic signals when less than 90% of the vehicles are autonomous, making AIM ineffective for a large portion of the transition period. This paper introduces a new protocol denoted Hybrid Autonomous Intersection Management (H-AIM), that is applicable as long as AIM is applicable and the infrastructure is able to sense approaching vehicles. Our experiments show that this protocol can decrease traffic delay for autonomous vehicles even at 1% technology penetration rate.

Keywords: Autonomous Intersection Management
Autonomous vehicles · Multiagent systems

1 Introduction

Autonomous driving capabilities are becoming increasingly common on vehicles. Such capabilities present opportunities for developing safer, cleaner and more efficient road networks. Looking towards a future when most vehicles are autonomous and connected, Dresner and Stone proposed a novel intersection control protocol denoted Autonomous Intersection Management (AIM) [5]. AIM was shown to lead to significant traffic delay reductions when compared to traditional traffic signals.

Connected and autonomous vehicles (CAVs), with the help of advanced sensing devices, are more accurate and predictable compared to *human operated vehicles* (HVs). By relying on the fine and accurate control of CAVs along with communication capabilities, the AIM protocol coordinates multiple vehicles to cross an intersection simultaneously.

© Springer International Publishing AG 2017
G. Sukthankar and J. A. Rodriguez-Aguilar (Eds.): AAMAS 2017 Best Papers,
LNAI 10642, pp. 151–167, 2017.
https://doi.org/10.1007/978-3-319-71682-4_10

The AIM protocol defines two types of autonomous agents: intersection managers, one per intersection, and driver agents, one per vehicle. Intersection managers are responsible for directing the vehicles through the intersections, while the driver agents are responsible for controlling the CAV to which they are assigned.

To improve the throughput and efficiency of the system, the driver agents "call ahead" to the intersection manager and request a path reservation (space-time sequence) within the intersection. The intersection manager then determines whether or not this request can be met by checking whether it conflicts with any previously approved reservation or a potential HV. HVs are assumed to occupy all trajectories that are allowed by the traffic signal i.e., are given a green light. If the intersection manager approves a driver agent's request, the driver agent must follow the assigned path through the intersection. On the other hand, if the intersection manager rejects a driver agent's request, the driver agent may not pass through the intersection but may attempt to request a new reservation.

AIM, assuming 100% of the vehicles are CAVs, was shown to reduce the delay imposed on vehicles by orders of magnitude compared to traffic signals [6]. On the other hand, AIM was shown to be not better than traffic signals when more than 10% of the vehicles are HVs [5].

Given that experts speculate that 90% CAV penetration will not occur anytime before 2045 [3], this paper suggests a new protocol denoted *Hybrid AIM* (H-AIM) that is suitable for the transition period. Unlike AIM, H-AIM assumes sensing of approaching vehicles which allows it to identify approaching HVs. This assumption is reasonable given technological advances allowing vehicle detection using video cameras [4], radar [9], and inductive loop detectors [8]. If no HV is observed on a given lane, then trajectories originating from that lane are assumed to not be occupied by HVs, allowing AVs more flexibility in obtaining reservations.

A single lane entering a four-way intersection can allow three different turning possibilities (turn left, continue straight, turn right) or any combination of the three. The performance of H-AIM is sensitive to the assignment of allowed turns. This paper studies the effect of assigning different turning options to different lanes and different vehicle types (HVs, CAVs).

The main contributions of this paper are:

1. Defining the H-AIM protocol.
2. Presenting a comprehensive empirical study showing that H-AIM improves over traditional traffic signals even for as low as 1% CAV penetration. To the best of our knowledge H-AIM is the first protocol that is shown to be beneficial for low CAV penetration rates. This attribute makes H-AIM relevant for the long transition period expected to take place.
3. Presenting guidelines, potentially useful for practitioners, for assigning allowed turning options from each incoming lane to both autonomous and human operated vehicles such that different traffic measurements are optimized.

2   Background

The work presented in this paper builds on top of the FCFS+Signals policy which is part of the Autonomous Intersection Management (AIM) protocol [5]. This section provides a short overview of both AIM and the FCFS+Signals policy.

2.1   Autonomous Intersection Management

AIM is a reservation-based protocol in which vehicles request to reserve trajectories crossing an intersection. The AIM protocol assumes that computer-controlled vehicles attempt to obtain a right of passage through the intersection by sending a reservation request message to the *intersection manager* (IM). When using a "First Come, First Served" (FCFS) policy, the IM approves reservation requests that do not conflict with any previously approved reservation or potential HVs. In brief, the protocol proceeds as follows.

1. An approaching CAV, *cav*, sends a message to the IM requesting a reservation. The *request-reservation message* contains data such as the vehicle's size, predicted arrival time, velocity, acceleration, and arrival and departure lanes.
2. The IM processes the request message by simulating the trajectory of *cav* through the intersection, the simulated trajectory is denoted by *path(cav)*.
3. If *path(cav)* does not conflict with any previously approved reservations or potential HVs then the IM issues a new reservation based on *path(cav)* and sends an *approve message* containing the new reservation back to *cav*.
4. If *path(cav)* does conflict with a previously approved reservations or potential HVs then the IM sends a *reject message* to *cav* which, after a predefined time period, may request a new reservation.
5. After receiving an approve message, it is the responsibility of *cav* to arrive at, and travel through, the intersection as specified in *path(cav)* (within a range of error tolerance).
6. A CAV may not enter the intersection unless it successfully obtained a reservation.
7. Upon leaving the intersection, the CAV informs the IM that its passage through the intersection was successful.

The AIM protocol does not rely on communication capabilities between vehicles (V2V) only between vehicles and the IM (V2I). The protocol is robust to communication failures: if a message is lost, either by the IM or by the CAV, the system's efficiency might be reduced, but safety is not compromised. Safety is guaranteed also when considering a mixed scenario where both HVs and CAVs are present. For such cases Dresner and Stone [5] introduced the FCFS+Signals policy.

2.2   FCFS+Signals

The FCFS+Signal policy [5] is a combination between AIM and traditional traffic signals. Whenever the traffic signal is green for a given lane, all vehicles arriving at that lane have the right of passage (excluding unprotected left turns). However, when the traffic signal shows a red light, only CAVs which were granted a reservation may drive through the intersection.

Since the protocol is not assumed to know the location and trajectory of HVs, such vehicles are assumed to occupy all trajectories that are approved by the traffic signal i.e., have a green light. In this paper we define such trajectories as *green trajectories*. Figure 1 shows an example of green trajectories across an intersection (both the solid and dashed lines represent green trajectories). Note that green trajectories are dynamically changing; once the signal changes, the green trajectories will also change. The signal's timing is assumed to be known so the protocol is able to predict green trajectories in advance.

FCFS+Signals prohibits CAVs from obtaining reservations that conflict with green trajectories. In our example from Fig. 1 all reservation requests will be automatically denied except those made by CAVs arriving from the south and those arriving from the North or East and request to turn right.[1]

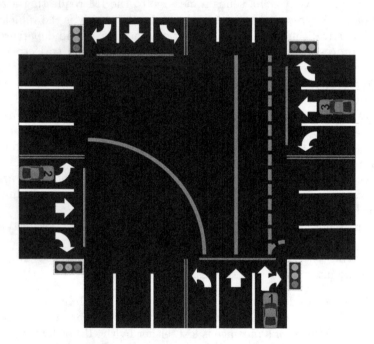

Fig. 1. Four-way intersection. Green light for all lanes originating from the South while all other lanes have a red light. Green trajectories marked with a solid or dashed green lines across the intersection. Active green trajectories marked only by dashed green lines. (Color figure online)

---

[1] This paper assumes driving on the right side of the road. However, the ideas can trivially be generalized to a left side driving policy.

2.3   Experimental Results

Dresner and Stone [5] reported *average delay* for a mixture of CAVs and HVs obtained from the AIM simulator running the FCFS+Signals policy. Delay is defined as the increase in travel time for a vehicle caused by red traffic signals or other vehicles. For CAV penetration of 90% and below, FCFS+Signals yielded a mild improvement. The improvement is attributed to CAVs that make right turns on red lights. If HVs are assumed to be able to turn right on red lights (as in the USA) or turning right has a designated lane bypassing the intersection, then there may be no improvement at all.

For AV penetration greater than 90% the *one-lane signal policy* was suggested which yielded a significant reduction in average delay. In the one-lane signal policy, right of passage for HVs (i.e., green light) is given to a single lane at a time instead of an entire road (all lanes arriving from the same direction). The one-lane signal policy results in a significant reduction in green trajectories at the cost of increased delay for HVs. As a result, the one-lane signal policy proved to be inefficient when considering high HV percentage (more than 10%).

3   Intersection Management Protocol for Mixed Traffic

CAVs are expected to penetrate the automobile market gradually over many years. Reaching 90% AV penetration rates will probably not happen in the near future [3]. Hence, a new intersection management protocol is required for managing traffic that is comprised mostly of HVs.

3.1   Assumptions and Desiderata

The new intersection management protocol should provide the following:

- Reduce the average delay suffered by vehicles crossing the intersection. Reduced delay translates into increased social welfare of the passengers.
- Reduce queue length on incoming lanes. Once the vehicle queue is longer than the length of the incoming link, a phenomenon known as queue spillback occurs [1]. Queue spillbacks have a negative cascading effect and should be avoided as much as possible [10].
- Increase throughput. Higher intersection throughput helps reduce congestion accumulated on links leading to the intersection.
- Provide a relative advantage to CAVs over HVs so as to incentivize drivers to transition to CAVs which are assumed to be safer [7] and more efficient [11].

In contrast to FCFS+Signals we make the following assumptions:

- Humans may turn right on red light if the path is clear. This is a common case in the USA. An alternative assumption is that a right turning lane follows a trajectory outside of the intersection (right turn that bypasses the intersection yields an effect similar to turning right on red).
- A sensor (loop detector, camera or radar) is able to detect approaching vehicles on each lane (sensing speed and heading is not assumed).

3.2 Hybrid AIM

We now present the Hybrid-AIM (H-AIM) protocol for mixed traffic intersection management. Similar to FCFS+Signals, H-AIM grants reservation in a FCFS order. However, while FCFS+Signals automatically rejects reservation requests that conflict with green trajectories, H-AIM rejects reservation requests that conflict with *active green trajectories*. Define an active green trajectory as a green trajectory with a HV present on it or on its incoming lane. Figure 1 illustrates active green trajectories shown as dashed green lines across the intersection (notice vehicle #1 on the incoming lane).

Active green trajectories are a subset of the green trajectories making H-AIM at least as efficient as FCFS+Signals; there can be no reservation that is approved by FCFS+Signals and denied by H-AIM. The other way around, on the other hand, is possible. As an example consider the setting depicted in Fig. 1. Assume vehicle #2 is a CAV and is heading North. Under the FCFS+Signals policy vehicle #2 would be automatically denied a reservation as it crosses a green trajectory. H-AIM on the other hand, would consider such a reservation as it doesn't cross an active green trajectory.

Note that the existence of a CAV on an incoming lane does not incur an active green trajectory. This requires the system to be able to identify whether an approaching vehicle is of type CAV or HV. For doing so we suggest the following procedure:

1. v = the number of vehicles detected on a given lane, l.
2. r = the number of reservation requests from unique vehicles seeking to enter the intersection from lane l. Reservations are considered only if the specified exit time is greater than the current time.
3. If $v > r$ then assume a human vehicle on lane l.

Note that the above procedure is safe in the sense that it will never misidentify a HV as a CAV. In the case of faulty communication this procedure might misidentify a CAV as a HV but this does not pose a safety issue. It might, however, hurt efficiency since a green trajectory might, mistakenly, be considered active.

Safety can be compromised if HVs are allowed to change lanes in close proximity to the intersection. For this reason HVs must be prohibited from changing lanes within detection range.

4   Reducing the Number of Green Trajectories

Green trajectories can limit CAVs from obtaining reservations. As such, CAVs benefit from reducing the number of green trajectories to a minimum. On the other hand, HVs cannot cross the intersection unless traveling on a green trajectory. Thus, HVs generally benefit from an increased number of green trajectories.

Dresner and Stone [5] presented the one-lane signal policy (see Sect. 2.3). This policy results in green trajectories that originate from a single lane at a time, which, in turn, leads to a significant reduction in the number of green trajectories. On the other hand, the one-lane signal policy was shown to have a dramatic negative effect on HVs.

We suggest a more conservative approach for reducing the number of green trajectories, which restricts turning options for HVs. Revisiting Fig. 1, assume vehicle #3 is autonomous and is heading west. When applying H-AIM Vehicle #3 is automatically denied a reservation since the requested reservation crosses an active green trajectory. Currently, the lane on which Vehicle #1 approaches the intersection allows crossing the intersection by continuing straight or turning right. If the turning policy on that lane is changed to "right only", the dashed straight green trajectory will no longer exist allowing vehicle #3 to obtain a reservation.

4.1   Turning Assignment Policy

As was shown in the previous section, the effectiveness of a managed intersection is affected by the allowed turning options in each lane. When considering a three-lane, four-way intersection, each incoming lane has between one and three turning options from the set {left, straight, right}. The turning assignment policy assigns each incoming lane with allowed turns.

For this study we consider four representative turn assignment policies that are depicted in Fig. 2. The policies are ordered and labeled according to degrees of freedom. Define degree of freedom for a lane as the number of turning options minus one. Define degree of freedom for a policy as the summation of degrees of freedom of all lanes.

A restrictive turning policy is one that has a low degree of freedom which, in turn, translates to fewer green trajectories. Policy 0 is an extreme case representing the most restrictive turning policy (0 degrees of freedom). Policy 4 is an extreme case of a liberal turning policy.

Define *safe turning policy* as one in which trajectories originated from the same road never cross each other. Turning policy 4 is not safe while 0, 2a and 2b are. Define *safe turning policy combination* as two policies in which no trajectory from one policy crosses any trajectory from the other when both originate from the same road. {0, 4} is a safe turning policy combination (even though 4 is not a safe policy on it's own). {2a, 4} is not a safe turning policy combination. A turning policy combination is considered when assigning one turning policy for HVs and a different one to CAVs.

For safety reasons we don't consider assigning HVs an unsafe policy. During our empirical study, we observed that assigning unsafe policy combinations for CAVs and HVs is counterproductive and should be avoided. Figure 3 demonstrates the inefficiency that stems from an unsafe turning policy combination. The figure presents a single road approaching a four-way intersection. CAVs are assigned the turning policy shown on the top level (checkerboard texture) while HVs are assigned the bottom turning policy (plain texture). Vehicle #1 is autonomous, it is located in the middle lane and would like to turn right. Assuming a green light for this incoming road and that HVs are arriving on the rightmost lane, vehicle #1 will not be able to obtain a reservation as it crosses an active green trajectory. Vehicle #1 will thus be stuck and will jam all vehicles behind it despite having a green light.

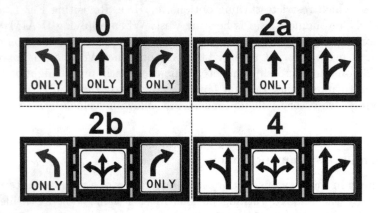

Fig. 2. 4 different turning assignment policies for a 3 lane road approaching a four way intersection.

5   Empirical Study

This section presents results from a comprehensive empirical study. The goals of these experiments are two-fold:

- Study the effectiveness of H-AIM for mixed traffic with an emphasis on low CAV ratios.
- Indicate which turning policy should be assigned to HVs and CAVs in different CAV penetration and traffic levels.

Unless stated otherwise, our experiments used settings identical to those presented by Dresner and Stone [5]:

- Speed limit set to 25 m/s
- CAV may communicate with the IM starting at a distance of 200 m, which at 25 m/s (approximately 56 miles/h) is 8 s before reaching the intersection.
- One simulated hour per instance. Results present the average over 20 instances per setting.

In line with our desiderata (presented in Sect. 3.1), we present average results for the following measurements:

- **Average delay** - see definition in Sect. 2.3.

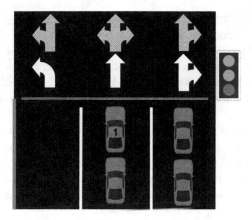

Fig. 3. An unsafe policy combination. Top policy (checkerboard texture) for AVs, bottom policy (plain texture) for HVs. (Color figure online)

- **Maximal queue length** - the maximal number of vehicles that simultaneously accommodate a single incoming lane. Note that 32 vehicles is the maximal queue length for any lane in the simulator, no new vehicles will be generated on a lane as long as this limit is reached. When high traffic volumes are considered, the maximal queue length is often reached and queue spillbacks occur. In such cases it is hard to compare different policies as they all return similar results making the maximal queue length measurement less valuable. Hence, we report maximal queue length only for low traffic levels.
- **Throughput** - the number of vehicles that passed the intersection in one hour. When low traffic volumes are considered the maximal throughput is often reached since all approaching vehicles eventually cross the intersection. At high traffic volumes, when queue spillbacks occur, throughput can give evidence on the severity of spillbacks i.e., the magnitude in which the spillbacks block new vehicles from entering the system. Hence, we report throughput only for high traffic levels.

The experiments presented in this section were obtained using the AIM4 simulator (http://www.cs.utexas.edu/~aim/). Several adaptations were required in order to run these experiments.

5.1 Modifications to the AIM Simulator

Below is a list of changes introduced to the AIM simulator (on top of the original specifications [5]) for running our experiments.

- vehicles are spawned with equal probability on all roads, and are generated via a Poisson process which is controlled by the probability that a vehicle will be generated at each time step. Each vehicle is randomly assigned a type (HV or CAV) and destination. Given the assigned destination a vehicle is placed on an incoming lane from which it can continue to its destination (the incoming lane must allow turning to the vehicle's destination). If several such lanes exist it will be placed on the lane with the least number of vehicles currently on it. For instance, consider Fig. 1, a vehicle arriving at the intersection from the South that is heading North would be assigned the middle lane since the left lane does not allow continuing North and the right lane already has one vehicle.

– CAVs are not granted reservations entering the intersection more than 3.5 s in the future. We add this constraint in order to allow the approaching vehicle detector enough time to detect all approaching HVs.
– A reservation is not necessarily denied if it conflicts with a green trajectory.
– A reservation is necessarily denied if it conflicts with an active green trajectory.

Table 1. Six-phase traffic signal timing. Green and yellow duration are given in seconds. Asterisk next to a phase number means that left turns are allowed during that phase.

Phase	Origin	Green	Yellow
1	East-West	30	0
2*	East	15	3
3*	North	15	0
4	North-South	30	0
5*	South	15	3
6*	West	15	0

5.2 Four-Way Intersection

Following Dresner and Stone [5] we start by presenting results from simulating a four way intersection with three lanes in each of the incoming roads (similar to the intersection presented in Fig. 1). 0.2 of the vehicles turn right at the intersection, 0.2 turn left and 0.6 continue straight regardless of the incoming road and vehicle type.[2] A six-phase traffic signal timing was used (the signal timing is presented in Table 1).

Recall that under our assumption that HVs can turn right on red, the FCFS+Signals protocol has no advantage over traditional traffic signals (unless using the one-lane signal policy, see Sect. 2.3). Since FCFS+Signals using the one-lane signal policy was stated to be helpful when considering 90% HVs and more, it is not relevant to our current study which focuses on early CAV adoption

---

[2] Dresner and Stone [5] do not report the turning ratios for their mixed traffic experiment. Our turning ratio was chosen since it results in a good balance between the incoming queues when 100% of the vehicles are HV.

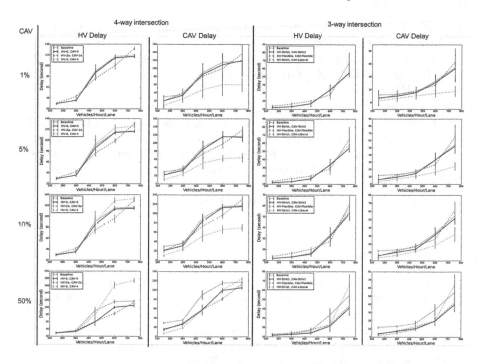

Fig. 4. Average delays for different traffic levels, CAV percentages, vehicle types, and intersection types. Each graph plots the average delay with 95% error intervals as a function of traffic level for three different turning policy combinations as well as the baseline (AIM).

stages. As a result the baseline for our experiments is the case where all vehicles yield to traffic signals while using turning policy 0 (similar to the turning policy used in [5]).

Figure 4 presents eight graphs for the four-way intersection case (left two columns). Each graph presents average delay in seconds (y-axis) versus traffic level in number of vehicles per hour per lane (x-axis). The data is presented for both HVs (first column) and CAVs (second column) and for different CAV penetration levels (with an emphasis on low CAV penetration levels - 1%, 5%, 10%, 50%).

Each graph compares three different safe turning policy combinations based on the policies presented in Fig. 2. Note that results for turning policy 2b are not presented. Using the specified experimental settings, policy 2b was inferior to the other policies across all measurements hence it is omitted.

When examining HVs' delay, the results teach us that for low traffic levels (≤ 300) and very high traffic levels (> 700) policy {HV-2a, CAV-2a} is inferior, while for traffic levels in the range (400–700) it is superior. For medium and high traffic levels (> 400) policy {HV-0, CAV-4} is inferior. In 50% CAV penetration levels, policy {HV-0, CAV-4} is particularly inferior. When examining CAVs' delay, we see a clear benefit for policy {HV-0, CAV-4} across all traffic levels and

CAV penetration levels except 50% penetration with high traffic levels (> 450). The advantage of this policy is due to its reduction of green trajectories as explained in Sect. 4.1. Looking at delays of both HV and CAV we see that H-AIM with the base policy {HV-0, CAV-0} was superior to the baseline (FCFS+Signals with policy {HV-0, CAV-0}).

Table 2 presents maximal queue length and throughput for the four-way intersection scenario. At low traffic levels (150, 300) we report maximal queue length. On the other hand, at high traffic levels (450, 600, 750) we report throughput (see Sect. 5 for reasoning). Similar to Fig. 4, results are presented for different traffic levels and different CAV penetration levels. We observe that avoiding congestion (minimizing queue length or maximizing throughput) is best achieved using policy {HV-2a, CAV-2a} regardless of the CAV penetration and traffic levels.

5.3   Three-Way Intersection

Next we present results from simulating a three way intersection with two lanes in each of the incoming roads (similar to the intersection presented in Fig. 5). 0.6 of the vehicles originating from the East or West continue straight while the rest (0.4) turn (either right or left depending on the incoming road). 0.5 of vehicles originating from the south turn right and the rest (0.5) left. We used a three-phase traffic signal timing presented in Table 3.

Figure 5 depicts three representative turning policies (Strict, Flexible, Liberal). Since a three-way intersection is not symmetrical, each turning policy is broken into three policies (one per origin road: West/East/South). We chose these three policies as they resemble the ones used in the four-way intersection experiment. "Strict" is the most restrictive policy, similar to policy 0 in the four-way case. "Flexible" has the highest degree of freedom among the safe policies, similar to policies 2a and 2b. "Liberal" has the maximal degrees of freedom overall, resembling policy 4. The baseline for our experiments is the case where 100% of the vehicles are HVs (i.e., all vehicles yield to traffic signals) using turning policy "Strict".

Figure 4 presents eight graphs for the three-way intersection case (right two columns). Each graph compares three different safe turning policies combinations based on the policies shown in Fig. 5. The results show a picture which is somewhat similar to the one drawn from the four-way intersection scenario. When considering HVs' delay, policy {HV-"Flexible", CAV-"Flexible"} is superior for intermediate traffic levels (600) with the exception of 50% CAV penetration levels where {HV-"Strict", CAV-"Strict"} proved most beneficial. Unlike the four-way intersection scenario, policy {HV-"Strict", CAV-"Liberal"} is never significantly inferior to other policies with the exception of 50% CAV penetration levels with traffic level of 750.

Similar to the four-way intersection scenario, when examining CAVs delay, we see a clear benefit for policy {HV-"Strict", CAV-"Liberal"} across all traffic levels and CAV penetration levels except 50% penetration with very high traffic levels (750). Similar to the four-way intersection scenario, H-AIM with base policy

Table 2. Results for a four-way intersection using different turning assignment policies for each vehicle type (HV, CAV) and different CAV penetration levels (CAV).
Reporting maximal queue length for low traffic volumes and throughput for high traffic volumes.

CAV	HV-0, CAV-0	HV-2a, CAV-2a	HV-0, CAV-4
Maximal queue			
150 Vehicles/Hour/Lane			
Base	13.2	**8.4**	12.8
1%	13.6	**8.5**	13.2
5%	13.3	**8.4**	12.6
10%	13.6	**8.7**	12.1
50%	12.4	**7.9**	8.4
300 Vehicles/Hour/Lane			
Base	21.8	**16.5**	21.5
1%	22.1	**16.3**	20.8
5%	21.7	**15.3**	21.6
10%	21.2	**16.0**	20.1
50%	20.8	**14.2**	14.7
CAV	Throughput		
450 Vehicles/Hour/Lane			
Base	4,621	**5,034**	4,630
1%	4,625	**5,039**	4,617
5%	4,639	**5,039**	4,647
10%	4,672	**5,057**	4,702
50%	4,865	**5,155**	5,118
600 Vehicles/Hour/Lane			
Base	4,989	**6,242**	4,983
1%	5,013	**6,239**	4,993
5%	5,029	**6,269**	5,027
10%	5,065	**6,309**	5,075
50%	5,367	**6,514**	5,804
750 Vehicles/Hour/Lane			
Base	5,314	**6,417**	5,328
1%	5,315	**6,429**	5,327
5%	5,361	**6,471**	5,414
10%	5,378	**6,520**	5,500
50%	5,718	**7,004**	5,972

Table 3. Three-phase traffic signal timing. Green and yellow duration are given in seconds. Asterisk next to a phase number means that left turns are allowed during that phase.

Phase	Origin	Green	Yellow
1	East-West	30	0
2*	East	15	3
3*	South	15	3

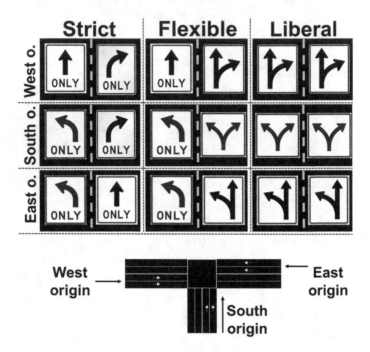

Fig. 5. 3 different turning assignment policies for a 2 lane road approaching a three way intersection.

{HV-"Strict", CAV-"Strict"} was superior to the baseline (FCFS+Signals with policy {HV-"Strict", CAV-"Strict"}) when examining delays over both HVs and CAVs. One exception is when considering HVs' delay at 1% CAV penetration, where H-AIM and the baseline performed similarly. Similar to Table 2, Table 4 presents maximal queue length and throughput but for the three-way intersection scenario. Again, we report maximal queue length for scenarios where queue spill back does not occur (traffic levels = {150, 300, 450}) else (traffic levels = {600, 750}) we report throughput. Similar to the four-way intersection scenario, we observe that avoiding congestion (minimizing queue length or maximizing throughput) is best achieved using policy {HV-"Flexible", CAV-"Flexible'}' with one exception at 50% CAV penetration level where {HV-"Strict", CAV-"Liberal"} was superior.

Table 4. Results for a three-way intersection using different turning assignment policies ("Strict" - S, "Flexible" - F, "Liberal" - L) for each vehicle type (HV, CAV) and different CAV penetration levels (CAV). Reporting maximal queue length for low traffic volumes and throughput for high traffic volumes.

CAV	HV-S, CAV-S	HV-F, CAV-F	HV-S, CAV-L
	Maximal queue		
	150 Vehicles/Hour/Lane		
Base	7.5	**6.7**	7.2
1%	7.5	**6.5**	7.6
5%	7.3	**7.0**	7.3
10%	7.2	**6.7**	6.9
50%	7.1	6.1	**5.1**
300 Vehicles/Hour/Lane			
Base	11.6	**10.9**	11.5
1%	11.3	**10.6**	11.6
5%	11.0	**10.8**	11.2
10%	11.0	**11.0**	11.3
50%	11.1	10.2	**8.5**
450 Vehicles/Hour/Lane			
Base	16.7	**15.2**	18.5
1%	17.1	**15.5**	16.9
5%	17.3	**14.6**	16.1
10%	17.0	**16.0**	16.3
50%	15.3	14.8	**11.4**
CAV	Throughput		
	600 Vehicles/Hour/Lane		
Base	3,239	**3,377**	3,253
1%	3,273	**3,388**	3,259
5%	3,275	**3,390**	3,301
10%	3,275	**3,391**	3,354
50%	3,355	3,407	**3,446**
750 Vehicles/Hour/Lane			
Base	3,754	**3,909**	3,774
1%	3,755	**3,907**	3,770
5%	3,793	**3,933**	3,862
10%	3,792	**3,941**	3,907
50%	3,942	**4,118**	3,975

5.4 Conclusions

Concluding the empirical study we provide the following guidelines:

- H-AIM is superior to FCFS+Signals (baseline) when considering average delay.
- When considering congestion reduction, H-AIM is not superior to FCFS+Signals until more than a 10% CAV technology penetration level is reached.
- If seeking to encourage CAV adoption at early stages (0%–10% CAV penetration levels), one should set turning policies that restrict HVs to the maximum (such as policy 0 and policy "Strict") while allowing maximal flexibility to CAVs (such as policy 4 and policy "Liberal").
- Our experiments showed that setting an unsafe turning policy combination is never worthwhile. These results are not presented in this paper.
- When seeking to reduce congestion, non-restrictive safe turning policies (such as policy 2a and policy "Flexible") should be set for both HVs and CAVs. Note that setting a non-restrictive policy for HVs gives little or no advantage to CAVs and thus, does not encourage CAV adoption.

6 Summary

Though the Autonomous Intersection Management (AIM) protocol was shown to be extremely efficient in coordinating Connected and Autonomous Vehicles (CAVs) traversing an intersection, it provides no improvement until 90% of the processed vehicles are CAV. This paper aims to enable efficient intersection management for early CAV penetration stages. To this end, we propose a modified AIM protocol denoted Hybrid-AIM (H-AIM). H-AIM is applicable under the assumption that vehicles approaching the intersection can be sensed (along with the assumptions required for AIM).

A comprehensive empirical study shows H-AIM to be superior to AIM when average delay imposed on vehicles is considered. Our study also gives guidelines as to how to assign turning options for each lane and vehicle type. Future work will study the effects of H-AIM when semi-autonomous vehicles [2] are considered and are assigned different turning policies. Future work will also examine restricting entire lanes to one vehicle type.

Acknowledgments. A portion of this work has taken place in the Learning Agents Research Group (LARG) at UT Austin. LARG research is supported in part by NSF (CNS-1330072, CNS-1305287, IIS-1637736, IIS-1651089), ONR (21C184-01), AFOSR (FA9550-14-1-0087), Raytheon, Toyota, AT&T, and Lockheed Martin. The authors would also like to thank the Texas Department of Transportation for supporting this research under project 0-6838, Bringing Smart Transport to Texans: Ensuring the Benefits of a Connected and Autonomous Transport System in Texas. Peter Stone serves on the Board of Directors of, Cogitai, Inc. The terms of this arrangement have been reviewed and approved by the University of Texas at Austin in accordance with its policy on objectivity in research.

References

1. Abu-Lebdeh, G., Benekohal, R.: Development of traffic control and queue management procedures for oversaturated arterials. Transp. Res. Record J. Transp. Res. Board **1603**, 119–127 (1997)
2. Au, T.-C., Zhang, S., Stone, S.: Autonomous Intersection Management for semi-autonomous vehicles. In: Handbook of Transportation. Taylor & Francis Group, Routledge (2015)
3. Bansal, P., Kockelman, K.M.: Forecasting Americans' long-term adoption of connected and autonomous vehicle technologies. In: Transportation Research Board 95th Annual Meeting, no. 16–1871 (2016)
4. Coifman, B., Beymer, D., McLauchlan, P., Malik, J.: A real-time computer vision system for vehicle tracking and traffic surveillance. Transp. Res. Part C Emerging Technol. **6**(4), 271–288 (1998)
5. Dresner, K., Stone, P.: A multiagent approach to Autonomous Intersection Management. J. Artif. Intell. Res. **31**, 591–656 (2008)
6. Fajardo, D., Au, T.-C., Waller, S., Stone, P., Yang, D.: Automated intersection control: Performance of future innovation versus current traffic signal control. Transp. Res. Record J. Transp. Res. Board **2259**, 223–232 (2011)
7. Furda, A., Vlacic, L.: Enabling safe autonomous driving in real-world city traffic using multiple criteria decision making. IEEE Intell. Transp. Syst. Mag. **3**(1), 4–17 (2011)
8. Gajda, J., Sroka, R., Stencel, M., Wajda, A., Zeglen, T.: A vehicle classification based on inductive loop detectors. In: Proceedings of the 18th IEEE Instrumentation and Measurement Technology Conference, IMTC 2001, vol. 1, pp. 460–464. IEEE (2001)
9. Hasch, J., Topak, E., Schnabel, R., Zwick, T., Weigel, R., Waldschmidt, C.: Millimeter-wave technology for automotive radar sensors in the 77 GHz frequency band. IEEE Trans. Microw. Theory Tech. **60**(3), 845–860 (2012)
10. Liu, Y., Chang, G.-L.: An arterial signal optimization model for intersections experiencing queue spillback and lane blockage. Transp. Res. Part C Emerg. Technol. **19**(1), 130–144 (2011)
11. Regele, R.: Using ontology-based traffic models for more efficient decision making of autonomous vehicles. In: Fourth International Conference on Autonomic and Autonomous Systems (ICAS 2008), pp. 94–99. IEEE (2008)

Evaluating Ad Hoc Teamwork Performance in Drop-In Player Challenges

Patrick MacAlpine$^{(\boxtimes)}$ and Peter Stone

Department of Computer Science, The University of Texas at Austin, Austin, USA
{patmac,pstone}@cs.utexas.edu

Abstract. Ad hoc teamwork has been introduced as a general challenge for AI and especially multiagent systems [16]. The goal is to enable autonomous agents to band together with previously unknown teammates towards a common goal: collaboration without pre-coordination. A long-term vision for ad hoc teamwork is to enable robots or other autonomous agents to exhibit the sort of flexibility and adaptability on complex tasks that people do, for example when they play games of "pick-up" basketball or soccer. As a testbed for ad hoc teamwork, autonomous robots have played in pick-up soccer games, called "drop-in player challenges", at the international RoboCup competition. An open question is how best to evaluate ad hoc teamwork performance—how well agents are able to coordinate and collaborate with unknown teammates—of agents with different skill levels and abilities competing in drop-in player challenges. This paper presents new metrics for assessing ad hoc teamwork performance, specifically attempting to isolate an agent's coordination and teamwork from its skill level, during drop-in player challenges. Additionally, the paper considers how to account for only a relatively small number of pick-up games being played when evaluating drop-in player challenge participants.

Keywords: Ad hoc teams · Multiagent systems · Teamwork
Robotics

1 Introduction

The increasing capabilities of robots and their decreasing costs is leading to increased numbers of robots acting in the world. As the number of robots grows, so will their need to cooperate with each other to accomplish shared tasks. Therefore, a significant amount of research has focused on multiagent teams. However, most existing techniques are inapplicable when the robots do not share a coordination protocol, a case that becomes more likely as the number of companies and research labs producing these robots grows. To deal with this variety of previously unseen teammates, robots can reason about *ad hoc teamwork* [16]. When participating as part of an ad hoc team, agents need to cooperate with

© Springer International Publishing AG 2017
G. Sukthankar and J. A. Rodriguez-Aguilar (Eds.): AAMAS 2017 Best Papers,
LNAI 10642, pp. 168–186, 2017.
https://doi.org/10.1007/978-3-319-71682-4_11

previously unknown teammates in order to accomplish a shared goal. Reasoning about these settings allows robots to be robust to the teammates they may encounter.

In [16], Stone et al. argue that ad hoc teamwork is "ultimately an empirical challenge." Therefore, a series of "drop-in player challenges" [5,6,14] have been held at the RoboCup competition,[1] a well established multi-robot competition. These challenges bring together real and simulated robots from teams from around the world to investigate the current ability of robots to cooperate with a variety of unknown teammates.

In each game of the challenges, robots are drawn from the participating teams and combined to form a new team. These robots are not informed of the identities of any of their teammates, but they are able to share a small amount of information using a limited standard communication protocol that is published in advance. These robots then have to quickly adapt to their teammates over the course of a single game and discover how to intelligently share the ball and select which roles to play.

Currently in drop-in player challenges, a metric used to evaluate participants is the average goal difference received by an agent across all games that an agent plays in. An agent's average goal difference is strongly correlated with how skilled an agent is, however, and is not necessarily a good way of evaluating an agent's *ad hoc teamwork performance*— how well agents are able to coordinate and collaborate with unknown teammates. Additionally, who an agent's teammates and opponents are during a particular drop-in player game strongly affects the game's result, and it may not be feasible to play enough games containing all possible combinations of agents on different ad hoc teams, thus the agent assignments to the ad hoc teams of the games that are played may bias an agent's average goal difference.

This paper presents new metrics for assessing ad hoc teamwork performance, specifically attempting to isolate an agent's coordination and teamwork from its skill level, during drop-in player challenges. Additionally, the paper considers how to account for only a relatively small number of games being played when evaluating drop-in player challenge participants.

The rest of the paper is structured as follows. A description of the the RoboCup 3D simulation domain used for this research is provided in Sect. 2. Section 3 explains the drop-in player challenge. Section 4 details our metric for evaluating ad hoc teamwork performance, and analysis of this metric is provided in Sect. 5. Section 6 discusses an extension to this metric when one can add agents with different skill levels, but the same level of teamwork, to a drop-in player challenge. How to account for a limited number of drop-in player games being played when evaluating ad hoc teamwork performance is presented in Sect. 7. A case study of the 2015 RoboCup 3D simulation drop-in player challenge demonstrating our work is analyzed in Sect. 8. Section 9 situates this work in literature, and Sect. 10 concludes.

---

[1] http://www.robocup.org/.

2   RoboCup Domain Description

Robot soccer has served as an excellent research domain for autonomous agents and multiagent systems over the past decade and a half. In this domain, teams of autonomous robots compete with each other in a complex, real-time, noisy and dynamic environment, in a setting that is both collaborative and adversarial. RoboCup includes several different leagues, each emphasizing different research challenges. For example, the humanoid robot league emphasizes hardware development and low-level skills, while the 2D simulation league emphasizes more high-level team strategy. In all cases, the agents are all fully autonomous.

The RoboCup 3D simulation environment—the setting for our work—is based on SimSpark,[2] a generic physical multiagent systems simulator. SimSpark uses the Open Dynamics Engine[3] (ODE) library for its realistic simulation of rigid body dynamics with collision detection and friction. ODE also provides support for the modeling of advanced motorized hinge joints used in the humanoid agents.

The robot agents in the simulation are homogeneous and are modeled after the Aldebaran Nao robot. The agents interact with the simulator by sending torque commands and receiving perceptual information. Each robot has 22° of freedom, each equipped with a perceptor and an effector. Joint perceptors provide the agent with noise-free angular measurements every simulation cycle (20 ms), while joint effectors allow the agent to specify the torque and direction in which to move a joint. Although there is no intentional noise in actuation, there is slight actuation noise that results from approximations in the physics engine and the need to constrain computations to be performed in real-time. Abstract visual information about the environment is given to an agent every third simulation cycle (60 ms) through noisy measurements of the distance and angle to objects within a restricted vision cone (120°). Agents are also outfitted with noisy accelerometer and gyroscope perceptors, as well as force resistance perceptors on the sole of each foot. Additionally, agents can communicate with each other every other simulation cycle (40 ms) by sending 20 byte messages.

Games consist of two 5 min halves of 11 versus 11 agents on a field size of 20 m in width by 30 m in length. Figure 1 shows a visualization of the simulated robot and the soccer field during a game.

3   Drop-In Player Challenge

For RoboCup 3D drop-in player challenges[4] each participating team contributes two drop-in field players to a game. Each drop-in player competes in full 10 min games (two 5 min halves) with both teammates and opponents consisting of

---

[2] http://simspark.sourceforge.net/.

[3] http://www.ode.org/.

[4] Full rules of the challenges can be found at http://www.cs.utexas.edu/~AustinVilla/sim/3dsimulation/2015_dropin_challenge/.

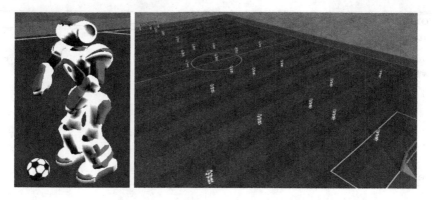

Fig. 1. A screenshot of the Nao-based humanoid robot (left), and a view of the soccer field during a 11 versus 11 game (right).

other drop-in field players. No goalies are used during the challenge to increase the probability of goals being scored.

Ad hoc teams are chosen by a greedy algorithm given in Algorithm 1 that attempts to even out the number of times agents from different participants in a challenge play with and against each other. In lines 6 and 7 of the algorithm agents are iteratively added to teams by getNextAgent() which uses the following ordered preferences to select agents that have:

1. Played fewer games.
2. Played against fewer of the opponents.
3. Played with fewer of the teammates.
4. Played a lower maximum number of games against any one opponent or with any one teammate.
5. Played a lower maximum number of games against any one opponent.
6. Played a lower maximum number of games with any one teammate.
7. Random.

Algorithm 1 terminates when all agents have played at least one game with and against all other agents.

Each drop-in player can communicate with its teammates using a simple protocol, —the use of the protocol is purely optional. The protocol communicates the following information:

- player's team
- player's uniform number
- player's current (x, y) position on the field
- (x, y) position of the ball
- time ball was last seen
- if player is currently fallen over

A C++ implementation of the protocol is provided to all participants.

Algorithm 1. Drop-In Team Agent Selection

Input: *Agents*

```
1: games = ∅
2: while not allAgentsHavePlayedWithAndAgainstEachOther() do
3:     team1 := ∅
4:     team2 := ∅
5:     for i := 1 to AGENTS_PER_TEAM do
6:         team1 ← getNextAgent(Agents \ {team1 ∪ team2})
7:         team2 ← getNextAgent(Agents \ {team1 ∪ team2})
8:     games ← {team1, team2}
9: return games
```

All normal game rules apply in this challenge. Each player is randomly assigned a uniform number from 2–11 at the start of a game. The challenge is scored by the average goal difference received by an agent across all games that an agent plays in.

4   Ad Hoc Teamwork Performance Metric

Since 2013 drop-in player challenges have been held at RoboCup in multiple robot soccer leagues including 3D simulation, 2D simulation, and the physical Nao robot Standard Platform League (SPL) [5,6,13–15]. Across these challenges there has been a high correlation between how well a team does in the challenge and how well a team performs in the main soccer competition. This correlation suggests it may be the case that better individual skills and ability—as opposed to teamwork—is a dominating factor when using average goal difference to rank challenge participants.

As drop-in player challenges are designed as a test bed for ad hoc teamwork, and the ability of an agent to interact with teammates without pre-coordination, ideally we would like to evaluate *ad hoc teamwork performance*—how well agents are able to coordinate and collaborate with unknown teammates. To measure this performance we need a way of isolating agents' ad hoc teamwork from their skill levels.

One way to infer an agent's skill level, relative to another agent, is to evaluate how agents perform in a drop-in player challenge when playing games with teams consisting entirely of their own agent. By playing two different agent teams against each other, and with each teams' members being of the same agent, we are able to directly measure the relative performance difference between the two agents. Although agents' skill levels may not be the only factor in the difference in performance between two teams—factors such as team coordination dynamics may affect performance as well—the teams' relative performance is used as a proxy for individual skills of its members. For agent team a playing agent team b we denote their skill difference, measured as the expected number of goals scored by agent team a minus the expected number of goals scored by agent team b, to be `relSkill`(a, b).

Given the `relSkill` value for all agent pairs, which can be measured by having all agents play each other in a round robin style tournament, we can estimate the expected goal difference of any mixed agent team drop-in player game by summing and then averaging the `relSkill` values of all agent pairs on opposing teams. Equation 1 shows the estimated score between two mixed agent teams A and B.

$$\texttt{score}(A, B) = \frac{1}{|A||B|} \sum_{a \in A, b \in B} \texttt{relSkill}(a, b) \tag{1}$$

Next, to determine the overall skill of an agent relative to all other agents, we compute the average goal difference across all possible $\left(\binom{N}{K} * \binom{N-K}{K}\right)/2$ drop-in player mixed team game permutations for an agent, where N is the total number of agents and K is the number of agents per team, using the estimated goal difference of each game from Eq. 1. We denote this value measuring the average goal difference (AGD) across all games for agent a as `skillAGD`(a). Instead of explicitly computing the score for all game permutations, we can simplify computation as shown in the following example to compute `skillAGD`(a) for a drop-in player challenge with agents $\{a, b, c, d\}$ and two agents on each team.

First determine the `score` of all drop-in game permutations involving agent a (rS used as shorthand for `relSkill`):

$$\texttt{score}\left(\{a, b\}, \{c, d\}\right) = \frac{\texttt{rS}(a, c) + \texttt{rS}(a, d) + \texttt{rS}(b, c) + \texttt{rS}(b, d)}{4}$$

$$\texttt{score}\left(\{a, c\}, \{b, d\}\right) = \frac{\texttt{rS}(a, b) + \texttt{rS}(a, d) + \texttt{rS}(c, b) + \texttt{rS}(c, d)}{4}$$

$$\texttt{score}\left(\{a, d\}, \{b, c\}\right) = \frac{\texttt{rS}(a, b) + \texttt{rS}(a, c) + \texttt{rS}(d, b) + \texttt{rS}(d, c)}{4}$$

Averaging all scores to get `skillAGD`(a), and as

$$\texttt{rS}(a, b) = -\texttt{rS}(b, a),$$

this simplifies to

$$\texttt{skillAGD}(a) = \frac{\texttt{rS}(a, b) + \texttt{rS}(a, c) + \texttt{rS}(a, d)}{6}.$$

Based on `relSkill` values canceling each other out when averaging over all drop-in game permutations, as shown in the above example, Eq. 2 provides a simplified form for estimating an agent's skill.

$$\texttt{skillAGD}(a) = \frac{1}{K(N-1)} \sum_{b \in Agents \setminus a} \texttt{relSkill}(a, b) \tag{2}$$

To evaluate agents' ad hoc teamwork we also need a measure of how well they do when playing in mixed team drop-in player games. Let `dropinAGD`(a) be the actual, measured average goal difference for agent a across all mixed team permutations of drop-in player games. Given an agent's `skillAGD`—estimated indirectly from `relSkill` values—and `dropinAGD`—measured directly—values, we compute a metric `teamworkAGD` for measuring an agent's teamwork. An agent's `teamworkAGD` value is computed by subtracting an agent's skill from it's measured performance in drop-in player games as shown in Eq. 3.

$$\text{teamworkAGD}(a) = \text{dropinAGD}(a) - \text{skillAGD}(a) \tag{3}$$

The `teamworkAGD` value serves to help remove the bias of an agent's skill from its measured averaged goal difference during drop-in player challenges, and in doing so provides a metric to isolate ad hoc teamwork performance.

5   Ad Hoc Teamwork Performance Metric Evaluation

To evaluate the `teamworkAGD` ad hoc teamwork performance metric presented in Sect. 4, we need to be able to create agents with different known skill levels and teamwork such that an agent's skill level is independent of its teamwork. Once we have agents with known differences in skill level and teamwork relative to each other, it is possible to check if the `teamworkAGD` metric is able to isolate agents' ad hoc teamwork from their skill levels during a drop-in player challenge. For our analysis, we designed a RoboCup 3D simulation drop-in player challenge with ten agents each having one of five skill levels and either poor or non-poor teamwork—there is a single agent for every combination of skill level and teamwork type—as follows.

We first created five drop-in player agents with different skill levels determined by how fast an agent is allowed to walk—the maximum walking speed is the only difference between the agents. While walking speed is only one factor for determining an agent's skill level—other factors such as how far an agent can kick the ball and how fast it can get up after falling are important too—by varying their maximum walking speed we ensure agents' overall skill levels differ significantly. The five agents, from highest to lowest skill level, were allowed to walk up to the following maximum walking speeds: 100%, 90%, 80%, 70%, 60%. We then played a round robin tournament with each of the five agents playing 100 games against each other. During these games members of each team consisted of all the same agent. Results from these games of the `relSkill` values of agents with different skill levels are shown in Table 1.

From the values in Table 1 we then compute the agents' skills relative to each other (`skillAGD`) using Eq. 2. When doing so we model the drop-in player challenge as being between ten participants consisting of two agents from each of the five skill levels. We also assume that the average goal difference between two agents of the same skill level is 0.[5] Agents' skill values are shown in Table 2.

---

[5] Empirically we have found that the average goal difference when one team plays itself approaches 0 across many games.

Table 1. Average goal difference of agents with different skill levels when playing 100 games against each other. A positive goal difference means that the row agent is winning. The number at the end of the agents' names refers to their maximum walk speed percentages.

	Agent60	Agent70	Agent80	Agent90
Agent100	1.73	1.36	0.78	0.24
Agent90	1.32	0.94	0.45	
Agent80	0.71	0.52		
Agent70	0.16			

Table 2. Skill values (`skillAGD`) for agents with different skill levels. The number at the end of the agents' names refers to their maximum walk speed percentages.

Agent	skillAGD
Agent100	0.183
Agent90	0.110
Agent80	0.000
Agent70	−0.118
Agent60	− 0.174

The default strategy for each of our drop-in player agents is for an agent to go to the ball if it is the closest member of its team to the ball. Once at the ball, an agent then attempts to kick or dribble the ball toward the opponent's goal. If the agent is not the closest to the ball, it waits at a position two meters behind the ball in a supporting position.

To create agents with poor teamwork, we made modified versions of each of the five different skill level agents such that the modified versions will still go to the ball if an unknown teammate—an agent that is not the exact same type—is closer or even already at the ball. These modified agents, which we refer to as "PT agents" for poor teamwork, can interfere with their unknown teammates and impede progress of the team as a whole. The only teammates they will not interfere with are known agent teammates—agents of the same type with the same maximum walking speed and poor teamwork attribute.

We played a drop-in player challenge with all ten agent types. The total number of possible drop-in team combinations is $(\binom{10}{5} * \binom{5}{5})/2 = 126$. Each combination was played ten times, resulting in a total of 1260 games. Data from these games showing each agent's `dropinAGD`, as well as the agents' `skillAGD` and computed `teamworkAGD`, are shown in Table 3. Note that a poor teamwork agent has the same `skillAGD` as the non-poor teamwork agents with the same walking speed—both agents behave identically when playing on a team consisting of all their own agents.

Table 3. Skill value, drop-in player tournament average goal difference, and ad hoc teamwork performance metric for different agents sorted by `teamworkAGD`.

Agent	skillAGD	dropinAGD	teamworkAGD
Agent70	−0.118	0.017	0.135
Agent60	−0.174	−0.055	0.119
Agent80	0.000	0.087	0.087
Agent100	0.183	0.204	0.021
Agent90	0.110	0.123	0.013
PTAgent60	−0.174	−0.196	−0.022
PTAgent70	−0.118	−0.169	−0.051
PTAgent100	0.183	0.109	−0.074
PTAgent80	0.000	−0.101	−0.101
PTAgent90	0.110	−0.018	−0.128

While the data in Table 3 shows a direct correlation of agents with higher skill levels having higher `dropinAGD` values, the `teamworkAGD` values rank all normal agents above poor teamwork agents. As `teamworkAGD` is able to discern between agents with different levels of teamwork, despite the agents having different levels of skill, `teamworkAGD` is a viable metric for analyzing ad hoc teamwork performance. However, there is a trend for agents with lower `skillAGD` values to have higher `teamworkAGD` values. We discuss and account for this trend in the next section.

6   Normalized Ad Hoc Teamwork Performance Metric

Part of the reason `teamworkAGD` in Table 3 is able to separate the agents with poor teamwork independent of an agent's skill level is due to agents with the same teamwork having similar values of `teamworkAGD`. Empirically we have noticed that is not always the case that teams with the same teamwork have similar `teamworkAGD` values. When skill levels between agents are more spread out, there is a trend for agents with lower skill levels to have higher values for `teamworkAGD`. This trend can be seen in Table 4 containing data from a drop-in player challenge with agents having maximum walking speeds between 100% and 40% of the possible maximum walking speed.

With the trend of agents with lower `skillAGD` having higher values for `teamworkAGD`, the poor teamwork PTAgent50 agent in Table 4 has a higher `teamworkAGD` than several of the non-poor teamwork agents.

To account for agents with the same teamwork, but different skill levels, we can normalize these agents' `teamworkAGD` values to 0. We define the value added to each of these agents' `teamworkAGD` values to set them to 0 as the agents' `normOffset` values. Thus for a set of multiple agents A with the same teamwork,

Table 4. Skill value, drop-in player tournament average goal difference, and ad hoc teamwork performance metric for different agents sorted by `teamworkAGD`.

Agent	`skillAGD`	`dropinAGD`	`teamworkAGD`
Agent40	−0.710	−0.270	0.440
Agent50	−0.226	−0.129	0.097
Agent55	−0.142	−0.081	0.061
Agent100	0.412	0.416	0.004
PTAgent50	−0.226	−0.230	−0.004
Agent90	0.296	0.259	−0.037
Agent70	0.028	−0.005	−0.033
Agent85	0.245	0.176	−0.069
PTAgent70	0.028	−0.179	−0.207
PTAgent90	0.296	0.043	−0.253

and for every agent $a \in A$, we let `normOffset`$(a) = -$`teamworkAGD`(a). This normalization produces a `normTeamworkAGD` value as shown in Eq. 4.

$$\text{normTeamworkAGD}(a) = \text{teamworkAGD}(a) + \text{normOffset}(a) \qquad (4)$$

While `normTeamworkAGD` will give the same value of 0 for agents that we know to have the same teamwork, we want to estimate `normOffset`, and then compute `normTeamworkAGD`, for agents that we do not necessarily know about their teamwork. To estimate `normOffset` values we first plot the `normOffset` values relative to `teamworkAGD` values for the agents with the same teamwork, and then fit a curve through these points. To intersect each point, we do a least squares fit to a $n - 1$ degree polynomial, where n is the number of points we are fitting the curve to. Then, to estimate any agent's `normOffset` value, we choose the point on this curve corresponding to the agent's `skillAGD`. A curve generated by the `normOffset` values normalizing `teamworkAGD` to 0 for Agent100, Agent85, Agent70, Agent55, and Agent40 from Table 4 is shown in Fig. 2.

Table 5 shows `normOffset` and `normTeamworkAGD` values for the agents in Table 4. The `normOffset` values for agents with 50% and 90% speeds are estimated. Considering that `normTeamworkAGD` is able to discern between agents with different levels of teamwork, it is a useful metric for analyzing ad hoc teamwork performance when agents with the same teamwork have larger differences in their `teamworkAGD` values. To compute `normTeamworkAGD`, however, a set of agents with the same teamwork, but different skill levels, must be included in a drop-in player challenge.

7   Drop-In Player Game Prediction

Computing `dropinAGD` requires results from all possible agent to team assignment permutations of drop-in player games. The number of games grows

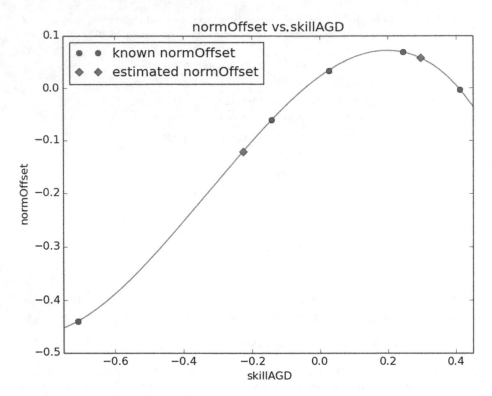

Fig. 2. Curve of `normOffset` vs. `skillAGD` based on `normOffset` values normalizing `teamworkAGD` to 0 for Agent 100, Agent 85, Agent 70, Agent 55, and Agent 40 from Table 4. Both data points used to generate the curve (blue dots) and points used to estimate `normOffset` for agents walking at 50% and 90% speeds (red diamonds) are shown. (Color figure online)

Table 5. `teamworkAGD`, `normOffset`, and `normTeamworkAGD` values for the agents in Table 4 sorted by `normTeamworkAGD`.

Agent	teamworkAGD	normOffset	normTeamworkAGD
Agent90	−0.037	0.057	0.020
Agent55	0.061	−0.061	0.000
Agent40	0.440	−0.440	0.000
Agent100	0.004	−0.004	0.000
Agent70	−0.033	0.033	0.000
Agent85	−0.069	0.069	0.000
Agent50	0.097	−0.121	−0.024
PTAgent50	−0.004	−0.121	−0.125
PTAgent70	−0.207	0.033	−0.174
PTAgent90	−0.253	0.057	−0.196

factorially as this is $\left(\binom{N}{K} * \binom{N-K}{K}\right)/2$ drop-in player games, where N is the total number of agents and K is the number of agents per team. Playing all permutations of drop-in player games may not be tractable or feasible, especially when drop-in player competitions involve physical robots [5,6].

To account for fewer numbers of drop-in player games being played, a prediction model can be built, based on data from previously played drop-in player games, to predict the scores of games that have not been played. Combining data from both the scores of games played and predicted games then allows for `dropinAGD` to be estimated.

One way to predict the scores of drop-in player games is to model them as a linear system of equations. More specifically, we can represent a drop-in player game as a linear equation with strength coefficients for individual agents, cooperative teammate coefficients for pairs of agents on the same team, and adversarial opponent coefficients for pairs of agents on opposing teams.

Given two drop-in player teams A and B, `score`(A, B) is modeled as the sum of strength coefficients S,

$$\sum_{a \in Agents} S_a * \begin{cases} 1 & \text{if } a \in A \\ -1 & \text{if } a \in B \\ 0 & \text{otherwise} \end{cases}$$

teammate coefficients T,

$$\sum_{a \in Agents, b \in Agents, a < b} T_{a,b} * \begin{cases} 1 & \text{if } a \in A \text{ and } b \in A \\ -1 & \text{if } a \in B \text{ and } b \in B \\ 0 & \text{otherwise} \end{cases}$$

and opponent coefficients O,

$$\sum_{a \in Agents, b \in Agents, a < b} O_{a,b} * \begin{cases} 1 & \text{if } a \in A \text{ and } b \in B \\ -1 & \text{if } a \in B \text{ and } b \in A. \\ 0 & \text{otherwise} \end{cases}$$

There are N strength coefficients, and $\binom{N}{2}$ of both teammate and opponent coefficients, for a total of $N + 2\binom{N}{2}$ coefficients.

To solve for the coefficients in the system of linear equations least squares regression is used. There needs to be enough data from games such that every agent has played with and against every other agent, however, so that there is at least one instance of every coefficient being multiplied by a non-zero number. Using Algorithm 1, with 10 agents total and 5 agents per team, having every coefficient multiplied by a non-zero number requires only 5 games. Figure 3 shows how the number of games required to create a prediction model increases as the number of agents increase when using Algorithm 1. Although it is possible to create a prediction model with a minimum number of games, such a system will be very underdetermined and more games will result in better predictions.

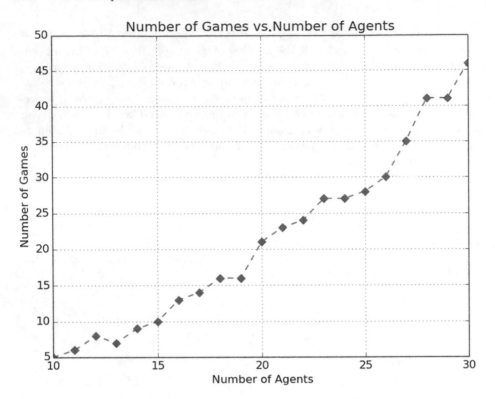

Fig. 3. The number of games required to play all agents with and against every other agent using Algorithm 1 as the number of agents increase. This data assumes there are five agents on each team.

As an example of our prediction model, Tables 6 and 7 show predicted values of `dropinADG` created from game scores generated by prediction models built from half the game data—data from 630 games—used to compute `dropinADG` values in Tables 3 and 4 respectively. More specifically, data from games encompassing half of all possible agent to team assignment permutations of drop-in player games—the first 63 out of 126 possible unique team permutations generated by letting Algorithm 1 continue to run even after all teams have played with and against each other—was used to build the prediction models.

The majority of the predicted `dropinAGD` values in Tables 6 and 7 are closer to the true `dropinAGD` values than that of their counterpart $\frac{1}{2}$ `dropinAGD` values computed directly from the games used to build the prediction models. Furthermore, the predicted `dropinAGD` values reduce the mean squared error relative to the $\frac{1}{2}$ `dropinAGD` values: from 6.405×10^{-4} to 3.212×10^{-4} and from 3.076×10^{-3} to 9.068×10^{-4} for Tables 6 and 7 respectively.

Table 6. The `dropinAGD` values from Table 3 (computed from all 1260 games) compared to both `dropinAGD` values from half the games played used to compute the data in Table 3 ($\frac{1}{2}$ `dropinAGD` with 630 games), and predicted `dropinAGD` values generated from a prediction model built from the game data used to compute $\frac{1}{2}$ `dropinAGD` (Pred. `dropinAGD` with 630 games). The difference (error) from the true `dropinAGD` values for both half the games played and predicted `dropinAGD` are shown in parentheses.

Agent	dropinAGD 1260 games	$\frac{1}{2}$ dropinAGD 630 games	Pred. dropinAGD 630 games
Agent100	0.204	0.194 (0.010)	0.223 (0.019)
Agent90	0.123	0.133 (0.010)	0.122 (0.001)
PTAgent100	0.109	0.114 (0.005)	0.117 (0.008)
Agent80	0.087	0.121 (0.034)	0.095 (0.008)
Agent70	0.017	0.006 (0.011)	0.021 (0.004)
PTAgent90	−0.018	−0.022 (0.004)	−0.019 (0.001)
Agent60	−0.055	−0.105 (0.050)	−0.094 (0.039)
PTAgent80	−0.101	−0.060 (0.041)	−0.073 (0.028)
PTAgent70	−0.169	−0.194 (0.025)	−0.181 (0.012)
PTAgent60	−0.196	−0.187 (0.009)	−0.212 (0.016)

Table 7. The `dropinAGD` values from Table 4 (computed from all 1260 games) compared to both `dropinAGD` values from half the games played used to compute the data in Table 4 ($\frac{1}{2}$ `dropinAGD` with 630 games), and predicted `dropinAGD` values generated from a prediction model built from the game data used to compute $\frac{1}{2}$ `dropinAGD` (Pred. `dropinAGD` with 630 games). The difference (error) from the true `dropinAGD` values for both half the games played and predicted `dropinAGD` are shown in parentheses.

Agent	dropinAGD 1260 games	$\frac{1}{2}$ dropinAGD 630 games	Pred. dropinAGD 630 games
Agent100	0.416	0.454 (0.038)	0.436 (0.020)
Agent90	0.259	0.356 (0.097)	0.296 (0.037)
Agent85	0.176	0.203 (0.027)	0.201 (0.025)
PTAgent90	0.043	0.105 (0.062)	0.048 (0.005)
Agent70	−0.005	−0.019 (0.014)	−0.016 (0.011)
Agent55	−0.081	−0.168 (0.087)	−0.132 (0.051)
Agent50	−0.129	−0.121 (0.008)	−0.098 (0.031)
PTAgent70	−0.179	−0.241 (0.062)	−0.173 (0.006)
PTAgent50	−0.230	−0.238 (0.008)	−0.241 (0.011)
Agent40	−0.270	−0.330 (0.060)	−0.323 (0.053)

8 Case Study: RoboCup 2015 Drop-In Player Challenge

Table 8 shows the results of computing normTeamworkAGD values for the ten released binaries of the 2015 RoboCup 3D simulation drop-in player challenge [15] participants. In doing so we added five agents with different skill levels but the same teamwork to the challenge: Agent100, Agent80, Agent65, Agent50, and Agent30. These agents, chosen specifically to have skillAGD values that span across the range of the 2015 RoboCup 3D simulation drop-in player challenge participants, are the same as the drop-in player agents used in our previous experiments—with the number at the end of the agents' names referring to their maximum walk speed percentages—except now the agents are made slightly more competitive by having them communicate to their known

Table 8. Computed values from released binaries of the 2015 RoboCup 3D simulation drop-in player challenge sorted by normTeamworkAGD. Values for skillAGD were computed from every agent playing 100 games against each of the other agents with teams consisting of all the same agent. Predicted dropinAGD (Pred. dropinAGD) values were computed using a prediction model built from the results of playing 1000 drop-in player games—only a very small partial amount of all 378,378 possible agent assignments for drop-in player games. These predicted dropinAGD values were then used in the computation of teamworkAGD, normOffset, and normTeamworkAGD values.

Agent	skillAGD	dropinAGD 1000 games	Pred. dropinAGD 1000 games	teamwork AGD	norm Offset	norm TeamworkAGD
UTAustin Villa	0.932	1.184	1.178	0.246	0.129	0.375
FCPortugal	0.384	0.228	0.262	−0.122	0.267	0.145
magma Offenburg	0.038	−0.069	−0.047	−0.085	0.139	0.054
Agent100	1.095	1.004	1.031	−0.064	0.064	0
Agent80	0.772	0.586	0.577	−0.195	0.195	0
Agent65	0.355	0.085	0.091	−0.264	0.264	0
Agent50	−0.278	−0.151	−0.129	0.149	−0.149	0
Agent30	−1.456	−0.432	−0.437	1.019	−1.019	0
BahiaRT	0.328	0.044	−0.029	−0.357	0.260	−0.097
RoboCanes	0.178	−0.207	−0.199	−0.377	0.216	−0.161
FUT-K	0.520	−0.027	0.029	−0.491	0.263	−0.228
Apollo3D	−0.533	−0.486	−0.506	0.027	−0.465	−0.438
HfutEngine 3D	−1.124	−0.468	−0.470	0.654	−1.100	−0.446
CIT3D	−0.574	−0.581	−0.589	−0.015	−0.519	−0.534
Nexus3D	−0.676	−0.713	−0.763	−0.087	−0.653	−0.740

teammates (those of the exact same agent type) where they are kicking the ball. Once an agent hears from a teammate the location its teammate is kicking the ball to, the agent then runs toward that location in anticipation of the ball being kicked there.

As there are 15 agents in the challenge, which would require $\left(\binom{15}{5} * \binom{10}{5}\right)/2 =$ 378,378 possible agent assignments for drop-in player games, we only played 1000 games—the first 1000 team permutations generated by letting Algorithm 1 continue to run even after all teams have played with and against each other— and then built a prediction model from the results of these games to compute predicted `dropinAGD` values for all agents. Using a prediction model is the only way for us to compute `dropinAGD`, and in turn `normTeamworkAGD`, given the large increase in the number of games needed to compute `dropinAGD` when adding five extra agents. The curve used to estimate `normOffset` values, and generated by the `normOffset` values normalizing `teamworkAGD` to 0 for Agent100, Agent80, Agent65, Agent50, and Agent30 from Table 8, is shown in Fig. 4.

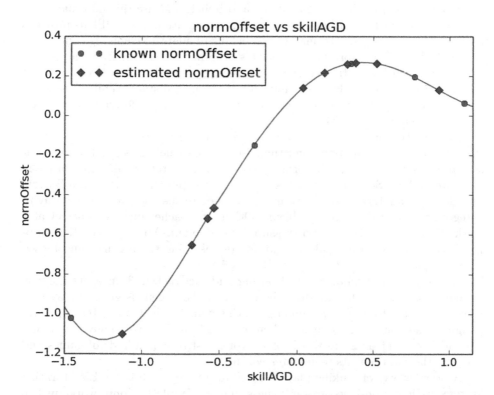

Fig. 4. Curve of `normOffset` vs `skillAGD` based on `normOffset` values normalizing `teamworkAGD` to 0 for Agent100, Agent80, Agent65, Agent50, and Agent30 from Table 8. Both data points used to generate the curve (blue dots) and points used to estimate `normOffset` (red diamonds) are shown. (Color figure online)

When analyzing the data in Table 8 we empirically find that most of the agents with lower `teamworkAGD` values interfere with their teammates when going to the ball. On the other hand, UTAustinVilla—the agent with the highest `teamworkAGD` value—purposely avoids running into teammates, and also checks to ensure it will not collide with other agents before attempting to kick the ball on its team's kickoffs [14].

9 Related Work

Multiagent teamwork is a well studied topic, with most work tackling the problem of creating standards for coordinating and communicating. One such algorithm is STEAM [17], in which team members build up a partial hierarchy of joint actions and monitor the progress of their plans. STEAM is designed to communicate selectively, reducing the amount of communication required to coordinate the team. In [7], Grosz and Kraus present a reformulation of the SharedPlans, in which agents communicate their intents and beliefs and use this information to reason about how to coordinate joint actions. In addition, SharedPlans provides a process for revising agents' intents and beliefs to adapt to changing conditions. In the TAEMS framework [9], the focus is on how the task environment affects agents and their interactions with one another. Specifically, agents reason about what information is available for updating their mental state. While these algorithms have been shown to be effective, they require that the teammates share their coordination framework.

On the other hand, ad hoc teamwork focuses on the case where the agents do not share a coordination algorithm. In [12], Liemhetcharat and Veloso reason about selecting agents to form ad hoc teams. Barrett et al. [2] empirically evaluate an MCTS-based ad hoc team agent in the pursuit domain, and Barrett and Stone [1] analyze existing research on ad hoc teams and propose one way to categorize ad hoc teamwork problems. Other approaches include Jones et al.'s work [10] on ad hoc teams in a treasure hunt domain. A more theoretical approach is Wu et al.'s work [18] into ad hoc teams using stage games and biased adaptive play.

In the domain of robot soccer, Bowling and McCracken [3] measure the performance of a few ad hoc agents, where each ad hoc agent is given a playbook that differs from that of its teammates. In this domain, the teammates implicitly assign the ad hoc agent a role, and then react to it as they would any teammate. The ad hoc agent analyzes which plays work best over hundreds of games and predicts the roles that its teammates will play.

A popular way of ranking players based on relative skill is the Elo [4] rating system originally designed to rank chess players. While Elo only works in two player games, the TrueSkill [8] rating system allows for ranking players in games with multiple player teams. These ranking systems do not attempt to decouple a player's skill from its teamwork performance, and we are unaware of any such previously existing metrics that decouple skill and teamwork in an ad hoc teamwork setting.

An alternative and potentially promising way of estimating scores of drop-in player games is Liemhetcharat and Luo's adversarial synergy graph model [11] which has been used to estimate the scores of basketball games based on player lineups.

10   Conclusions

Drop-in player challenges serve as an exciting testbed for ad hoc teamwork, in which agents must adapt to a variety of new teammates without pre-coordination. These challenges provided an opportunity to evaluate agents' abilities to cooperate with new teammates to accomplish goals in complex tasks. They also served to encourage the participants in the challenges to reason about teamwork and what is actually necessary to coordinate a team.

This paper presents new metrics for assessing ad hoc teamwork performance, specifically attempting to isolate an agent's coordination and teamwork from its skill level, during drop-in player challenges. Additionally, the paper offers a prediction model for the scores of drop-in player games. This prediction model allows for smaller numbers of drop-in games being played when evaluating drop-in player challenge participants. When combined these contributions make it easier to study and perform research on ad hoc teamwork.

Acknowledgments. This work has taken place in the Learning Agents Research Group (LARG) at UT Austin. LARG research is supported in part by NSF (CNS-1330072, CNS-1305287, IIS-1637736, IIS-1651089), ONR (21C184-01), AFOSR (FA9550-14-1-0087), Raytheon, Toyota, AT&T, and Lockheed Martin. Peter Stone serves on the Board of Directors of, Cogitai, Inc. The terms of this arrangement have been reviewed and approved by the University of Texas at Austin in accordance with its policy on objectivity in research.

References

1. Barrett, S., Stone, P.: An analysis framework for ad hoc teamwork tasks. In: AAMAS 2012, June 2012
2. Barrett, S., Stone, P., Kraus, S.: Empirical evaluation of ad hoc teamwork in the pursuit domain. In: AAMAS 2011, May 2011
3. Bowling, M., McCracken, P.: Coordination and adaptation in impromptu teams. In: AAAI (2005)
4. Elo, A.: The Rating of Chess Players, Past and Present. Arco Publishing, New York (1978)
5. Genter, K., Laue, T., Stone, P.: Benchmarking robot cooperation without pre-coordination in the robocup standard platform league drop-in player competition. In: Proceedings of the 2015 IEEE/RSJ International Conference on Intelligent Robots and Systems (IROS 2015), September 2015
6. Genter, K., Laue, T., Stone, P.: Three years of the robocup standard platform league drop-in player competition: creating and maintaining a large scale ad hoc teamwork robotics competition. Auton. Agents Multi-Agent Syst. (JAAMAS) 31(4), 790–820 (2017). Springer

7. Grosz, B., Kraus, S.: Collaborative plans for complex group actions. Artif. Intell. **86**, 269–368 (1996)
8. Herbrich, R., Minka, T., Graepel, T.: Trueskill^{TM}: a bayesian skill rating system. In: Proceedings of the 19th International Conference on Neural Information Processing Systems, pp. 569–576. MIT Press (2006)
9. Horling, B., Lesser, V., Vincent, R., Wagner, T., Raja, A., Zhang, S., Decker, K., Garvey, A.: The TAEMS White Paper, January 1999
10. Jones, E., Browning, B., Dias, M.B., Argall, B., Veloso, M.M., Stentz, A.T.: Dynamically formed heterogeneous robot teams performing tightly-coordinated tasks. In: ICRA, pp. 570–575, May 2006
11. Liemhetcharat, S., Luo, Y.: Applying the synergy graph model to human basketball. In: Proceedings of the 2015 International Conference on Autonomous Agents and Multiagent Systems, pp. 1695–1696 (2015)
12. Liemhetcharat, S., Veloso, M.: Modeling mutual capabilities in heterogeneous teams for role assignment. In: IROS 2011, pp. 3638–3644 (2011)
13. MacAlpine, P., Depinet, M., Liang, J., Stone, P.: UT Austin Villa: RoboCup 2014 3D simulation league competition and technical challenge champions. In: Bianchi, R.A.C., Akin, H.L., Ramamoorthy, S., Sugiura, K. (eds.) RoboCup 2014. LNCS (LNAI), vol. 8992, pp. 33–46. Springer, Cham (2015). https://doi.org/10.1007/978-3-319-18615-3_3
14. MacAlpine, P., Genter, K., Barrett, S., Stone, P.: The RoboCup 2013 drop-in player challenges: experiments in ad hoc teamwork. In: Proceedings of the IEEE/RSJ International Conference on Intelligent Robots and Systems (IROS), September 2014
15. MacAlpine, P., Hanna, J., Liang, J., Stone, P.: UT Austin Villa: RoboCup 2015 3D simulation league competition and technical challenges champions. In: Almeida, L., Ji, J., Steinbauer, G., Luke, S. (eds.) RoboCup 2015. LNCS (LNAI), vol. 9513, pp. 118–131. Springer, Cham (2015). https://doi.org/10.1007/978-3-319-29339-4_10
16. Stone, P., Kaminka, G.A., Kraus, S., Rosenschein, J.S.: Ad hoc autonomous agent teams: collaboration without pre-coordination. In: AAAI 2010, July 2010
17. Tambe, M.: Towards flexible teamwork. J. Artif. Intell. Res. **7**, 81–124 (1997)
18. Wu, F., Zilberstein, S., Chen, X.: Online planning for ad hoc autonomous agent teams. In: IJCAI (2011)

Convention Emergence in Partially Observable Topologies

James Marchant[(⊠)] and Nathan Griffiths

Department of Computer Science, University of Warwick, Coventry, UK
{james,nathan}@dcs.warwick.ac.uk

Abstract. In multi-agent systems it is often desirable for agents to adhere to standards of behaviour that minimise clashes and wasting of (limited) resources. In situations where it is not possible or desirable to dictate these standards globally or via centralised control, convention emergence offers a lightweight and rapid alternative. Placing fixed strategy agents within a population, whose interactions are constrained by an underlying network, has been shown to facilitate faster convention emergence with some degree of control. Placing these fixed strategy agents at topologically influential locations (such as high-degree nodes) increases their effectiveness. However, finding such influential locations often assumes that the whole network is visible or that it is feasible to inspect the whole network in a computationally practical time, a fact not guaranteed in many real-world scenarios. We present an algorithm, PO-PLACE, that finds influential nodes given a finite number of network observations. We show that PO-PLACE finds sets of nodes with similar reach and influence to the set of high-degree nodes and we then compare the performance of PO-PLACE to degree placement for convention emergence in several real-world topologies.

Keywords: Convention emergence · Partial observability
Local information

1 Introduction

Coordinating the actions of independent agents within a multi-agent system (MAS) increases efficiency within the system. Incompatible action choices made during interactions can cause clashes, which may incur resource costs and limit the overall effectiveness of the system. Establishing protocols of interaction, such as which action to choose in a given situation, minimises such clashes and helps to maximise the potential of the system.

However, it is not always possible to dictate such rules and protocols in a top-down manner. In multi-agent systems, with agents controlled by multiple parties or systems which lack a centralised control mechanism, it is often infeasible to establish this level of *a priori* coordination. Additionally, for systems where the

© Springer International Publishing AG 2017
G. Sukthankar and J. A. Rodriguez-Aguilar (Eds.): AAMAS 2017 Best Papers,
LNAI 10642, pp. 187–202, 2017.
https://doi.org/10.1007/978-3-319-71682-4_12

range of choices available to agents is large or has no evident optimal selection, it may be undesirable to enforce rules of this nature.

Convention emergence allows a system to deal with these problems in a decentralised, online manner. A *convention* represents a socially-adopted expected behaviour amongst agents, for instance the correct course of action in a given scenario. Convention emergence has been shown to be possible in both static and dynamic networks with minimal requirements, needing only rational agents that are able to learn [6,13,22].

Fixed strategy (FS) agents are those that continue to choose the same action regardless of the behaviour of others around them or the results of their actions. Placing such agents within a system has been shown to affect the direction and speed of convention emergence, with small numbers of FS agents eliciting change in much larger populations [19]. In systems constrained by an underlying network topology, placing such agents by heuristics based on network features such as degree magnifies this effect [8,9].

Previous work on convention emergence often assumes that the topology constraining agent interactions is fully observable, allowing highly influential locations to be found easily [11,17,19,21]. However, in many real-world applications such information is not always readily available. This can be due to factors such as the problem size or external limitations such as restricted access to network information or a network's API as is the case with Twitter or Facebook.

In this paper we explore the effect of the restrictions placed on FS agent placement in partially observable topologies. We propose an algorithm, PO-PLACE, to find influential locations within such topologies given a highly limited number of network queries. We show the effectiveness of PO-PLACE at finding approximations of the highest degree locations for several real-world topologies under a number of restrictions on available information. We then apply PO-PLACE to select FS agents within these networks and examine the effect on convention emergence compared to placing with full topological knowledge. This approach allows an interested third party, with limited access to the system, to find the appropriate locations to target their influence efforts.

The remainder of this paper is arranged as follows. In Sect. 2 we explore related work on convention emergence and local information strategies for finding influential nodes. Section 3 describes the algorithm and its design, whilst Sect. 4 describes the network datasets and experimental setup. Section 5 contains the analysis and discussion of the results and Sect. 6 concludes the paper.

2   Related Work

Ensuring coordination in MAS allows increased system efficiency and conventions are a lightweight method of doing so. Conventions place 'soft constraints' on agent choices by encouraging mutually beneficial behaviour by adherence to the convention. Unlike *norms* there is no explicit punishment for going against the convention but doing so is likely to incur a cost to the agent due to increased clashes often represented as a negative interaction payoff [10,18]. Conventions can thus be described as "an equilibrium everyone expects in interactions that have more than one equilibrium" [24]. Agents adhering to the convention expect others to behave in a certain way and, because of this, can act efficiently when this expectation is met. Conventions have been shown to emerge unaided from local agent interactions in systems [6,11,20,22] and require no additional or assumed agent capabilities to enable punishment (as is the case for norms). The only assumptions necessary for conventions to emerge are that agents are rational and have the capability to learn from their interactions. Numerous works have shown that rapid and robust convention emergence occurs with these minimal assumptions [9,19,22].

'Social learning' has been proposed as a way for agents to converge on a convention where agents monitor payoffs they receive from their choices when interacting with others and use a simplified Q-Learning algorithm to inform future decisions [19]. The payoffs directly quantify the notion of an action clash costing resources and convention emergence can occur without explicit memory of the interaction. However, the work does not consider a connecting topology that limits agent interactions. In many application domains such a topology is likely, whether it be a social network or a more explicit communication network and can have a large effect on the nature of convention emergence [5,21].

Despite lacking a connecting topology, Sen and Airiau's work introduces the concept of fixed strategy (FS) agents, those agents which always choose the same action regardless of the current situation or convention, as a way to influence convention emergence. They show that a small number of such agents is able to manipulate the convention emergence within a much larger population. Griffiths and Anand [9] expand on this by considering FS agents in a network topology. In their model, all agents are situated as nodes within the network and interactions are limited to neighbours. They showed that *where* FS agents are placed is a key factor in their effectiveness. Placing the FS agents at influential locations such as nodes of high degree or betweenness centrality offers substantially better performance than random placement. This was explored further by Franks *et al.* [7,8] who included more advanced placement metrics such as eigencentrality.

This previous work assumes full visibility of the network topology to inform FS placement. Indeed, little work on partial observability for convention emergence has been done. This paper expands the state of the art by considering the effect of restricted observations on the ability to robustly and efficiently place FS agents in static, real-world topologies. Convention destabilisation [14] and dynamic topologies [13] will be investigated in future work.

Related work exists in the fields of graph algorithms and influence spread, the latter sharing many qualities with convention emergence. For instance, Brautbar and Kearns present a novel model [2], *Jump and Crawl*, motivated by operations commonly available in networks such as Facebook. Their model consists of two aspects: *Jump* which moves to a randomly selected node in the network and *Crawl* which searches all neighbours of the selected node for high-degree nodes. They provide bounds for many different types of network but, for an arbitrary network, finding the highest degree node approaches $O(n \log n)$, a large factor for even medium-sized networks.

The influence maximisation problem [3,4] attempts to find a selection of nodes such that the spread of influence (often modelled as single chance 'cascades') from them is maximised. As in this paper, Mihara et al. [15] assume the network is initially unknown and show that influence maximisation effectiveness of 60–90% with 1–10% network observation is achievable. This work also uses a 'growing fringe' approach with priority based on degree estimation. As influence maximisation and convention emergence are similar in aim, this indicates that results are achievable under partial observability constraints.

Whilst many of these approaches are similar in application they differ in that our investigation focuses on the often encountered scenario of limited, finite observations. Making optimal use of these is paramount and so necessitates a different set of considerations.

3   Placement Strategy

In this paper, the partial observability problem for networks can be described as any scenario where a network's topology is initially unknown and is revealed incrementally within a local neighbourhood of nodes already explored [1]. As a solution to the partial observability problem for FS agent selection we propose a heuristic algorithm, PO-PLACE. This section describes the function of the algorithm as well as the justification for the design choices.

3.1   Partial Observability Algorithm

The placement strategy is presented in Algorithms 1 and 2 and has the following aim: Given a network, $G = (V, E)$, a desired number of locations, n, and a limited number of observations, o, find a selection of nodes $\{v_1, ..., v_n\} \subset V$ such that $deg\text{-}sum = deg(v_1) + ... + deg(v_n)$ is maximised. We define an observation as a query that retrieves the list of neighbours, $N(u)$ for a given node, u. This functionality is frequently available in real-world network APIs (such as Twitter or Facebook) and so we assume that such information is available. This assumption is later relaxed to allow the algorithm to explore situations with only limited neighbour information. We assume that the set of nodes, V, is known but the set of edges, E, (and hence neighbours and degree of a node) is not. Finding the highest degree nodes is desirable since fixed strategy agent placement by degree consistently produces effective convention emergence [8,9,13,14] but without requiring computationally expensive metrics such as betweenness centrality. The degree of nodes can be entirely derived from local information and, as such, is an applicable heuristic within partially observable networks.

The algorithm begins by creating an empty set, S, to monitor which nodes have already been explored and an empty mapping, N, that maps a node v to $N(v)$, its set of neighbours. As we only consider static topologies in this paper, by storing this information we can avoid using observations redundantly.

Algorithm 1. Partial Observability Placement

```
 1: procedure PO-PLACE(G, n, o, s, p, f)
 2:        Create empty node set, S
 3:        Create empty mapping, N
 4:        o_rem ← o

 5:        while o_rem > 0 ∧ |S| < |V| do
 6:            Select v u.a.r from {V \ S}
 7:            if o_rem mod s ≠ 0 then
 8:                o_local ← min(⌈o/s⌉, o_rem)
 9:            else
10:                o_local ← min(⌊o/s⌋, o_rem)
11:            end if
12:            o_rem ← o_rem − o_local
13:            o_unused ← Traverse(G, o_local, v, p, f, S, N)
14:            o_rem ← o_rem + o_unused
15:        end while

16:        return n highest-degree nodes in S
17: end procedure
```

Many of the other approaches [1,15] to finding high-degree nodes select a random starting node and then 'grow' outwards, selecting the highest degree nodes from the neighbourhood surrounding those already explored. However, this is not desirable in FS agent placement since, with limited observations, it is likely to produce a single cluster of well-explored nodes. Selecting from this cluster will then mean that all FS agents are close together, making some of their influence redundant. Instead, we build on the notion of *Jump and Crawl* [2]. We explore a local area up to a defined amount and then 'jump' to another location and explore around this new point. This helps to minimise the risk of overlap between high-degree nodes, as well as ensuring that a bad initial random selection does not hinder the final selection.

To facilitate this, we introduce a parameter, s, which dictates the minimum number of separate local area explorations that will take place. The observations are split, as evenly as possible, between each of these explorations with the earlier ones receiving any spare observations (this is achieved between Lines 7 and 11 of Algorithm 1). This subset of observations is then passed to the local area traversal which is presented in Algorithm 2. If any observations are unused by the local area traversal (for instance if it finds a local maxima) they are returned to the pool of available observations and used in later, additional local traversals.

Algorithm 2, TRAVERSE, describes the local area traversals. It is passed both S and N, to avoid redundant exploration, as well as the initial start node of the local area, v. It is also passed its own local limit of observations and two parameters from outside, p and f, which are explained below. It maintains a max-priority queue to determine which node(s) it should next explore by highest degree and begins by adding v to this queue. Throughout Algorithm 2, observation of a node's neighbour list is stored in N to avoid additional queries. The algorithm then performs the following, until either the queue is empty or all assigned observations have been used up:

Algorithm 2. Local area traversal algorithm

```
 1: procedure TRAVERSE(G, o, v, p, f, S, N)
 2:      Create max-priority queue, Q
 3:      count ← 0
 4:      if v not in N then
 5:          N[v] ← N(v)
 6:          Add v to S
 7:          count ← count + 1
 8:      end if
 9:      Add (v, |N[v]|) to Q

10:      while |Q| > 0 ∧ count < o do
11:          Fringe ← top min(f, |Q|) elements of Q
12:          for all u in Fringe do
13:              Avail ← {N[u] \ S}
14:              num ← min(|Avail|, max(f, ⌊p × |Avail|⌋))
15:              Chosen ← u.a.r select num members of Avail
16:              for all w in Chosen do
17:                  N[w] ← N(w)
18:                  Add w to S
19:                  count ← count + 1
20:                  if count = o then
21:                      return 0
22:                  end if
23:                  Add (w, |N[w]|) to Q
24:              end for
25:          end for
26:      end while

27:      return o − count
28: end procedure
```

1. Take the top f nodes from the queue (or all elements, if fewer).
2. For each of these nodes, find the set of unexplored nodes in its neighbours.
3. Choose a proportion, p, of these (or up to f if this proportion would be less than f).
4. Add these nodes to the queue after finding their neighbours.

Parameter f is the 'fringe size', the number of nodes that are expanded simultaneously before their neighbours are queued. This acts as a control over how 'breadth-first' or 'depth-first' the local traversal approach will be. Parameter p is the proportion of the node's neighbours that should be queried. This allows the algorithm to simulate situations where a node's full neighbour list is either not fully available (for instance, an API that only returns a subset) or where doing so incurs additional cost. In the latter case we seek to explore the effect that only querying p proportion of neighbours has on the performance of PO-PLACE. Whilst it will reduce the effectiveness, establishing the extent of this reduction, and whether the results are still close enough to degree placement, allows PO-PLACE to be effective over a wider range of scenarios.

4   Experimental Setup

This section defines the real-world topologies and the experimental setup used for analysis of PO-PLACE. We then describe the model of convention emergence used to study the efficacy of PO-PLACE for FS placement selection.

4.1 Networks

We make use of three real-world networks from the Stanford SNAP datasets [12]. These datasets represent a number of different methods of social interaction and, as such, each have different features allowing a wide-ranging look at the effectiveness of PO-PLACE. The three datasets chosen are: CA-CondMat, the collaboration network of the arXiv COND-MAT (Condensed Matter Physics) category; Email-Enron, the email communications between workers at Enron; and Ego-Twitter, a crawl of Twitter follow relationships from public sources (for our purposes we ignore the directed nature of the edges). These datasets are used frequently in both convention emergence and influence spread research [3, 7, 16, 23] as performance benchmarks.

For the purposes of monitoring convention emergence in these networks, we only want to examine a single, connected component. As such, all 3 networks were reduced to their largest weakly connected component (WCC). Additionally, any self-loops (edges from a node to itself) were removed as such edges artificially inflate a node's degree whilst not increasing its ability to influence others. Table 1 shows the number of nodes and edges in each network and the number of nodes and edges (without self-loops) in their largest WCC.

Table 1. Original and modified network sizes

	Network		Largest WCC	
	$|V|$	$|E|$	$|V|$	$|E|$
CA-CondMat	23,133	93,497	21,363	91,286
Enron-Email	36,692	183,831	33,696	180,811
Twitter	81,306	1,768,149	81,306	1,342,296

4.2 Experimental Setup

We performed simulations of PO-PLACE on the real-world networks described above. We varied both the number of nodes ($n = 5$ to $n = 30$) being requested as well as the number of observations provided ($o = 500$ to $o = 5000$ [$o = 3500$ for CondMat]). To establish an upper bound and allow comparison a full-observability degree placement was also performed for each of the networks with the same range of values. Each set of parameters was averaged over 30 runs.

For convention emergence, a population of agents is situated in the topologies. Each timestep, each agent chooses one of its neighbours u.a.r to play the 10-action coordination game [19] receiving positive or negative payoffs depending on whether their choices match. Agents use a simplified Q-Learning algorithm to learn the most beneficial choice. We utilise the 10-action game as used by [14] to avoid the issues of small convention spaces raised in Sect. 2 and to allow comparison to previous work. They have a chance to randomly choose their action ($p_{explore} = 0.25$) or else choose the most beneficial one. FS agents replace the agents at the chosen locations and always choose their predetermined action.

5 Results and Discussion

In this section we present the analysis of PO-PLACE and compare it to the upper bound from degree placement. We explore the effects of the various parameters on PO-PLACE at different levels of observation. We then use these findings as insight to compare the performance of PO-PLACE to degree for convention emergence when used to place FS agents into the chosen networks.

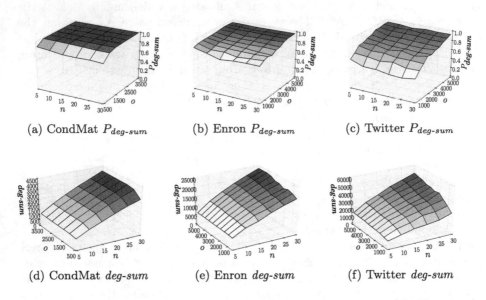

Fig. 1. $P_{deg\text{-}sum}$ and *deg-sum* performance of PO-PLACE for varying n (# of locations) and o (# of observations) in the real-world networks.

5.1 PO-PLACE Output

We begin by looking at the isolated algorithm output, comparing it to the output generated by a degree placement scheme. As the aim of PO-PLACE is to maximise *deg-sum* this is our primary metric by which to evaluate PO-PLACE. The highest *deg-sum* possible in each network is that of the set of highest degree nodes. Establishing this as an upper bound allows evaluating the performance of PO-PLACE by comparing the *deg-sum* of its output as a proportion of that of the pure degree network. We denote this as $P_{deg\text{-}sum}$.

Whilst *deg-sum* describes the maximum reach of the nodes selected, another useful metric is the size of the 1-hop neighbourhood of those nodes. This can be defined as: 1-HOP$(L, G) = \{v \in \{V \setminus L\} | \exists(u, v) \in E \wedge u \in L\}$ where L is the set of nodes selected for placement and $G = (V, E)$ is the network. That is, the 1-Hop neighbourhood is the set of nodes that are connected to a member of S but are not in S themselves. The 1-Hop neighbourhood offers a slightly different measure of influence by discounting nodes that are connected to multiple members of S. Whilst normally tied closely to *deg-sum* a noticeable disparity indicates that the selected nodes are likely to be clustered close to one another, which is undesirable. As with *deg-sum* we concern ourselves with the proportionate behaviour of 1-HOP size, $P_{|1\text{-}HOP|}$.

The final metric we use to evaluate the performance is based on the Jaccard Index which measures similarity between two sets. The Jaccard Index is defined as $J(A, B) = |A \cap B|/|A \cup B|$. However, in our instance, one of the sets is static. We are trying to approximate that set with the other (i.e. a one-way similarity), whilst the Jaccard Index is looking at the two-way similarity between them. Instead we want to measure how close the selection of PO-PLACE is to the baseline, and so we define a distance measure, D_{Base}, thus: $D_{Base}(L, Base) = |L \cap Base|/|Base|$. That is, the fraction that elements of L make up of the baseline set, $Base$. This metric enables evaluation of how close the actual node selection of PO-PLACE is to that of degree placement, whilst the previous two measure the selection's features.

These metrics offer insight into the influence and reach of the nodes selected by PO-PLACE as well as allowing a direct comparison to degree-based placement with full observability. Thus they should be good predictors of the performance of PO-PLACE in the convention emergence setting.

Varying Observations. We begin by considering the base case of the algorithm where $s = p = f = 1$. This allows us to study the effect of varying the number of observations and provides a lower bound on the expected performance of PO-PLACE. With these settings, PO-PLACE closely resembles the algorithms presented by Borgs et al. [1] and Mihara et al. [15].

We examine the effects of varying both the number of observations available (o) as well as the number of locations requested (n) in all three networks. For all networks, n was varied between 5 and 30 in increments of 5 and o was varied from 500 observations up to 3500 (for CondMat) or 5000 (for Enron and Twitter). The results are presented in Fig. 1.

As can be seen in Fig. 1, all networks respond well, even with minimal numbers of observations. Even at $o = 500$, the degree sum of the nodes selected by PO-PLACE is often a substantial proportion of the optimal one. The performance varies across the three networks, with placement in CondMat doing best where it varies from 90% ($\pm5\%$) at $n = 5$ to 83% ($\pm5\%$) at $n = 30$. The algorithm similarly performs well in Enron, though to a lesser extent. The performance in Twitter is noticeably worse, varying from 61% to 48% with larger standard deviations for both. This is to be expected, as 500 observations represents a substantially smaller proportion of the population in Twitter than it does in CondMat or Enron (0.61%, 2.34% and 1.48% respectively). Even with this, the percentage achieved in Twitter with such limitations substantially outperforms the naïve solution of using all observations at random locations (16% ($\pm6\%$) for $n = 5$, $o = 500$, averaged over 100 runs).

Performance rapidly increases with the number of observations. For $n = 30$, the worst performing value of n, in both CondMat and Twitter $P_{deg\text{-}sum}$ exceeds 90% at round 5% network observation ($o = 1000$ for CondMat and $o = 5000$ for Twitter) and Enron exceeds 90% at around 10% observation ($o = 3500$). Figure 1 also shows that the relationship between $P_{deg\text{-}sum}$ and increasing o is one of diminishing returns, with improvements in $P_{deg\text{-}sum}$ most noticeable at lower values of o. This is to be expected, the relative increase in o is smaller at higher values, but dictates that increasing the effectiveness of PO-PLACE at low values of o will have the most benefit. Additionally, in each network, the difference in performance across the values of n becomes less noticeable at higher o. Thus, any increased performance from PO-PLACE will be most noticeable early on.

The other metrics we use to evaluate PO-PLACE show similar behaviour to $P_{deg\text{-}sum}$, increasing rapidly with the number of observations. Figure 2 shows a representative example of the three metrics' variation with o for the Twitter network when requesting 20 locations. The shaded regions represent the standard deviations. As can be seen, both the *deg-sum* and 1-HOP proportions increase rapidly up until $o = 2000$ and then any further gains occur over longer spans. The standard deviations for each of these decrease as well, from approximately 15% at $o = 500$ down to around 5% at $o = 5000$. This indicates that, not only is PO-PLACE finding sets of nodes with higher degree, it is doing so consistently at higher numbers of observations, a finding that is repeated across all networks and values of n. $P_{|1\text{-}HOP|}$ is consistently at the same level, if not better than, $P_{deg\text{-}sum}$. Whilst the two should be well-correlated, this shows that PO-PLACE is not simply choosing nodes close to one another and, indeed, is often choosing nodes that have a better neighbourhood size than the *deg-sum* would indicate.

The performance of PO-PLACE when evaluated by D_{Base} is noticeably different than the other two metrics and offers an interesting insight. The same pattern of diminishing returns is not present and D_{Base} continues to increase with additional observations. Note that, although both the degree sum and neighbourhood size are comparable to that of pure degree placement, the low values of D_{Base} indicate that the nodes selected are not the same as the actual highest degree nodes. Section 5.2 evaluates whether this difference has a noticeable effect on convention emergence or if the reach and influence indicated by high *deg-sum* and 1-HOP scores is the best indicator of success as hypothesised.

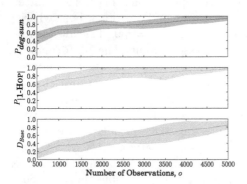

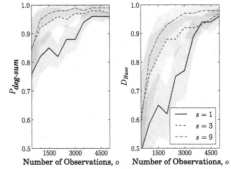

Fig. 2. Metric performances of PO-PLACE for the Twitter network, $n = 20$.

Fig. 3. Effect of varying s on $P_{deg\text{-}sum}$ and D_{Base}. Enron network, $n = 30$.

Varying Concurrent Searches. Having established a baseline for PO-PLACE and explored the effects of limited observations we now explore the variants of the algorithm. As noted in the prior section, at low values of o the *deg-sum* performance of PO-PLACE is consistently lower, with performance in the Twitter network as low as 48%. With very limited observations, making the best use of them is paramount. In Sect. 3 we hypothesised that splitting the available observations between multiple locations in the network and exploring them in parallel may offer improvements over crawling from a singular location.

To test this hypothesis, we varied s from 1 to 9 to determine the effect that these concurrent searches would have. Figure 3 shows a typical case in the Enron network for $n = 30$. Shaded areas represent the errors of each plot. The left-hand graph shows the effect on $P_{deg\text{-}sum}$ of varying the number of concurrent searches, splitting the observations between them. As can be seen, adding concurrent starting points has an immediate and noticeable effect, especially at low numbers of observations. At $o = 500$ the proportion achieved by *deg-sum* is 10% higher when additional starting locations are introduced and this difference becomes even more noticeable as o increases. Indeed, for most values of o, adding additional starting locations had significant benefits in both the Enron and Twitter networks, with the benefits become less marked at high o where $P_{deg\text{-}sum}$ approaches 1.0 unaided. Whilst there is a noticeable drop-off in effectiveness after initial parallelisation ($s = 5$ and $s = 7$, not included in the results to aid readability, offer little improvement over $s = 3$ for example) the effect at low values of s is substantial as can be seen. Concurrent starting points enable saturation of the algorithm's effectiveness at much lower values of o and not only increase $P_{deg\text{-}sum}$ and $P_{|1\text{-}\text{HOP}|}$ (not pictured) but, as shown in Fig. 3, cause marked improvement in D_{Base} as well, indicating that this change facilitates much better approximation of the degree placement.

However, it should be noted that this pattern is not consistent. In the Cond-Mat network, increasing s had little effect and in a few settings was actually detrimental. This indicates that there is perhaps an underlying feature of the CondMat topology that benefits from localised crawling and will be an area of future study. The results of CondMat in Fig. 1a lend additional weight to this hypothesis, with behaviour that is substantially different than the other two topologies despite being of comparable size to Enron. Overall though, increasing s by even a small amount is likely to benefit the performance of PO-PLACE.

Partial Neighbour Lists. In many settings, retrieving the whole of an agent's neighbour list may also be impossible. Whether this is due to a technical limitation (only being able to retrieve a certain percentage of information) or because such information is not publicly available and is instead reserved for 'premium' or 'subscribed' users of such a network, ensuring that PO-PLACE is robust to such issues is a necessity to make it widely viable.

To simulate these restrictions, and measure their effect on the performance of PO-PLACE, the parameter, p, controls the proportion of an agent's neighbours that may be explored. Results until this point have assumed that the full

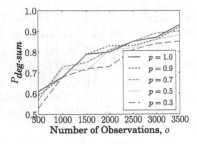
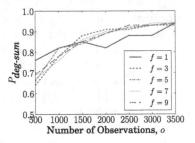

Fig. 4. Effect of varying p on $P_{deg\text{-}sum}$. Twitter network, $n = 5$.

Fig. 5. Effect of varying f on $P_{deg\text{-}sum}$. Enron network, $n = 30$.

neighbour list for any agent is available upon request (i.e. $p = 1.0$). p is varied between 0.3 and 0.9 to determine the impact of this limitation. Representative results are shown in Fig. 4 for the Twitter network and $n = 5$ but are applicable across all networks and values of o and n.

The results in Fig. 4 show that different values of p have minimal effect on the performance of PO-PLACE. For all values of p, $P_{deg\text{-}sum}$ is comparable. Performing a 95% confidence interval Welch's t-test against the $p = 1.0$ results at each point, only $p = 0.3$ ($o = 1500, 2000, 3500$) and $p = 0.5$ ($o = 1500, 3500$) are significantly worse. This pattern of minimal difference is repeated in all networks, with none seemingly more susceptible or affected by partial neighbour lists. We conclude that PO-PLACE is robust to receiving only partial information of this nature and is primarily unaffected by such limitations.

Breadth-First vs. Depth-First Expansion. Finally, we turn our attention to the concept of breadth-first vs depth-first expansion in PO-PLACE. That is, when crawling the local area, should additional current area expansion be performed before considering new additions (breadth-first) or purely iteratively (depth-first). Where there is locally a clearly defined degree gradient we expect the latter to perform better. However, depth-first expansion also risks expending all the observations whilst exploring a suboptimal, locally maximal path.

Parameter f allows study of this by controlling how many of the current highest degree nodes that PO-PLACE is aware of are expanded concurrently. Experiments up until now have had $f = 1$ (depth-first). We now vary f from 1 to 9. Figure 5 presents these findings in the Enron network for $n = 30$. As with the previous results, it is our finding that the patterns here are replicated throughout the different topologies and values of n.

Similar to the findings when varying p, varying f has little absolute impact on the capabilities of PO-PLACE. However, using a 95% confidence interval Welch's t-test, all but $f = 9$ are statistically significantly worse at $o = 500$. This is likely due to the limited observations being focused too locally. All are significantly better between $o = 2000$ and $o = 3000$ but there is little gain in selecting values of f beyond 3 as the performance of PO-PLACE is almost identical. Overall, PO-PLACE seems to gain little from considering the local area more thoroughly before further expansion. Whether this is intrinsic in the design or a facet of the topologies being explored is ongoing work.

5.2   Convention Emergence Under Partial Observability

Having explored the performance of PO-PLACE under different topologies and types of partially observability, we now examine how PO-PLACE compares to degree placement for FS agents in convention emergence in static networks. Having established ranges of parameters that offer the best performance improvements for each topology, these will be utilised to compare the algorithm to degree placement. Additionally, basic settings (small numbers of observations, no concurrent placements) provide a baseline comparison of PO-PLACE.

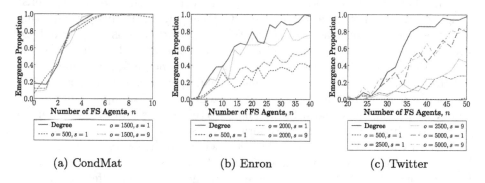

(a) CondMat   (b) Enron   (c) Twitter

Fig. 6. Comparison of PO-PLACE and Degree FS agent placement for convention emergence in real-world topologies. The y-axis indicates the proportion of runs where the desired strategy emerged as the convention.

A convention has emerged when the population has converged to have one action as the dominant choice of agents in the network. Most work considers this to be the case when the convention reaches 90% dominance [9,13,19]. However, much of this work utilises synthetic networks rather than real-world topologies and populations that are substantially smaller than those we consider. Preliminary experiments show that the topologies are relatively resistant to convention emergence, requiring both high numbers of FS agents as well as substantial time. As we are concerned with a comparison of the performance of PO-PLACE against pure degree placement we wish to find settings that are guaranteed to repeatedly experience convention emergence. As such, we consider a convention to have emerged when the 80% Kittock criteria is met, $K_{80\%}$ [11]. That is, a convention has emerged when 80% of the population, when not exploring, would choose the same action. This indicates a high level of dominance of the desired action and allows more robust comparisons. We find that such a threshold is reliably reached, if it is likely to be reached at all, within 10000 iterations for the CondMat and Twitter networks and within 15000 iterations for the Enron network. As such, we measure the proportion of runs that have converged to the desired strategy within these time-frames across all networks.

The results are presented in Fig. 6. We utilise PO-PLACE and Degree Placement to allocate FS agents across a range found to exhibits noticeable changes in convention emergence rates with the parameters indicated. The values of o chosen within each topology are such that the number of observations is, at most, approximately 5% of the agent population. All runs are performed 50 times and the proportion of runs that produce the desired convention (strategy chosen u.a.r at time $t = 0$ and assigned to all FS agents) is measured.

Figure 6a shows the results for the CondMat topology. As was expected, due to the behaviour of CondMat in the PO-PLACE experimentation, all of the chosen parameters produce comparable results to the pure degree placement. Even at the worst performing parameters ($o = 500, s = 1$) there is no discernible difference between the performance of degree placement and PO-PLACE, whilst at higher number of observations (where PO-PLACE was entirely approximating the highest-degree nodes as seen in Fig. 1a) the performance is as expected. Of note is the fact that, whilst it resulted in worse output of PO-PLACE in the prior section, increasing s does not noticeably affect the performance here.

Within the other networks the difference in performance is more noticeable but still indicates that PO-PLACE is generating close approximation of the degree placement. In both Enron and Twitter (Figs. 6b and c) the minimal observation situation performs substantially worse than degree placement, particularly in the Twitter network. However, when given observations of 5% of the network, PO-PLACE performs noticeably better. Whilst it still falls behind the performance of degree placement in both networks the difference is substantially smaller with PO-PLACE performing around 50–70% as effectively on average as degree placement in both networks (0.52 ± 0.08 in Enron, 0.69 ± 0.18 in Twitter). However, when we increase s, as was found in Sect. 5.1, it improves this substantially to 0.82 ± 0.15 average effectiveness compared to degree placement in Enron and, less substantially, to 0.79 ± 0.3 in the Twitter network. We quantify these values by comparing the emergence proportions of PO-PLACE and degree at each value of n and calculating the ratio between them which we then average. We discount values where either placement is achieving less than a 0.1 emergence proportion to avoid noisy results influencing the measure. As 0.1 is the expected emergence proportion of our desired strategy in a convention emergence we do not influence, we believe discounting values below this allows a more accurate comparison between the two algorithms. In the Twitter network, we also consider $o = 2500$ as the effect of increased s was more pronounced for this value during Sect. 5.1. Whilst there is a noticeable improvement at higher n the average compared effectiveness differs only marginally: 0.24 ± 0.06 for $s = 1$ and 0.3 ± 0.11 for $s = 9$. In the Twitter network, o is the dominant factor.

Overall, we have shown that even when only observing a small portion of the underlying topology, and strategically using these observations to maximise their effect, it is possible to achieve comparable performance to degree placement with full network visibility using PO-PLACE.

6   Conclusion

Finding influential positions within a network topology to maximise the effectiveness of fixed strategy agents is an ongoing area of research in convention emergence. The problem has many facets and variations that make it difficult to find an optimal yet general approach. In many cases, placing the fixed strategy agents at high degree nodes provides effective convention emergence with little computational overhead. Finding high-degree nodes in a network is trivial when the network is fully observable. In many domains, this may not always be possible. Technical limitations such as memory constraints or incomplete information and usage limitations such as finite API calls mean that often a network topology may only be *partially observable*. Finding effective placement for FS agents with these restrictions adds another level of complexity.

In this paper we presented a placement algorithm, PO-PLACE, that is designed for use in partially observable topologies. It uses finite observations to find sets of high-degree nodes and approximates the set of nodes that would be selected given full observability.

With small proportions of the network being observable, PO-PLACE can locate nodes with similar reach and influence as degree placement. We evaluate the performance in three real-world topologies and show that the addition of concurrent searches and splitting of observations improves the performance of the algorithm across all metrics. With 1–10% observation the algorithm is able to find sets of nodes with >90% of the reach and influence of degree placement.

Finally, we showed that PO-PLACE performs comparably to degree placement when used to facilitate convention emergence using fixed strategy agents whilst only observing 5% of a network topology. We found that the additional aspects of PO-PLACE benefit the placement mechanism and demonstrated that convention emergence is easily facilitated in partially observable networks.

References

1. Borgs, C., Brautbar, M., Chayes, J., Khanna, S., Lucier, B.: The power of local information in social networks. In: Goldberg, P.W. (ed.) WINE 2012. LNCS, vol. 7695, pp. 406–419. Springer, Heidelberg (2012). https://doi.org/10.1007/978-3-642-35311-6_30
2. Brautbar, M., Kearns, M.J.: Local algorithms for finding interesting individuals in large networks. In: Proceedings of Innovation in Computer Science, pp. 188–199 (2010)
3. Chen, D.-B., Xiao, R., Zeng, A.: Predicting the evolution of spreading on complex networks. Sci. Rep. **4**, 6108 (2014)
4. Chen, W., Wang, Y., Yang, S.: Efficient influence maximization in social networks. In: Proceedings of the 15th ACM SIGKDD International Conference on Knowledge Discovery and Data Mining, pp. 199–208 (2009)
5. Delgado, J.: Emergence of social conventions in complex networks. Artif. Intell. **141**(1–2), 171–185 (2002)
6. Delgado, J., Pujol, J.M., Sangüesa, R.: Emergence of coordination in scale-free networks. Web Intell. Agent Syst. **1**(2), 131–138 (2003)

7. Franks, H., Griffiths, N., Anand, S.S.: Learning agent influence in MAS with complex social networks. Auton. Agents Multi-Agent Syst. **28**(5), 836–866 (2014)
8. Franks, H., Griffiths, N., Jhumka, A.: Manipulating convention emergence using influencer agents. Auton. Agents Multi-Agent Syst. **26**(3), 315–353 (2013)
9. Griffiths, N., Anand, S.S.: The impact of social placement of non-learning agents on convention emergence. In: Proceedings of the 11th International Conference on Autonomous Agents and Multiagent Systems, pp. 1367–1368 (2012)
10. Kandori, M.: Social norms and community enforcement. Rev. Econ. Stud. **59**(1), 63–80 (1992)
11. Kittock, J.: Emergent conventions and the structure of multi-agent systems. In: Lectures in Complex Systems: Proceedings of the 1993 Complex Systems Summer School, pp. 507–521 (1995)
12. Leskovec, J., Krevl, A.: SNAP Datasets: Stanford large network dataset collection, June 2014. http://snap.stanford.edu/data
13. Marchant, J., Griffiths, N., Leeke, M.: Convention emergence and influence in dynamic topologies. In: Proceedings of the 2015 International Conference on Autonomous Agents and Multiagent Systems, pp. 1785–1786 (2015)
14. Marchant, J., Griffiths, N., Leeke, M., Franks, H.: Destabilising conventions using temporary interventions. In: Ghose, A., Oren, N., Telang, P., Thangarajah, J. (eds.) COIN 2014. LNCS (LNAI), vol. 9372, pp. 148–163. Springer, Cham (2015). https://doi.org/10.1007/978-3-319-25420-3_10
15. Mihara, S., Tsugawa, S., Ohsaki, H.: Influence maximization problem for unknown social networks. In: Proceedings of the IEEE/ACM International Conference on Advances in Social Networks Analysis and Mining, pp. 1539–1546 (2015)
16. Pei, S., Tang, S., Zheng, Z.: Detecting the influence of spreading in social networks with excitable sensor networks. PloS One **10**(5), e0124848 (2015)
17. Salazar, N., Rodriguez-Aguilar, J.A., Arcos, J.L.: Robust coordination in large convention spaces. AI Commun. **23**(4), 357–372 (2010)
18. Savarimuthu, B.T.R., Arulanandam, R., Purvis, M.: Aspects of active norm learning and the effect of lying on norm emergence in agent societies. In: Kinny, D., Hsu, J.Y., Governatori, G., Ghose, A.K. (eds.) PRIMA 2011. LNCS (LNAI), vol. 7047, pp. 36–50. Springer, Heidelberg (2011). https://doi.org/10.1007/978-3-642-25044-6_6
19. Sen, S., Airiau, S.: Emergence of norms through social learning. In: Proceedings of the 20th International Joint Conference on Artifical Intelligence, pp. 1507–1512 (2007)
20. Villatoro, D., Sabater-Mir, J., Sen, S.: Social instruments for robust convention emergence. In: Proceedings of the 22nd International Joint Conference on Artificial Intelligence, pp. 420–425 (2011)
21. Villatoro, D., Sen, S., Sabater-Mir, J.: Topology and memory effect on convention emergence. In: Proceedings of the 2009 IEEE/WIC/ACM International Joint Conference on Web Intelligence and Intelligent Agent Technology, pp. 233–240 (2009)
22. Walker, A., Wooldridge, M.: Understanding the emergence of conventions in multi-agent systems. In: International Conference on Multi-Agent Systems, pp. 384–389 (1995)
23. Wang, X., Su, Y., Zhao, C., Yi, D.: Effective identification of multiple influential spreaders by DegreePunishment. Phys. A: Stat. Mech. Appl. **461**, 238–247 (2016)
24. Young, H.P.: The economics of convention. J. Econ. Perspect. **10**(2), 105–122 (1996)

Reasoning About Opportunistic Propensity in Multi-agent Systems

Jieting Luo[1](✉), John-Jules Meyer[1], and Max Knobbout[2]

[1] Utrecht University, Utrecht, The Netherlands
{J.Luo,J.J.C.Meyer}@uu.nl
[2] Triple, Alkmaar, The Netherlands
m.knobbout@wearetriple.com

Abstract. Opportunism is a behavior that takes advantage of knowledge asymmetry and results in promoting agents' own value and demoting others' value. We want to eliminate such selfish behavior in multi-agent systems, as it has undesirable results for the participating agents. In order for monitoring and eliminating mechanisms to be put in place, it is needed to know in which context agents will or are likely to perform opportunistic behavior. In this paper, we develop a framework to reason about agents' opportunistic propensity. Opportunistic propensity refers to the potential for an agent to perform opportunistic behavior. In particular, agents in the system are assumed to have their own value systems and knowledge. With value systems, we define agents' state preferences. Based on their value systems and incomplete knowledge about the state, they choose one of their rational alternatives, which might be opportunistic behavior. We then characterize the situations where agents will or will not perform opportunistic behavior and prove the computational complexity of predicting opportunism.

Keywords: Opportunism · Propensity · Logic · Reasoning
Decision theory

1 Introduction

Let us first consider an example scenario. A seller sells a cup to a buyer and it is known by the seller beforehand that the cup is actually broken. The buyer buys the cup without knowing it is broken. The seller exploits the knowledge asymmetry about the transaction to achieve his own gain at the expense of the buyer. Such behavior which is intentionally performed by the seller was named opportunistic behavior (or opportunism) by economist Williamson [19]. Opportunism

G. Sukthankar and J. A. Rodriguez-Aguilar (Eds.): AAMAS 2017 Best Papers,
LNAI 10642, pp. 203–221, 2017.
https://doi.org/10.1007/978-3-319-71682-4_13

is a selfish behavior that takes advantage of relevant knowledge asymmetry and results in promoting one's own value and demoting others' value [9]. In the context of multi-agent systems, it is normal that knowledge is distributed among participating agents in the system, which creates the ability for the agents to behave opportunistically. We want to eliminate such a selfish behavior, as it has undesirable results for other agents in the system. Evidently, not every agent is likely to be opportunistic. In social science, ever since the theory about opportunism was proposed by Williamson in economics, it has gained a large amount of criticism due to over-assuming that all economic players are opportunistic. [5] highlights the challenge on how to predict opportunism *ex ante* and introduces a cultural perspective to better specify the assumptions of opportunism. In multi-agent systems, we also need to investigate the interesting issues about opportunistic propensity so that the appropriate amount of monitoring [10] and eliminating mechanisms can be put in place.

Based on decision theory, an agent's decision on what to do depends on the agent's ability and preferences. If we apply it to opportunistic behavior, an agent will perform opportunistic behavior when he can do it and he prefers doing it. Those are the two issues that we consider in this paper without discussing any normative issues. Based on this assumption, we develop a model of transition systems in which agents are assumed to have their own knowledge and value systems, which are related to the ability and the desire of being opportunistic respectively. Our framework can be used to predict and specify when an agent will perform opportunistic behavior, such as which kinds of agents are likely to perform opportunistic behavior and under what circumstances. A monitoring mechanism for opportunism benefits from this result as monitoring devices may be set up in the occasions where opportunism will potentially occur. We can also design mechanisms for eliminating opportunism based on the understanding of how agents decide to behave opportunistically. Besides, our framework can be used by autonomous agents to decide whether to participate in the system, as their actions might potentially be regarded as opportunistic behavior given their knowledge and value systems.

In this paper, we introduce a framework to reason about agents' opportunistic propensity. Opportunistic propensity refers to the potential for an agent to perform opportunistic behavior. More precisely, agents in the system are assumed to have their own value systems and knowledge. We specify an agent's value system as a strict total order over a set of values, which are encoded within our logical language. Using value systems, we define agents' state preferences. Moreover, agents have partial knowledge about the true state where they are residing. Based on their value systems and incomplete knowledge, they choose one of their rational alternatives, which might be opportunistic. We thus provide a natural bridge between logical reasoning and decision making, which is used for reasoning about opportunistic propensity. We then characterize the situations where agents will or will not perform opportunistic behavior and prove the computational complexity of predicting opportunism.

2 Framework

We use Kripke structures as our basic semantic models of multi-agent systems. A Kripke structure is a directed graph whose nodes represent the possible states of the system and whose edges represent accessibility relations. Within those edges, equivalence relation $\mathcal{K}(\cdot) \subseteq S \times S$ represents agents' epistemic relation, while relation $\mathcal{R} \subseteq S \times Act \times S$ captures the possible transitions of the system that are caused by agents' actions. We use s_0 to denote the initial state of the system. It is important to note that, because in this paper we only consider opportunistic behavior as an action performed by a hypothetical agent, we do not model concurrent actions labeled with agents so that every possible transition of the system is caused by an action instead of joint actions (see e.g., [2,16] for related models). For simplification, we assume that the actions in our model are deterministic. We use $\Phi = \{p, q, ...\}$ of atomic propositional variables to express the properties of states S. A valuation function π maps each state to a set of properties that hold in the corresponding state. Formally,

Definition 2.1. *Let $\Phi = \{p, q, ...\}$ be a finite set of atomic propositional variables. A Kripke structure over Φ is a tuple $\mathcal{T} = (Agt, S, Act, \pi, \mathcal{K}, \mathcal{R}, s_0)$ where e.g.*

- *$Agt = \{1, ..., n\}$ is a finite set of agents;*
- *S is a finite set of states;*
- *Act is a finite set of actions;*
- *$\pi : S \rightarrow \mathcal{P}(\Phi)$ is a valuation function mapping a state to a set of propositions that are considered to hold in that state;*
- *$\mathcal{K} : Agt \rightarrow 2^{S \times S}$ is a function mapping an agent in Agt to a reflexive, transitive and symmetric binary relation between states; that is, given an agent i, for all $s \in S$ we have $s\mathcal{K}(i)s$; for all $s, t, u \in S$ $s\mathcal{K}(i)t$ and $t\mathcal{K}(i)u$ imply that $s\mathcal{K}(i)u$; and for all $s, t \in S$ $s\mathcal{K}(i)t$ implies $t\mathcal{K}(i)s$; $s\mathcal{K}(i)s'$ is interpreted as state s' is epistemically accessible from state s for agent i. For convenience, we use $\mathcal{K}(i, s) = \{s' \mid s\mathcal{K}(i)s'\}$ to denote the set of epistemically accessible states from state s;*
- *$\mathcal{R} \subseteq S \times Act \times S$ is a relation between states with actions, which we refer to as the transition relation labeled with an action; we require that for all $s \in S$ there exists an action $a \in Act$ and one state $s' \in S$ such that $(s, a, s') \in \mathcal{R}$, and we ensure this by including a stuttering action sta that does not change the state, that is, $(s, sta, s) \in \mathcal{R}$; we restrict actions to be deterministic, that is, if $(s, a, s') \in \mathcal{R}$ and $(s, a, s'') \in \mathcal{R}$, then $s' = s''$; since actions are deterministic, sometimes we denote state s' as $s\langle a \rangle$ for which it holds that $(s, a, s\langle a \rangle) \in \mathcal{R}$. For convenience, we use $Ac(s) = \{a \mid \exists s' \in S : (s, a, s') \in \mathcal{R}\}$ to denote the available actions in state s.*
- *$s_0 \in S$ denotes the initial state.*

Now we define the language we use. The language $\mathcal{L}_{\mathrm{KA}}$, propositional logic extended with knowledge and action modalities, is generated by the following grammar:

$$\varphi ::= p \mid \neg\varphi \mid \varphi_1 \vee \varphi_2 \mid K_i\varphi \mid \langle a \rangle\varphi \quad (i \in Agt, a \in Act)$$

The semantics of $\mathcal{L}_{\mathrm{KA}}$ are defined with respect to the satisfaction relation $\models$. Given a Kripke structure $\mathcal{T}$ and a state s in $\mathcal{T}$, a formula φ of the language can be evaluated as follows:

- $\mathcal{T}, s \models p$ iff $p \in \pi(s)$;
- $\mathcal{T}, s \models \neg\varphi$ iff $\mathcal{T}, s \not\models \varphi$;
- $\mathcal{T}, s \models \varphi_1 \vee \varphi_2$ iff $\mathcal{T}, s \models \varphi_1$ or $\mathcal{T}, s \models \varphi_2$;
- $\mathcal{T}, s \models K_i\varphi$ iff for all t such that $s\mathcal{K}(i)t$, $\mathcal{T}, t \models \varphi$;
- $\mathcal{T}, s \models \langle a \rangle\varphi$ iff there exists s' such that $(s, a, s') \in \mathcal{R}$ and $\mathcal{T}, s' \models \varphi$;

Other classical logic connectives (e.g., "$\wedge$", "$\rightarrow$") are assumed to be defined as abbreviations by using $\neg$ and $\vee$ in the conventional manner. As is standard, we write $\mathcal{T} \models \varphi$ if $\mathcal{T}, s \models \varphi$ for all $s \in S$, and $\models \varphi$ if $\mathcal{T} \models \varphi$ for all Kripke structures $\mathcal{T}$.

In this paper, in addition of the $\mathcal{K}$-relation being S5, we also place restrictions of *no-forgetting* and *no-learning* based on Moore's work [11] for the simplification of our framework. It is defined as follows: given a state s in S, if there exists s' such that $s\langle a \rangle\mathcal{K}(i)s'$ holds, then there is a s'' such that $s\mathcal{K}(i)s''$ and $s' = s''\langle a \rangle$ hold; if there exists s' and s'' such that $s\mathcal{K}(i)s'$ and $s'' = s'\langle a \rangle$ hold, then $s\langle a \rangle\mathcal{K}(i)s''$. Following this restriction, we have $\models K_i(\langle a \rangle\varphi) \leftrightarrow \langle a \rangle K_i\varphi$. The *no-forgetting* principle says that if after performing action a agent i considers a state s' possible, then before performing action a agent i already considered possible that action a would lead to this state. In other words, if an agent has knowledge about the effect of an action, he will not forget about it after performing the action. The *no-learning* principle says that all the possible states resulting from the performance of action a in agent i's possible states before action a are indeed his possible states after action a. In other words, the agent will not gain extra knowledge about the effect of an action after performing the action.

3   Value System and Rational Alternative

Agents in the system are assumed to have their own value systems and knowledge. Based on their value systems and incomplete knowledge about the state, agents choose their rational alternatives for the next action they will perform.

3.1   Value System

Given several (possibly opportunistic) actions available to an agent, it is up to the agent's decision to perform opportunistic behavior. Basic decision theory applied to intelligent agents relies on three things: agents know what actions they can carry out, the effects of each action and agents' preference over the effects [13]. In this paper, the effects of each action are expressed by our logical language, and we will specify agents' abilities and preferences in this section. It is worth noting that we only study a single action being opportunistic in this paper, so we will apply basic decision theory for one-shot (one-time) decision problems, which concern the situations where a decision is experienced only once.

One important feature of opportunism is that it promotes agents' own value but demotes others' value. In this section, we define agents' value systems, as it is the standard of agents' consideration about the performance of opportunistic behavior. A value can be seen as an abstract standard according to which agents have their preferences over states. For instance, if we have a value denoting *equality*, we prefer the states where equal sharing or equal rewarding hold. Related work about values can be found in [12,17].

Because of the abstract feature of a value, it is usually interpreted in more detail as a state property, which is represented as a $\mathcal{L}_{\text{KA}}$ formula. The most basic value we can construct is simply a proposition p, which represents the value of achieving p. More complex values can be interpreted such as of the form $\langle a \rangle \varphi \wedge \langle a' \rangle \neg \varphi$, which represents the value that there is an option in the future to either achieve φ or $\neg \varphi$. Such a value corresponds to *freedom of choice*. A formula of a value can also be in the form of $K\varphi$, meaning that it is valuable to *achieve knowledge*. In this paper we denote values with v, and it is important to remember that v is an element from the language $\mathcal{L}_{\text{KA}}$. However, not every formula from $\mathcal{L}_{\text{KA}}$ can be intuitively classified as a value.

We argue that agents can always compare any two values, as we can consider two equivalent values as one value. In other words, every element in the set of values is comparable to each other and none of them is logically equivalent to each other. Therefore, we define a value system as a strict total order over a set of values, representing the degree of importance, which are inspired by the goal structure and the preference in [1,4].

Definition 3.1 (Value System). *A value system $V = (\text{Val}, \prec)$ is a tuple consisting of a finite set* $\text{Val} = \{v, ..., v'\} \subseteq \mathcal{L}_{\text{KA}}$ *of values together with a strict total ordering $\prec$ over* Val. *When $v \prec v'$, we say that value v' is more important than value v.*

We also use a natural number indexing notation to extract the value of a value system, so if V gives rise to the ordering $v \prec v' \prec \ldots$ then $V[0] = v$, $V[1] = v'$, and so on. Value promotion and demotion along a state transition can be defined as follows:

Definition 3.2 (Value Promotion and Demotion). *Given a value v and an action a, we define the following shorthand formulas:*

$$\text{promoted}(v, a) := \neg v \wedge \langle a \rangle v$$
$$\text{demoted}(v, a) := v \wedge \langle a \rangle \neg v$$

We say that a value v is promoted along the state transition (s, a, s') if and only if $s \models \text{promoted}(v, a)$, and we say that v is demoted along this transition if and only if $s \models \text{demoted}(v, a)$.

An agent's value v gets promoted along the state transition (s, a, s') if and only if v doesn't hold in state s and holds in state s'; an agent's value v gets demoted along the state transition (s, a, s') if and only if v holds in state s and doesn't hold in state s'. Note that in principle an agent is not always aware that his or her value gets demoted or promoted, i.e. it might be the case where $s \models$ promoted(v, a) but agent i does not know this, i.e. $s \models \neg(K_i \text{ promoted}(v, a))$.

Now we can define a multi-agent system as a Kripke structure together with agents' value systems, representing their basis of practical reasoning. We also assume that value systems are common knowledge in the system to simplify the model. Formally, a multi-agent system $\mathcal{M}$ is an $(n+1)$-tuple: $\mathcal{M} = (\mathcal{T}, V_1, ..., V_n)$, where $\mathcal{T}$ is a Kripke structure, and for each agent i in $\mathcal{T}$, V_i is a value system.

We now define agents' preferences over two states in terms of values, which will be used for modeling the effect of opportunism. We first define a function highest(i, s, s') that maps a value system and two different states to the most preferred value that changes when going from state s to s' from the perspective of agent i. In other words, it returns the value that changes which the agent most cares about, i.e. the most important change between these states for the agent.

Definition 3.3 (Highest Value). *Given a multi-agent system $\mathcal{M}$, an agent i and two states s and s', function* highest $: Agt \times S \times S \to Val$ *is defined as follows:*

$$\text{highest}(i, s, s')_{\mathcal{M}} := V_i[min\{j \mid \forall k > j : \mathcal{M}, s \models V_i[k] \Leftrightarrow \mathcal{M}, s' \models V_i[k]\}]$$

We write highest(i, s, s') *for short if $\mathcal{M}$ is clear from context.*

Note that if no values change between s and s', we have that highest$(i, s, s') = V_i[0]$, i.e. the function returns the agents least preferred value. Moreover, it is not hard to see that highest$(i, s, s') = $ highest(i, s', s), meaning that the function is symmetric for the two state arguments.

With this function we can easily define agents' preference over two states. We use a binary relation "$\precsim$" over states to represent agents' preferences.

Definition 3.4 (State Preferences). *Given a multi-agent system $\mathcal{M}$, an agent i and two states s and s', agent i weakly prefers state s' to state s, denoted as $s \precsim_i^{\mathcal{M}} s'$, iff*

$$\mathcal{M}, s \models \text{highest}(i, s, s') \Rightarrow \mathcal{M}, s' \models \text{highest}(i, s, s')$$

We write $s \precsim_i s'$ for short if $\mathcal{M}$ is clear from context. Moreover, we write $S \precsim_i S'$ for sets of states S and S' whenever $\forall s \in S, \forall s' \in S' : s \precsim s'$.

As is standard, we also define $s \sim_i s'$ to mean $s \precsim_i s'$ and $s' \precsim_i s$, and $s \prec_i s'$ to mean $s \precsim_i s'$ and $s \not\precsim_i s'$. The intuitive meaning of the definition of $s \precsim_i s'$ is that agent i weakly prefers state s' to s if and only if the agent's most important value does not get demoted (either stays the same or gets promoted). In other words, agent i weakly prefers state s' to s: if highest(i, s, s') holds in state s, then it must also hold in state s', and if highest(i, s, s') does not hold in state s, then it does matter whether it holds in state s' or not. Clearly there is a correspondence between state preferences and promotion or demotion of values, which we can make formal with the following proposition.

Proposition 3.1. *Given a model $\mathcal{M}$ with agent i, state s and available action a in s. Let $v^* = \text{highest}(i, s, s\langle a\rangle)$. We have:*

$$s \prec_i s\langle a\rangle \Leftrightarrow \mathcal{M}, s \models \text{promoted}(v^*, a)$$
$$s \succ_i s\langle a\rangle \Leftrightarrow \mathcal{M}, s \models \text{demoted}(v^*, a)$$
$$s \sim_i s\langle a\rangle \Leftrightarrow \mathcal{M}, s \models \neg(\text{demoted}(v^*, a) \vee \text{promoted}(v^*, a))$$

Proof. Firstly we prove the third one. We define $s \sim_i s\langle a\rangle$ to mean $s \precsim_i s\langle a\rangle$ and $s\langle a\rangle \precsim_i s$. $s \precsim_i s\langle a\rangle$ means that value v^* doesn't get demoted when going from s to $s\langle a\rangle$, and $s\langle a\rangle \precsim_i s$ means that value v^* doesn't get demoted when going from $s\langle a\rangle$ to s. Hence, value v^* doesn't get promoted or demoted (stays the same) by action a. Secondly we prove the first one. We define $s \prec_i s\langle a\rangle$ to mean $s \precsim_i s\langle a\rangle$ and $s \not\precsim_i s\langle a\rangle$. $s \precsim_i s\langle a\rangle$ means that value v^* doesn't get demoted when going from s to $s\langle a\rangle$, and $s \not\precsim_i s'$ means that either value v^* gets promoted or demoted by action a. Hence, value v^* gets promoted by action a. We can prove the second one in a similar way.

Additionally, apart from the fact that $s \prec_i s\langle a\rangle$ implies that the highest changed value gets promoted, we also have that no other value which is more preferred gets demoted or promoted. We have the result that the $\precsim_i$ relation obeys the standard properties we expect from a preference relation.

Proposition 3.2 (Properties of State Preferences). *Given an agent i, his preferences over states "$\precsim_i$" are*

- *Reflexive:* $\forall s \in S : s \precsim_i s$;
- *Transitive:* $\forall s, s', s'' \in S :$ *if* $s \precsim_i s'$ *and* $s' \precsim_i s''$, *then* $s \precsim_i s''$.

Proof. The proof follows Definition 3.4 directly. In order to prove $\precsim_i$ is reflexive, we have to prove that for any arbitrary state s we have $s \precsim_i s$. From Definitions 3.3 and 3.4 we know $\text{highest}(i, s, s') = V_i[0]$ when $s = s'$, and for any arbitrary state s we always have $\mathcal{M}, s \models V_i[0]$ implies $\mathcal{M}, s \models V_i[0]$. Therefore, $s \precsim_i s$ and we can conclude that $\precsim_i$ is reflexive.

In order to prove transitivity, we have to prove $\mathcal{M}, s \models v^*$ implies $\mathcal{M}, s'' \models v^*$, where $v^* = \text{highest}(i, s, s'')$. It can be the case where v^* stays the same in state s and s'' or the case where $\mathcal{M}, s \models \neg v^*$ and $\mathcal{M}, s'' \models \neg v^*$. For the first case, when $s \sim s'$ and $s' \sim s''$, meaning that all the values stay the same when going from s to s' and from s' to s'', it is also the case when going from s to s''. We now consider the case where $\mathcal{M}, s \models \neg v^*$ and $\mathcal{M}, s'' \models \neg v^*$. Firstly, we denote $\text{highest}(i, s, s')$ as u^* and $\text{highest}(i, s', s'')$ as w^*. It can either be that $u^* \sim_i w^*$, $u^* \prec_i w^*$ or $u^* \succ_i w^*$. If $u^* \sim_i w^*$, we can conclude that $u^* \sim_i w^* \sim_i v^*$, hence the implication holds. We now distinguish between the cases where $u^* \prec_i w^*$ or $u^* \succ_i w^*$.

- If $u^* \prec_i w^*$, we know that w^* is the highest value that changes and gets promoted when going from s' to s'', but stays the same between s and s'. Hence, we can conclude that $\mathcal{M}, s \models \neg w^*$ and $\mathcal{M}, s'' \models w^*$, and that $w^* = v^*$ (i.e., w^* is the highest value that changes between s and s''). Hence we have $\mathcal{M}, s \models v^*$ implies $\mathcal{M}, s'' \models v^*$.

- If $u^* \succ_i w^*$, we know that u^* is the highest value that changes and gets promoted when going from s to s', but stays the same between s' and s''. Hence, we can conclude that $\mathcal{M}, s \models \neg u^*$ and $\mathcal{M}, s'' \models u^*$, and that $u^* = v^*$ (i.e. v^* is the highest value that changes between s and s''). Hence we have $\mathcal{M}, s \models v^*$ implies $\mathcal{M}, s'' \models v^*$.

In our system, we only look at the value change that is most cared about to deduce state preferences. Certainly, there are other ways of deriving these preferences from a value system. Instead of only considering the value change that is cared about in the state an rational alternative transition, it is also possible to take into account all the value changes in the state transition. For opportunism, what we want to stress is that opportunistic agents ignore (rather than consider less) other agents' interest, which has a lower index in the agent's value system. In order to align with this aspect, we use the highest value approach in this paper.

3.2   Rational Alternatives

Since we have already defined values and value systems as agents' standards for decision-making, we can start to apply decision theory to reason about agents' decision-making. Given a state in the system, there are several actions available to an agent, and he has to choose one in order to go to the next state. We can see the consideration here as a one-shot decision making. In decision theory, if agents only act for one step, a rational agent should choose an action with the highest (expected) utility without reference to the utility of other agents [13]. Within our framework, this means that a rational agent will always choose an rational alternative based on his value system.

Before choosing an action to perform, an agent must think about which actions are available to him. We have already seen that for a given state s, the set of available actions is $Ac(s)$. However, since an agent only has partial knowledge about the state, we argue that the actions that an agent knows to be available is only part of the actions that are physically available to him in a state. For example, an agent can call a person if he knows the person's phone number; without this knowledge, he is not able to do it, even though he is holding a phone. Recall that the set of states that agent i considers as being the actual state in state s is the set $\mathcal{K}(i, s)$. Given an agent's partial knowledge about a state as a precondition, he knows what actions he can perform in that state, which is the intersection of the sets of actions physically available in the states in this knowledge set.

Definition 3.5 (Subjectively Available Actions). *Given an agent i and a state s, agent i's subjectively available actions are the set:*

$$Ac(i, s) = \bigcap_{s' \in \mathcal{K}(i,s)} Ac(s').$$

Because a stuttering action sta is always included in $Ac(s)$ for any state s, we have that $sta \in Ac(i, s)$ for any agent i. When only sta is in $Ac(i, s)$, we say that the agent cannot do anything because of his limited knowledge. Obviously an agent's subjectively available actions is always part of his physically available actions ($Ac(i, s) \subseteq Ac(s)$). Based on rationality assumptions, he will choose an action based on his partial knowledge of the current state and the next state. Given a state s and an action a, an agent considers the next possible states as the set $\mathcal{K}(i, s\langle a \rangle)$. For another action a', the set of possible states is $\mathcal{K}(i, s\langle a' \rangle)$. The question now becomes: How do we compare these two possible set of states? Clearly, when we have $\mathcal{K}(i, s\langle a \rangle) \precsim_i \mathcal{K}(i, s\langle a' \rangle)$, meaning that all alternatives of performing action a' are at least as desirable as all alternatives of choosing action a, it is always better to choose action a'. However, in some cases it might be that some alternatives of action a are better than some alternatives of action a' and vice-versa. In this case, an agent cannot decisively conclude which of the actions is optimal. This approach has natural ties to game theory in the context of (non-)dominated strategies [7]. This leads us to the following definition:

Definition 3.6 (Rational Alternatives). *Given a state s, an agent i and two actions $a, a' \in Ac(i, s)$, we say that action a is* dominated *by action a' for agent i in state s iff $\mathcal{K}(i, s\langle a \rangle) \precsim_i \mathcal{K}(i, s\langle a' \rangle)$. The set of* rational alternatives *for agent i in state s is given by the function $a_i^* : S \to 2^{Act}$, which is defined as follows:*

$$a_i^*(s) = \{a \in Ac(i, s) \mid \neg \exists a' \in Ac(i, s) : a \neq a' \text{ and}$$
$$a' \text{ dominates } a \text{ for agent } i \text{ in state } s\}.$$

The set $a_i^*(s)$ are all the actions for agent i in state s which are available to him and are not dominated by another action which is available to him. In other words, it contains all the actions which are rational alternatives for agent i. Since it is always the case that $Ac(i, s)$ is non-empty because of the stuttering action sta, and since it is always the case that there is one action which is non-dominated by another action and $Ac(i, s)$ is finite, we conclude that $a_i^*(s)$ is non-empty. We can see that the actions that are available to an agent not only depend on the physical state, but also depend on his knowledge about the state. The more he knows, the better he can judge what his rational alternative is. In other words, agents try to make a best choice based on their value systems and incomplete knowledge about the state. The following proposition shows how agents remove actions with our approach.

Proposition 3.3. *Given a state s, an agent i and two actions $a, a' \in Ac(i, s)$, action a is dominated by action a' iff*

$$\neg \exists s', s'' \in \mathcal{K}(i, s) : s'\langle a \rangle \succ s''\langle a' \rangle.$$

Proof. $\exists s', s'' \in \mathcal{K}(i, s) : s'\langle a \rangle \succ s''\langle a' \rangle$ is equivalent to $\mathcal{K}(i, s\langle a \rangle) \not\precsim_i \mathcal{K}(i, s\langle a' \rangle)$, because $s'\langle a \rangle \in \mathcal{K}(i, s\langle a \rangle)$ and $s''\langle a' \rangle \in \mathcal{K}(i, s\langle a' \rangle)$. And $\mathcal{K}(i, s\langle a \rangle) \not\precsim_i \mathcal{K}(i, s\langle a' \rangle)$ is equivalent to the fact that action a is non-dominated by action a'.

From this proposition we can see that agents remove all the options (actions) that are always bad to do, and there is no possibility to be better off by choosing a dominated action. The following proposition connects Definition 3.6 with state preferences.

Proposition 3.4. *Given a multi-agent system* $\mathcal{M}$, *a state* s *and an agent* i,

$$sta \notin a^*(s) \Rightarrow \forall a \in a^*(s) : s \precsim_i s\langle a \rangle.$$

Proof. We prove it by contradiction. If there exists an action $a \in a^*(s)$ such that agent i's value will get demoted by performing it, it will be dominated by the stuttering action sta, which can always keep agent i's values neutral, and sta might be in $a^*(s)$. Contradiction!

If the stuttering action sta is not in the set of rational alternatives for agent i, meaning that it is dominated by the actions in the set of rational alternatives, agent i can always at least keep his value neutral by performing any action in his rational alternatives. We will illustrate the above definitions and our approach through the following example.

Example 1. Figure 1 shows a transition system $\mathcal{M}$ for agent i. State s and s' are agent i's epistemic alternatives, that is, $\mathcal{K}(i, s) = \{s, s'\}$. Now consider the actions that are physically available and subjectively available to agent i. $Ac_i(s) = \{a_1, a_2, a_3, sta\}$, $Ac_i(s') = \{a_1, a_2, sta\}$. Because $Ac(i, s) = Ac_i(s) \cap Ac_i(s')$, agent i knows that only sta, a_1 and a_2 are available to him in state s.

Next we talk about agent i's rational alternatives in state s. Given agent i's value system $V_i = (u \prec v \prec w)$, and the following valuation: u, $\neg v$ and $\neg w$ hold in $\mathcal{K}(i, s)$, $\neg u$, $\neg v$ and w hold in $\mathcal{K}(i, s\langle a_1 \rangle)$, and u, v and $\neg w$ hold in $\mathcal{K}(i, s\langle a_2 \rangle)$, we then have the following state preferences: $\mathcal{K}(i, s) \prec \mathcal{K}(i, s\langle a_1 \rangle)$, $\mathcal{K}(i, s) \prec \mathcal{K}(i, s\langle a_2 \rangle)$ and $\mathcal{K}(i, s\langle a_2 \rangle) \prec \mathcal{K}(i, s\langle a_1 \rangle)$, meaning that action a_2 and the stuttering action sta are dominated by action a_1. Thus, we have $a_i^*(s) = \{a_1\}$.

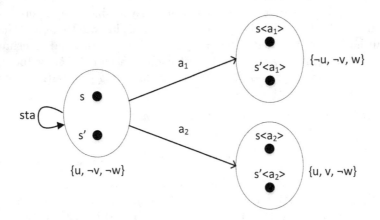

Fig. 1. A transition system $\mathcal{M}$ for agent i

4 Opportunism Propensity

Before reasoning about opportunistic propensity, we should first formally know what opportunistic propensity actually is. Opportunism is a social behavior that takes advantage of relevant knowledge asymmetry and results in promoting one's own value and demoting others' value [9]. It means that it is performed with the precondition of relevant knowledge asymmetry and the effect of promoting agents' own value and demoting others' value. Firstly, knowledge asymmetry is defined as follows.

Definition 4.1 (Knowledge Asymmetry). *Given two agents i and j, and a $\mathcal{L}_{\mathrm{KA}}$ formula ϕ, knowledge asymmetry about ϕ between agent i and j is the abbreviation:*

$$\mathrm{KnowAsym}(i, j, \phi) := K_i \phi \wedge \neg K_j \phi \wedge K_i(\neg K_j \phi)$$

It holds in a state where agent i knows ϕ while agent j does not know ϕ and this is also known by agent i. It can be the other way around for agent i and agent j. But we limit the definition to one case and omit the opposite case for simplicity. Now we can define opportunism.

Definition 4.2 (Opportunism Propensity). *Given a multi-agent system $\mathcal{M}$, a state s and two agents i and j, the assertion $\mathrm{Opportunism}(i, j, a)$ that action a performed by agent i is opportunistic behavior is defined as:*

$$\mathrm{Opportunism}(i, j, a) := \mathrm{KnowAsym}(i, j, \mathrm{promoted}(v^*, a) \wedge \mathrm{demoted}(w^*, a))$$

where $v^* = \mathrm{highest}(i, s, s\langle a \rangle)$ and $w^* = \mathrm{highest}(j, s, s\langle a \rangle)$.

This definition shows that if the precondition KnowAsym is satisfied in state s then the performance of action a will be opportunistic behavior. The asymmetric knowledge that agent i has is about the change of the truth value of v^* and w^* along the transition by action a, where v^* and w^* are the values that agent i and agent j most care about along the transition respectively. It follows that agent j is partially or completely not aware of it. Compared to the definition of opportunism in [9], Definition 4.2 focuses on the opportunistic propensity of an agent in a state, in the sense that the precondition of performing opportunistic behavior is modeled in an explicit way. As is stressed in [9], opportunistic behavior is performed by intent rather than by accident. In this paper, instead of explicitly modeling intention, we interpret it from agents' rationality that they always intentionally promote their own values. We can derive three propositions from the definition, which are useful in our next section.

Proposition 4.1 (Value Promotion and Demotion). *Given a multi-agent system $\mathcal{M}$ and an opportunistic behavior a performed by agent i to agent j in state s, action a will promote agent i's value but demote agent j's value, which can be formalized as*

$$\mathcal{M}, s \models \mathrm{Opportunism}(i, j, a) \quad \Rightarrow \quad s \prec_i s\langle a \rangle \ \text{and} \ s \succ_j s\langle a \rangle$$

Proof. From $\mathcal{M}, s \models \text{Opportunism}(i, j, a)$ we have: $\mathcal{M}, s \models K_i(\text{promoted}(v^*, a) \wedge \text{demoted}(w^*, a))$. And thus since all knowledge is true, we have that $\mathcal{M}, s \models \text{promoted}(v^*, a)$ and $\mathcal{M}, s \models \text{demoted}(w^*, a)$. Using the correspondence found in Proposition 3.1, we can conclude $s \prec_i s\langle a \rangle$ and $s \succ_j s\langle a \rangle$.

Proposition 4.2 (Different Value Systems). *Given a multi-agent system $\mathcal{M}$ and opportunistic behavior a performed by agent i to agent j in state s, agent i and agent j have different value systems, which can be formalized as*

$$\mathcal{M}, s \models \text{Opportunism}(i, j, a) \quad \Rightarrow \quad V_i \neq V_j$$

Proof. We prove it by contradiction. We denote $v^* = \text{highest}(i, s, s\langle a \rangle)$ and $w^* = \text{highest}(j, s, s\langle a \rangle)$, for which v^* and w^* are the property changes that agent i and agent j most care about in the state transition. If $V_i = V_j$, then $v^* = w^*$. However, because $\mathcal{M}, s \models K_i(\text{promoted}(v^*, i) \wedge \text{demoted}(w^*, j))$, and thus $\mathcal{M}, s \models K_i(\neg v^* \wedge w^*)$, and because knowledge is true, we have $\mathcal{M}, s \models \neg v^* \wedge w^*$. But, since $v^* = w^*$, we have $\mathcal{M}, s \models \neg v^* \wedge v^*$. Contradiction!

From this proposition we can see that agent i and agent j care about different things based on their value systems about the transition.

Proposition 4.3 (Inclusion). *Given a multi-agent system $\mathcal{M}$ and opportunistic behavior a performed by agent i to agent j in state s, agent j's knowledge set in state s is not a subset of agent i's and action a is available in agent i's knowledge set:*

$$\mathcal{M}, s \models \text{Opportunism}(i, j, a) \quad \Rightarrow \mathcal{K}(j, s) \nsubseteq \mathcal{K}(i, s) \text{ and } a \in Ac(i, s)$$

Proof. We can prove it by contradiction. Knowledge set is the set of states that an agent considers as possible in a given actual state. $\forall t \in \mathcal{K}(i, s)$, agent i considers state t as a possible state where he is residing. The same with $\mathcal{K}(j, s)$ for agent j. If $\mathcal{K}(j, s) \nsubseteq \mathcal{K}(i, s)$ is false, we have $\mathcal{K}(j, s) \subseteq \mathcal{K}(i, s)$ holds, which means that agent j knows more than or exactly the same as agent i. However, Definition 4.2 tells that agent i knows more about the transition by action a than agent j. So $\mathcal{K}(j, s) \subseteq \mathcal{K}(i, s)$ is false, meaning that $\mathcal{K}(j, s) \nsubseteq \mathcal{K}(i, s)$ holds. Further, because from $\mathcal{M}, s \models \text{Opportunism}(i, j, a)$ we have $\mathcal{M}, s \models K_i(\langle a \rangle v^* \wedge \langle a \rangle \neg w^*)$, by the semantics of $\langle a \rangle v^*$ and $\langle a \rangle \neg w^*$, for all $t \in \mathcal{K}(i, s)$ there exists $(t, a, s') \in R$. Thus, we have $a \in Ac(i, s)$.

These three propositions are three properties that we can derive based on Definition 4.2. The first one shows that opportunistic behavior results in value opposition for the agents involved; the second one tells that the two agents involved in the relationship evaluate the transition based on different value systems; the third one indicates the asymmetric knowledge that agent i has for behaving opportunistically. We will illustrate the above definitions through the example mentioned at the beginning of the paper.

Example 2. Figure 2 shows the example of selling a broken cup: The action selling a cup is denoted as *sell* and we use two value systems V_s and V_b for the seller and the buyer respectively. State s_1 is the seller's epistemic alternative, while state s_1 and s_2 are the buyer's epistemic alternatives. We also use a dash line circle to represent the buyer's knowledge $\mathcal{K}(b, s_1)$ (not the seller's). In this example, $\mathcal{K}(s, s_1) \subset \mathcal{K}(b, s_1)$. Moreover, $hm = $ highest$(s, s_1, s_1 \langle sell \rangle)$, $\neg hb = $ highest$(b, s_1, s_1 \langle sell \rangle)$, meaning that the seller only cares about if he gets money from the transition, while the buyer only cares about if he has a broken cup from the transition. We also have $\mathcal{M}, s_1 \models K_s(promoted(hm, sell) \land demoted(\neg hb, sell))$, meaning that the seller knows the transition will promote his own value while demote the buyer's value in state s_1. For the buyer, action *sell* is available in both state s_1 and s_2. However, hb doesn't hold in both s_1 and s_2, so he doesn't know if he has a broken cup or not. Therefore, there is knowledge asymmetry between the seller and the buyer about the value changes from s_1 to $s_1 \langle sell \rangle$. Action *sell* is potentially opportunistic behavior in state s_1.

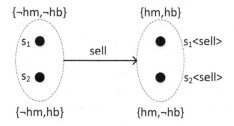

Fig. 2. Selling a broken cup

5 Reasoning About Opportunistic Propensity

In this section, we will characterize the contexts where agents will perform opportunistic behavior and where opportunism is impossible to happen and prove the computational complexity of predicting opportunism.

5.1 Having Opportunism

Agents will perform opportunistic behavior when they have the ability and the desire of doing it. The ability of performing opportunistic behavior can be interpreted by its precondition: it can be performed whenever its precondition is fulfilled. Agents have the desire to perform opportunistic behavior whenever it is a rational alternative. There are also relations between agents' ability and desire of performing an action. As rational agents, firstly we think about what actions we can perform given the limited knowledge we have about the state, and secondly we choose the action that may maximize our utilities based on our partial knowledge. This practical reasoning in decision theory can also be applied to reasoning about opportunistic propensity. Given the asymmetric knowledge an agent has, there are several (possibly opportunistic) actions available to him, and he may choose to perform the action which is a rational alternative to him, regardless of the result for the other agents. Based on this understanding, we have the following theorem, which implies agents' opportunistic propensity:

Theorem 5.1. *Given a multi-agent system $\mathcal{M}$, a state s, two agents i and j and an action a, agent i will perform action a to agent j as opportunistic behavior in state s:*

$$\exists a \in a_i^*(s) : \mathcal{M}, s \models \text{Opportunism}(i, j, a)$$

iff

1. $\forall t \in \mathcal{K}(i, s) : \mathcal{M}, t \models \text{promoted}(v^*, a) \wedge \text{demoted}(w^*, a)$, $\exists t \in \mathcal{K}(j, s) : \mathcal{M}, t \models \neg(\text{promoted}(v^*, a) \wedge \text{demoted}(w^*, a))$, *where* $v^* = \text{highest}(i, s, s\langle a \rangle)$ *and* $w^* = \text{highest}(j, s, s\langle a \rangle)$;
2. $s \prec_i s\langle a \rangle$ *and* $s \succ_j s\langle a \rangle$;
3. $\neg \exists a' \in Ac(i, s) : a \neq a'$ *and* a' *dominates* a.

Proof. **Forwards:** If action a is opportunistic behavior, we can immediately have statement 1 by the definition of Knowledge Set. Because action a is in agent i's rational alternatives in state s ($a \in a_i^*(s)$), by Definition 3.6, action a is not dominated by any action in $Ac(i, s)$. Also because action a is opportunistic, by Proposition 4.1 it results in promoting agent i's value but demoting agent j's value ($s \prec_i s\langle a \rangle$ and $s \succ_j s\langle a \rangle$). **Backwards:** Statement 1 means that there is knowledge asymmetry between agent i and agent j about the formula promoted$(v^*, a) \wedge$ demoted(w^*, a). From this we can see the knowledge asymmetry is the precondition of action a. If this precondition is satisfied, agent i can perform action a. Moreover, by statement 2, because action a promotes agent i's value but demotes agent j's value, we can conclude that action a is opportunistic behavior. By statement 3, because action a is not dominated by any action in $Ac(i, s)$, it is a rational alternative for agent i in state s to perform action a.

Given an opportunistic behavior a, in order to predict its performance, we should first check the asymmetric knowledge that agent i has for enabling its performance. Based on agent i's and agent j's value systems, we also check if it is not dominated by any actions in $Ac(i, s)$ and its performance can promote agent i's value but demote agent j's value. It is important to stress that Theorem 5.1 never states that agents will for sure perform opportunistic behavior if the three statements are satisfied. Instead, it shows opportunism is likely to happen because it is in agents' rational alternatives.

5.2   Not Having Opportunism

We need much information about the system as Theorem 5.1 states to predict opportunism, and it might be difficult to achieve all of them. Fortunately, sometimes it is already enough to know that opportunism is impossible to occur.

An example might be detecting opportunism: if we already know in which context agents cannot perform opportunistic behavior, there is no need to set up any monitoring mechanisms for opportunism in those contexts. The following propositions characterize them:

Proposition 5.1. *Given a multi-agent system* $\mathcal{M}$*, a state* s*, two agents* i *and* j *and an action* a*,*

$$\mathcal{K}(i, s) = \mathcal{K}(j, s) \Rightarrow \mathcal{M}, s \models \neg Opportunism(i, j, a).$$

Proof. When $\mathcal{K}(i, s) = \mathcal{K}(j, s)$ holds, which means that both agent i and agent j have the same knowledge. In this context, Statement 1 in Theorem 5.1 is not satisfied, so action a is not opportunistic behavior.

Proposition 5.2. *Given a multi-agent system* $\mathcal{M}$*, a state* s*, two agents* i *and* j *and an action* a*,*

$$V_i = V_j \Rightarrow \mathcal{M}, s \models \neg Opportunism(i, j, a).$$

Proof. If $V_i = V_j$ holds, which means that both agent i and agent j have the same value system, the values of both agents don't go opposite, that is, Statement 2 in Theorem 5.1 is not satisfied. So action a is not opportunistic behavior.

In this section, we specified the situation where agents will perform opportunistic behavior and characterized the contexts where opportunism is impossible to happen. This information is essential not only for the system designers to identify opportunistic propensity, but also for an agent to decide whether to participate in the system given his knowledge and value system, as his behavior might be regarded as opportunistic. Moreover, our approach can be used in practice. For instance, in the electronic market place, only the seller knows that the product is not good for the buyer before he ships it, and he can earn more money if he still claims that the product is good. In this context the seller can and wants to perform opportunistic behavior, selling the product, to the buyer according to Theorem 5.1. Monitoring and eliminating opportunism mechanism should be put there in order to demotivated such a behavior. However, if we can ensure that both the seller and the buyer are aware of the quality of the product before the seller ships it, it is impossible for him to get benefits from the buyer.

5.3   Computational Complexity

Theorem 5.1 shows that whether a given action will be performed by an agent as opportunistic behavior. More generally, we would like to know given a multi-agent system we design, whether there exists opportunistic behavior between agents and how difficult it is to check it. In this section, we will investigate this issue through proposing an algorithm. The decision problem associated with predicting opportunistic behavior is as follows:

PREDICTING OPPORTUNISM
_____

Given: Multi-agent system $\mathcal{M}$.

Question: Does there exist opportunistic behavior between agents for $\mathcal{M}$?

Theorem 5.2. *Given a multi-agent system $\mathcal{M}$, the problem that whether there exists opportunistic behavior between agents for $\mathcal{M}$ is in* **P**.

Proof. In order to prove it is a **P** problem, we need to find an algorithm that allows us to solve the decision problem in polynomial-time. We design Algorithm 1 for verifying opportunistic behavior in a multi-agent system $\mathcal{M}$ based on Theorem 5.1. The algorithm loops through all the possible transitions in the system, which has complexity $O(n)$, where $n = |\mathcal{R}|$. Notice that transitions are executed by hypothetical agents, meaning that the value systems we consider for the transition is assumed to be known once the transition is given. For each transition, it verifies the statements listed in Theorem 5.1 one by one. Line 21–24 is to verify whether there is no action a' that dominates action a. Based on the definition of dominance between actions, the algorithm has to perform the comparison $\mathcal{K}(i, s\langle a \rangle)$ with $\mathcal{K}(i, s\langle a' \rangle)$ for all a' in $Ac(i, s)$. If for all $s' \in \mathcal{K}(i, s\langle a \rangle)$ and for all $s'' \in \mathcal{K}(i, s\langle a' \rangle)$ we have $s' \prec s''$, then action a is dominated by action a'. Hence, the complexity of executing line 21–24 is $O(mk^2)$, where $m = |Ac(i, s)|$ and $k = |\mathcal{K}(i, s)|$. The computational complexity of the whole algorithm is $O(nmk^2)$, which implies that Algorithm 1 can check whether there exists opportunistic behavior between agents for a given multi-agent system in polynomial-time. Therefore, we can conclude that given a multi-agent system $\mathcal{M}$, the problem that whether there exists opportunistic behavior between agents is in **P**.

6   Related Work

The technical framework we used in this paper is a transition system extended with value systems. As standards for specifying preferences, people usually use goals rather than value (e.g. [6,14]) in logic-based formalization and utilities in decision theory and game theory (e.g. [15,18]) for the same purpose. Only some work in the area of argumentation reasons about agents' preferences and decision making by values (e.g. [3,12,17]). Goals are concrete and should be specified with time, place and objects, while value is relatively stable and not limited to be applied in a specific situation. Since state transitions are caused by the performance of actions, we can evaluate actions by whether our value is promoted or demoted in the state transition. For representing agents' evaluation on states, Keeney and Raiffa proposed Multi-Attribute Utility Theory (MAUT) in which states are described in terms of a set of attributes and the utilities of the states are calculated by the sum of the scores on each attribute based on agents' value system [8]. Apparently, not everything can be evaluated with numbers, which is one of the reasons why people consider using value systems as an alternative. A value system is like a box that allows us to define its content

Algorithm 1. Predicting Opportunism

```
 1: procedure HASKNOWASYM(S₁, S₂, π, φ) returns true or false
 2:     set g₁ ← true
 3:     set g₂ ← false
 4:     for each s ∈ S₁ do
 5:         if φ ∉ π(s) then
 6:             set g₁ ← false
 7:             break
 8:     for each s ∈ S₂ do
 9:         if ¬φ ∈ π(s) then
10:             set g₂ ← true
11:             break
12:     return g₁ ∧ g₂
13:
14: procedure PREDICTING(M) returns true or false
15:     set flag ← false
16:     for each (s, a, s⟨a⟩) ∈ R do
17:         set v* ← highest(i, s, s⟨a⟩)
18:         set w* ← highest(j, s, s⟨a⟩)
19:         if HASKNOWASYM(K(i, s), K(j, s), π, promoted(v*, a) ∧ demoted(w*, a))
     then
20:             if promoted(v*, a) ∧ demoted(w*, a) ∈ π(s) then
21:                 set h ← 0
22:                 for each a′ ∈ Ac(i, s) do
23:                     if a ≠ a′ and K(i, s⟨a⟩) ⪯ K(i, s⟨a′⟩) then
24:                         h++
25:                 if h == 0 then
26:                     set flag ← true
27:                     break
28:     return flag
```

as we need. In this paper, a value is modeled as a formula in our language and a value system is constructed as a total order over a set of values. Instead of calculating the utility of states, agents specify their preferences over states by evaluating the value change that they most care about.

We reason about agents' opportunistic propensity based on decision theory extended with knowledge and value systems, which correspond to some concepts from game theory. In game theory, agents can be situated in a game which is not fully observable, and the notion of information sets is introduced to represent the states that the agent cannot distinguish [7]. In this paper, we use a similar concept *knowledge set* to represent the set of states that the agent considers as possible. Based on the representation of uncertainty, we use the notion of dominance to compare two different actions: a dominated action is an action that is always bad to perform regardless of the uncertainty about the system, which is an approach bridging to (non-)dominated strategies in game theory. It is thus already seen that we can apply techniques from game theory based on the concept similarities to enrich the existing decision theory and enhance the reasoning capabilities on agents' opportunistic propensity.

7   Conclusion and Future Work

The investigation about opportunism is still new in the area of multi-agents systems. We ultimately aim at designing mechanisms to eliminate such selfish behavior in the system. In order to avoid over-assuming the performance of opportunism so that monitoring and eliminating mechanism can be put in place, we need to know in which context agents will or are likely to perform opportunistic behavior. In this paper, we argue that agents will behave opportunistically when they have the ability and the desire of doing it. With this idea, we developed a framework of multi-agent systems to reason about agents' opportunistic propensity without considering normative issues. Agents in the system were assumed to have their own value systems. Based on their value systems and incomplete knowledge about the state, agents chose one of their rational alternatives, which might be opportunistic behavior. With our framework and our definition of opportunism, we characterized the situations where agents will/will not perform opportunistic behavior and proved the computational complexity of predicting opportunism. Certainly there are multiple ways to extend our work. One interesting way is to enrich our formalization of value systems over different sets of values, and the enrichment might lead to the investigation about the compatibility of value systems and different results about opportunistic propensity. Another way is to consider normative issues in our framework in addition to the ability and the desire of being opportunistic.

Acknowledgments. The research is supported by China Scholarship Council. We would like to thank Allan van Hulst and anonymous reviewers for their helpful comments.

References

1. Ågotnes, T., van der Hoek, W., Wooldridge, M.: Normative system games. In: Proceedings of the 6th International Joint Conference on Autonomous Agents and Multiagent Systems, p. 129. ACM (2007)
2. Alur, R., Henzinger, T.A., Kupferman, O.: Alternating-time temporal logic. J. ACM **49**(5), 672–713 (2002)
3. Bench-Capon, T., Atkinson, K., McBurney, K.: Using argumentation to model agent decision making in economic experiments. Auton. Agents Multi-Agent Syst. **25**(1), 183–208 (2012)
4. Bulling, N., Dastani, M.: Norm-based mechanism design. Artif. Intell. **239**, 97–142 (2016)
5. Chen, C.C., Peng, M.W., Saparito, P.A.: Individualism, collectivism, and opportunism: a cultural perspective on transaction cost economics. J. Manag. **28**(4), 567–583 (2002)

6. Cohen, P.R., Levesque, H.J.: Intention is choice with commitment. Artif. Intell. **42**(2–3), 213–261 (1990)
7. Dixit, A.K., Nalebuff, B.: The Art of Strategy: A Game Theorist's Guide to Success in Business & Life. WW Norton & Company, New York (2008)
8. Keeney, R.L., Raiffa, H.: Decisions with Multiple Objectives: Preferences and Value Trade-Offs. Cambridge University Press, Cambridge (1993)
9. Luo, J., Meyer, J.-J.: A formal account of opportunism based on the situation calculus. AI Soc. **32**(4), 527–542 (2017). https://doi.org/10.1007/s00146-016-0665-4
10. Luo, J., Meyer, J.-J., Knobbout, M.: Towards a framework for detecting opportunism in multi-agent systems. In: ECAI 2016–22nd European Conference on Artificial Intelligence, pp. 1636–1637 (2016)
11. Moore, R.C.: Reasoning About Knowledge and Action. SRI International Menlo Park, CA (1980)
12. Pitt, J., Artikis, A.: The open agent society. retrospective and prospective views. Artif. Intell. Law **23**(3), 241–270 (2015)
13. Poole, D.L., Mackworth, A.K.: Artificial Intelligence: Foundations of Computational Agents. Cambridge University Press, Cambridge (2010)
14. Rao, A.S., Georgeff, M.P.: Modeling rational agents within a BDI-architecture. KR **91**, 473–484 (1991)
15. Steele, K., Orri Stefánsson, H.: 'Decision theory. In: Zalta, E.N. (ed.) The Stanford Encyclopedia of Philosophy. Metaphysics Research Lab, Stanford University, winter 2016 edn. (2016)
16. van der Hoek, W., Wooldridge, M.: Cooperation, knowledge, and time: Alternating-time temporal epistemic logic and its applications. Studia Logica **75**(1), 125–157 (2003)
17. Van der Weide, T.L.: Arguing to motivate decisions, Ph.D. dissertation (2011)
18. Von Neumann, J., Morgenstern, O.: Theory of Games and Economic Behavior. Princeton University Press, New Jersey (2007)
19. Williamson, O.E.: Markets and Hierarchies, Analysis and Antitrust Implications: A Study in the Economics of Internal Organization. Free Press, New York (1975)

Approaching Interactions in Agent-Based Modelling with an Affordance Perspective

Franziska Klügl[1]($\boxtimes$) and Sabine Timpf[2]

[1] School of Natural Science and Technology,
Örebro University, 70182 Örebro, Sweden
franziska.klugl@oru.se
[2] Institute for Geography, Augsburg University, 86135 Augsburg, Germany
sabine.timpf@geo.uni-augsburg.de

Abstract. Over the last years, the affordance concept has attracted more and more attention in agent-based simulation. Due to its grounding in cognitive science, we assume that it may help a modeller to capture possible interactions in the modelling phase as it can be used to clearly state under which circumstances an agent might execute a particular action with a particular environmental entity.

In this discussion paper we clarify the concept of affordance and introduce a light-weight formalization of the notions in a way appropriate for agent-based simulation modelling. We debate its suitability for capturing interaction compared to other approaches.

1 Introduction

A critical part of building an agent-based model is related to interactions between agents, as well as between agents and other objects in their environment. There is an inherent gap between formulating agents, their properties, individual goals and/or behaviour at the micro level and the overall intended outcome observable at a macro level. When running the simulation, the simulated agents – put together and into an environment –, eventually *generate* this aggregated outcome. Interaction hereby forms the element of the model that connects micro- and macro level. Yet, one cannot easily foresee who will actually interact with whom in the running simulation. Diverse methodologies for developing agent-based simulation models propose different solutions to produce some form of predictability of interactions, defining a systematic approach to formulations in the model.

In this contribution we aim at clarifying the concept of an affordance so that it becomes a helpful notion for general agent-based simulation model development. We suggest a formalization that – if embedded into an appropriate development methodology – can support a more reliable model development by explicitly representing potential interactions. Affordances have been seen as useful in so diverse areas such as Human-Computer Interaction and Virtual Reality,

© Springer International Publishing AG 2017
G. Sukthankar and J. A. Rodriguez-Aguilar (Eds.): AAMAS 2017 Best Papers,
LNAI 10642, pp. 222–238, 2017.
https://doi.org/10.1007/978-3-319-71682-4_14

Robotics or spatially explicit agent-based simulation. Our goal is to develop a concept that supports a modeller in capturing interactions, not in a way to be able to automatically reason about them before running a simulation, but in a way to make the modeller aware of the circumstances in which interactions happen or not. Interactions that occur during a simulation run need to be fully explainable. Analysis of interactions that actually happened during the simulation, shall supporting understanding and thus quality control of the simulation model. We argue that affordances – due to their grounding in cognitive science theory – form a natural basis for guiding a modeller.

In the following we first set the scene by discussing how interactions are handled when developing agent-based simulation models, this is followed by a discussion of related work on affordances in agent-based simulation. We then introduce our particular interpretation of the original affordance notion, define affordances for use during simulation runtime and affordance schemata to be specified during model definition. We illustrate how those notions can be applied in a small example. The contribution ends with a discussion of challenges not yet addressed and our future planned work.

2   Formulating Interactions

There are different perspectives that a modeller may consider when developing with an agent-based simulation model. Depending on the particular methodology applied, the set of perspectives is different. Yet, there is a core set containing first, a model of the agents, and second, a model of the simulated environment in which the agents are embedded. The third perspective aims at capturing the interactions between the those elements of the model. A fourth perspective deals with the simulation infrastructure containing information about scheduling updates, time model, et cetera.

The perspective of a single agent is well understood - using techniques and meta-models elaborated in diverse agent architectures such as rule-based systems or BDI agents. The environmental perspective received much attention over the last years, sometimes mixed with the infrastructural elements especially when handling the environments' update from the actions of the agents that should happen in parallel. Yet, the interaction perspective in a general sense appears to be neglected.

The Merriam-Webster dictionary shortly characterizes "interaction" as "mutual or reciprocal action or influence". In addition it distinguishes between two forms of interaction: (1) Interaction as communication and (2) interaction as mutual effect. The first form is often adopted in Agent-Oriented Software Development when specification of agent interaction is reduced to specification of protocols for exchanges of structured messages. The second form is more current. Considering interactions is basically a first step to connect the micro-level behaviour of agents to observations at the system or macro-level.

2.1   Interaction in AOSE

It is not surprising that formulating and dealing with interactions is at the heart of developing and analysing multi-agent systems. Basically from the beginning, researchers analysed conditions and circumstances for interaction of all kinds of agents – simple reactive to intelligent agents. J. Ferber [8] systematically analysed interaction situations.

Organization models were spotlighted as a mean to structure societies of agents. Hereby, the number of potential interaction partners is restricted to agents within the group or to agents having adopted particular roles. The basic idea behind those endeavours is to make system-level behaviour more predictable. Organizational notions hereby allow determining with whom to interact, while what actually happens during interaction is formulated in a protocols. Many meta-models were proposed for organizational models (such as [9] or [14]).

AUML [4] became the de facto standard for representing agent communication protocols as it provided more flexibility and a higher abstraction level than plain UML sequence diagrams at that time. Meanwhile, the corresponding UML2 diagrams offer similar features [5]. At a higher abstraction level interactions and relations between agents can also be represented using UML Use Case diagrams [7].

2.2   Interaction in Agent-Based Simulation

When specifying a particular behaviour with which an agent interacts or interferes with another entity, it essential to understand in which context the interaction will actually happen during runtime. This reads strange as one may assume that only what is given during modelling, is actually happening during simulation. However, this is just the case in models in which interaction situations are fully given - e.g. the above mentioned pre-defined organizations exactly provide such fully determined place. Yet this is not the case in general. In a simulation with a kind of stigmergic interaction, an agent modifies an environmental entity, another agent perceives the result and reacts to it. Interaction here consists of action and perception in a decoupled way. Who actually reacts to the modification is unclear, when the modeller determines the agent behaviour and potential interaction. Stigmergic interaction may be an extreme example, but similar situations happen in all cases in which the agent behaviour definition contains elements that are determined during runtime – when the agent interacts in a particular situations with the entities that are actually there. This makes it so difficult to handle interaction in agent-based simulation modelling.

Definition and simulation of interactions consequently forms a major source of errors eventually leading to extensive debugging and analysing. Thus, it is a highly critical element of a modelling methodology to get the interactions right as early as possible. Their proper specification, documentation and analysis is essential.

Over the years, several approaches have been published that suggest ways of explicitly handling interactions when creating a model. Like in the AOSE case, particular organization-level models have been proposed, such as using an institutional perspective as in the MAIA methodology [12]. Also integrating a model of social networks, such as discussed in [2] provides a structure who interacts with whom.

As in AOSE, UML Sequence diagrams that enable to formulate an interaction as a sequence of messages, can be used to specify interaction in agent-based simulation models [6]. For a higher abstraction level UML Use Case diagrams may be used. But, the flexible element may concern the interaction partner.

A basic framework for supporting the modelling of interaction is presented with the IODA approach [25]. In their methodology they propose to define inter-actions explicitly in a way separated for the actual agent models. This is also done when using explicit models of organizations, yet IODA is special as it directly couples interaction to agent action. The central element of IODA is a table that label how agents of one family interact with others. The label con-nected to a program or script as well as some form of condition. Hereby, Kubera et al. [24] also argue that everything can be an agent; so basically all actions can be phrased as interaction. IODA is particularly apt for reactive agents. It does not cover selection of interaction partners – all agents of a particular type within a specified distance may interact.

In this contribution, we want to explore if the concept of affordances helps to capture possible interactions in the modelling phase. Affordances also could be used to select interaction partners during a running simulation. Before we elaborate on our thoughts, we give an overview on what affordances actually are supposed to be as well as how they are currently used in agent-based simulation.

3   Notions and Usages of Affordances

3.1   The Concept of Affordance

The notion of affordance is at the core of ecological psychology, brought for-ward by Gibson [13]. Gibson defined affordances as action potentials provided by the environment: "The affordances of the environment are what it offers the animal, what it provides or furnishes, whether for good or ill". For example, a bench affords sitting to a human. The potential action of 'sitting' depends on properties of the bench, properties of the human, and on the current activ-ity the human is engaged in. Gibson put special emphasis on this reciprocity between animal and environment, insisting that affordances are neither objec-tive nor subjective. Thus, Stoffregen [32] defined affordances as "properties of the animal-environment system [...] that do not inhere in either the environment or the animal".

In the context of cognitive engineering Norman [27] determines the usability of environmental objects for a human carrying out a specific task by considering not only the affordances but the "perceivable affordances" of objects and the ease of perception for humans. Norman is dedicated the designing objects in such a way that their affordances become immediately perceivable by a person engaged in some task. Transferred to the context of modelling interactions a modeller needs to anticipate the affordances that will be needed within a specific action context and that the agent should be able to perceive.

Affordances in robotics reasoning [3] are used for enabling robots to handle unexpected situations. The moment an object is recognized as for example a mug, the robot can retrieve what actions it affords from an object-affordance database. Based on this the robot may adapt its plan to water plants using the mug instead of another container which is not available. This approach requires an extensive database which first needs to be assembled. Raubal and Moratz [30] developed a functional model for affordance based agent, aiming at enabling robots to perceive action-relevant properties of the environment. This research clarifies the notions but stays at an abstract level of formalization. Although they don't name it "affordances" but services, [11] present an idea that is related to both the robotics idea of affordances. They use an action planner to configure a learning scenario an educational game. Appropriate objects providing the services that are needed in the scenario are integrated into the scenario. This is not a simulation application per se, but has some relation.

In Geographic Information Science the affordance concept has been used extensively in order to model and understand human environmental perception and cognition. Jordan et al. [20] created an affordance-based model of place, discovering that the agent, the environment and the task of the agent need to be modelled in order to be able to determine affordances of places. Raubal [29] based his model of wayfinding in airports on an extended concept of affordances, including social and emotional aspects, thus enabling agents to interpret the meaning of environmental entities relevant to the task at hand. Jonietz and Timpf [17,18] interpret affordances as a higher-order property of an agent-environment system, which is determined by agent- and environment-related properties termed capabilities and dispositions at a lower level. As in the previous modelling approaches, affordances are interpreted as properties that may be modelled and not as something that emerges from the interaction between agent and environment. However, the affordance concept emphasizes the central role of action potentials and ties the afforded action and the respective environmental entities in a pragmatic sense [15].

3.2   Affordances in Agent-Based Simulation Modelling

During the last years the concept of affordances has become popular in agent-based simulation. Affordances were basically used to enable a modeller to formulate some element in the simulated environment that the agents could use for deciding about where to go next or what to do next.

There are a number of models that aim at reproducing how a human reasons about its environment for achieving more realism. These models are highly motivated by cognitive science. The basic assumption is that following hypotheses how humans really think, the model can achieve a higher degree of structural validity. Examples for those models are [28–30] or [17]. A formalisation focussing on affordances as an emergent property based on a detailed model of spatially explicit environment as well as actions and relations in that environment can be found in [1].

Other works interpret the notion of affordances more freely: Joo et al. [19] propose affordance-based Finite State Automata. They use affordance-effect pairs to structure the transitions between states of a simulated human. In an evacuation scenario, an agent follows a given route to the exit, but checks every step whether necessary affordances are fulfilled, using affordances to evaluate different local options.

Kapadia et al. ([21]) use "affordance fields" for representing the suitability of possible actions in a simulation of pedestrian steering and path-planning behaviour. An affordance is hereby a potential steering action. The affordance field is calculated from a combination of multiple fields filled with different kinds of perception data. The agent selects the action with the best value in the affordance field. A particular interesting approach is suggested by Ksontini et al. ([23]). They use affordances in traffic simulation denoting virtual lanes as an occupyable space. Agents reason about what behaviour is enabled by the environmental situation. The affordances offered by the environment are explicitly represented by those virtual objects that offer driving on them. [22] labelled environmental entities with "affordances" such as "provides medication" as counterparts of agent needs enabling the agents to flexibly search for interaction partners or destinations.

In these approaches, affordances are used as more as rules, for representing constraints or for identifying options. They serve as a tool for flexibly connecting an agent to interaction partners. There is no intention to advance the research in cognitive science.

3.3   Our Concept of Affordances

Affordances capture an emerging potential for interaction between an agent in a particular mind set intending to carry out a particular action and an environmental entity or ensemble of entities that the intended action involves. The entities need to have specific dispositions that can match up with the capabilities of the agent.

We use "affordance" as a kind of technical term capturing something that would be not be capturable otherwise. We do no claim to formalize the psychological, cognitive-science view on how humans actually reason about affordances. Our focus is on helping the modeller understand and think about interactions between agent and environment. Affordance shall make the potential for interaction between an agent and its environment explicit. So, we let the affordance stand per se for a potential interaction independent of how an agent selects its actions during simulation runtime. One can see it as a "shortcut" for representing what the agent perceives as relevant for selecting an entity as an interaction partner, without explicitly listing relevant features. In Gibson's original affordance idea there is no space for explicit selection between different affordances - the potential for action is directly linked to action in a Boolean fashion.

4 From Affordances to Interaction

Our aim is to deal with interactions in an explicit and flexible way using affordances. Therefore, we need to differentiate between an *affordance* emerging for a simulated agent while it "moves" through its environment, and between a representation that is defined by the modeller as some kind of declarative pattern from which the perception is generated. In principle, we assume that there is something like an explicit, declarative model that the modeller creates which is then interpreted or compiled for actually executing it during simulation runtime when the agent actually "lives". We call the run-time representations "affordance", and the modelling-time representation "affordance schema".

For running the actual Agent-Based Simulation, there must be some process to generate affordances. Theoretically, there is no emergence involved when a simulated agent perceives simulated affordances, as everything is defined by the modeller. Yet, from the point of view of the agent, an affordances may be in deed unexpected. For a modeller an affordance cannot "emerge" surprisingly.

4.1 Affordance and Affordance Schema

We define an affordance as a relation between a potential action and an environmental configuration. So, theoretically, it is neither a part of teh environment, nor a part of the agent, but connects both. The affordance becomes noticeable by an agent a at a particular time point t during simulation. $pAff_{a,t}$ are all affordances that the agent a can perceive at time t:

$$\langle a, act, x \rangle \in pAff_{a,t} \tag{1}$$

Such an affordance denotes the possibility of establishing a relation between an agent a and an entity x with respect to action act. x may serve as an interaction partner, if the given action is executed. With the perception of the affordance, the action becomes possible. Both, a and x are in a particular state at the time point t. We apply an extended view on "state" that goes beyond pure representation of kind-of metabolic values, but also contains activity, motivations and goals, beliefs, etc. We do not make assumptions on how this state looks like in a particular model. We also need to assume that the agent a has an explicit set of distinct, potential actions from which it selects one to perform in its environment.

As given in the previous section, an affordance links a potential action to an environmental *constellation*. Per se, such a constellation is not just a single entity in a particular state, but contains context. For example, a bench affords sitting-down just if the area to sit on is sufficiently stable (state of the bench entity). Selection is influenced by the context of the entity - whether it is below a tree casting shadow on it during sunny, too hot times or under a roof that protects it from rain on a rainy day. We assume that all information that qualifies an entity for offering a particular potential action is represented in its state; information that makes it more or less qualified in comparison to other entities affording the same potential action is determined by its context. Preferences or degrees of qualification are not considered in the classical affordance concept. In [16] this classical view is extended by elaborating gradation of affordances. Nevertheless, the selection of affordances depends on the particular way the agent reasons about affordances which should be independent from the actual affordance. An agent might follow the first affordance relation that it encounters or a random one or might evaluate different options for determining which one to prefer.

This formalization is different from [31] who see an affordance as an acquired relation between a combination of an environmental object with behaviour and an effect of this behaviour. The idea of acquiring knowledge about an affordance illustrates their robotics perspective.

Thus, we image an affordance as an explicit object (as a kind of data structure) during simulation runtime about which the agent reasons with respect to carrying it out or not. Thus, there is a need for an higher-level data structure or schema that enables to create such runtime affordance objects. Such a schema must be more than a class in the object-oriented sense from which affordance instances can be created. For being useful in modelling per se, the schema needs to contain more contextually relevant information and conditions under which the affordance actually "emerges". We define such an *affordance schema* in the following way:

$$\langle AType, act, EType, hContext, sContext \rangle \tag{2}$$

Such an *affordance schema* can be seen as a "pattern" that can be used to generate or determine affordances present in the agents environment[1]. An affordance schema is specified during modelling, but does not necessarily exist as an explicit data structure during simulation runtime. When a modeller specifies such affordance schemata, she explicitly writes down under which circumstances an interaction might happen between an agent of type *AType* performing action *act* with an object of type *EType*. The action in the affordance can be a more specific and parametrized version of the action given in the affordance schema. The actual action representation may depend on the applied agent architecture. The fourth and fifth elements *hContext* and *sContext* capture the circumstances under which an affordance $\langle a, act, x \rangle$ can be really created for a being a kind Of *AType* and x being a kind Of *EType*, offering the action. The difference between *hContext* and *sContext* is that the former contains hard conditions that enable the affordance - focussing on object and agent properties directly; the latter contains weaker conditions or even just criteria that make a particular constellation more favourable than others.

For example, in a park simulation (such as [33]), during a hot day, an agent a enters a park that is equipped with a currently broken bench $b1$ in the shadow, a clean bench $b2$ in the sun and a nice looking stone $st1$ under shady trees. The agent a entering the park is tired and searches for a place

---

[1] Our idea of an affordance schema is on a higher abstraction level than what W. Kuhn called "Image Schema" in [26]. He describes an environmental constellation using spatial categories and connects them to a process that they afford.

to sit down (action *sitDown*). Thus, *a* scans the environment using the affordance schema for *sitDown* and generates the following affordance objects: $\langle a, sitDown, b2 \rangle$ and $\langle a, sitDown, st1 \rangle$. It does not generate an affordance for *b1* because the bench is not apt for sitting on it due to the broken surface. Depending on some form of ontology capturing environmental entities such as benches or stones as entities with flat surfaces, the modeller has defined the following affordance schema as a pattern to describe potential interactions: $\langle Visitor, sitDown, ObjectWithFlatSurface, hConditions, sConditions \rangle$. Agent *a* is of type *Visitor* and requires for the action *sitDown* an entity of type *EntityWithFlatSurface* which is the superclass of benches, stones, etc. However, not every one of those entities affords the action for a *Visitor* agent. This is represented in *hConditions* which define under which circumstances the environmental entity *e* affords the action *sitDown*: {*stable(surface(e))*, *height(surface(e)) < 110 cm)*, ...}. The set of soft conditions may contain for example {*inShadow(e)*}. The conditions may also refer to other objects present in the vicinity of *e* affecting whether and how well the entity *e* actually can afford the given action.

It is important to stress that our definition of affordance and affordance schema does not contain a description of the effect of the action it refers to. The description as in the example just contained a label *sitDown*. What this means. In simulation, we are dealing with an environment for the agents' behaviour that is part of the model - that means fully defined by the modeller. With the environment the effect of actions is usually fully defined, even if a modeller follows the conceptually cleaner distinction between agent action and environmental reaction as described by [10].

Another essential aspect distinguishing the specification of affordances in agent-based simulation versus robotics (and also agent-oriented software development) is that actions in simulation may be defined at arbitrary levels of abstraction – adapted to the abstraction level of the environment. For example the action *sitDown* may not have a lower level correspondence when the agent executes it. There might be no going towards, arching joints, lowering backs, or whatever low-level commands are necessary to execute such an action. Abstraction levels might differ a lot between models describing the same phenomenon. This is also the reason why we would not expect to be able to create a set of "standard" affordances that can be used across many simulation applications.

Enabling a distinction between different environmental objects so that the agent may prefer one to another is NOT part of the original affordance idea. An affordance connects an environmental object to a potential action of the agent. How the agent reasons is not part of the affordance. Thus, conditions are intentionally only present on the affordance schema, i.e., the modelling level. The runtime affordance depends on the simulated agents' point of view within the simulated environment. But somehow during simulation, there must be a process generating the affordances that the agent then can select, etc. Thus, we need to discuss processes of how the affordance schemata generate affordances and determine the agents' actual behaviour and interaction.

4.2 Processes Around Affordances

In theory, an affordance *emerges* as a potential action for an actor with a particular motivation (goal, desire...). In a simulation, it needs to be determined either by the simulated agent itself or by some higher level process which may not be manifested as an actor in the simulation. In Fig. 1 an abstract view is visualized of how different elements of such a process can be connected.

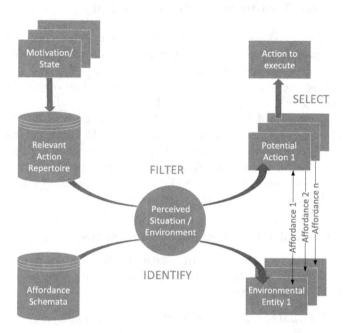

Fig. 1. Overview over processes related to affordance generation and usage.

Following Fig. 1 we need to elaborate partial processes relevant for creating interactive agent behaviour from explicitly defined affordances:

- Mechanism that connects agent goals to actions that are apt to achieve the agents' goal or satisfy its motivation. This process element is responsible for a pre-selection of actions from an action repertoire capturing what the agent is able to do in general. The selected actions need to be connected to the agents' motivational concepts and perceptions/beliefs in a classical way: the agent shall not select actions that it believes not to work in a particular environment, etc.
- Potential actions are filtered based on the environment checking whether the prerequisites for the actions are fulfilled or not. This is done by doing some kind of "pattern matching" of affordance schemata to perceived environment. This connects potential actions to environmental entities by generating (identifying) affordance relations.

– Having established a set of realizable actions with potential interaction partners, the agent can use the affordances connecting actions and entities to evaluate which of the combinations are the preferable ones. Such a preference relation between affordances (based on an evaluation of the context information given on the modelling level) is then used to select the action that is executed.

5   Illustrative Example: A Supermarket

Consider the following situation and process: The agent A has collected a number of items in a supermarket and moves towards the cash points to pay. The agent has sufficient money (in cash or on/with card). There exist two manned cash points as well as one self-payment counter machine: $\{CashierRight, Cashier-Middle, AutoCashier\}$. Of the two manned cash points, only $CashierRight$ is actually busy. $CashierMiddle$ misses the cashier agent. Each of the working cash points has a queue of agents waiting for their turn: In front of $CashierRight$ there are 4 persons queuing up, in front of $AutoCashier$ only another person is waiting for the current person to finish. So it is highly probably that A will be served earlier when queueing up at the electronic cash point. A strongly prefers to interact with humans, yet is under time pressure.

5.1   Description of Interaction with Affordances

When A approaches the cash point area with the intention of doing the action $Pay2Leave$, A perceives the three cashiers and immediate sees that only two are available for the intended action.

$$\langle A, Pay2Leave, CashierRight, Cond_{CashierRight,now} \rangle$$
$$\langle A, Pay2Leave, AutoCashier, Cond_{AutoCashier,now} \rangle$$

with $Cond_{CashierRight,now} = \{queue(CashierRight, 4), female(CashierRight), young(CashierRight), friendly(CashierRight)\}$ describing the configuration of the particular cashpoint at time now. The configuration of $AutoCashier$ is $Cond_{AutoCashier,now} = (queue(AutoCashier, 1))$.

There is no affordance for $CashierMiddle$ as it is actually not working due to the missing cashier. Both affordances have particular properties that describe the current configuration the agent evaluates for making a decision for one of the interaction partner $CashierRight$ or $AutoCashier$. The agent needs to evaluate whether it prefers to wait for the interaction with a nice human cashier or wants to go for the faster automated way. The $Pay2Leave$ action may have a particular implementation for each of the interaction partners specified as a communication protocol as given below.

While this describes a simulation run-time situation, the modeller defines affordances schemata to specify the interaction. In this example case, the relevant affordance schema may look like that:

$$\langle SHOPPER, Pay2Leave, CASHpOINT, Prereq, PrefCriteria \rangle$$

The affordance schema contains the following elements: first, a combination of agent type *SHOPPER* and a particular action/activity that the agent wants to do: *Pay2Leave*. Hereby, *Pay2Leave* $\in$ *ActionRepertoire(SHOPPER)*, that means the action must be part of the default – as defined on the class level – action repertoire of any shopper agent. *A* as an instance of a *SHOPPER* has this action in its action repertoire. *CASHPOINT* is an abstract class from which the classes of *MANNED−CASHPOINT* AND *AUTOMATED−CASHPOINT* are inheriting. Both types can afford the *Pay2Leave* action of the shopping agent. Yet, additional conditions must be fulfilled. These conditions and criteria depend on the concrete type of *CASHPOINT* (see Table 1).

Table 1. Prerequisites and Conditions in Cashier Scenario

AType	Conditions	Preference Criteria
MANNED − CASHPOINT	manned(C)	Queue, Friendliness...
AUTOMATED − CASHPOINT	functioning(C), available(A,card)	Queue

For achieving a fully functional model clearly a lot of elements are missing. We just focus on a small number of potential interactions. Additional interactions could be between *A* and diverse products that *A* wants to buy. Hereby each product affords to be taken and put into the cart. Before we continue discussing our approach, we have a look how the corresponding formulations would look like when using ways of specification as introduced in Sect. 2.

5.2   Description of Interaction with IODA or MAIA

In the following we intend to give a general impression of two rather extreme alternatives to formulate interactions: IODA [25] aiming more at simulation of emergent phenomena and MAIA [12] following an explicit organization-oriented approach. We do neither give full models, nor the does our description describe exactly the same part of the model. Thus, a lot of context is missing which would be necessary to precisely apply these two methodologies for developing agent-based simulations.

The central element of designing this scenario with IODA [25] is to define the interaction matrix as shown in Table 2.

The table specifies that elements of one agent "family" interact with an element of another. For example, a Customer agent may initiate an interaction with a Cashier agent, if the distance between them is lower than 3 units. "Pay&Pack" is hereby a label for a sequence of actions describing the actions of the involved entities during the interaction. The model specifies what happens during an interaction, under which circumstances the interaction is triggered and what type of agents are involved. What is actually done is represented as a sequence

Table 2. Raw Interaction Matrix in the supermarket scenario following IODA [25]. In following the steps of the overall methodology, interactions would be detailed and selection process is specified, etc.

Source	Target			
	Shelf	Customer	Employee	Cashier
Customer	TakeGoods (d = 0)	WaitBehind (d = 2)	AskForHelp (d = 2)	Pay& Pack (d = 3)
Employee	ReFill (d = 0)			
Cashier		RequestPayment (d = 3)		

of actions executed by the contributing agents. How the actual interaction is selected is determined by the actual agent architecture. Initially, Kubera et al. assumed reactive agents, that means agents that more or less directly connect perception to action without reasoning about explicit representations of agent goals.

As introduced above, MAIA [12] forms a framework for agent-based simulation based on the formalization of a particular organizational model. It is especially apt for social models.

In a simulation reproducing how humans behave in a supermarket, one may assume two types of agents: Customers as individual actors and the supermarket as a composite actor, bringing together all its employees that temporally take over a particular role, such as ReFiller or Cashier. Agents may have particular attributes, such as contents of the shopping chart or entries on the shopping list. The roles have an associated objective, such as acquire and pay all items on the shopping list. There may be dependencies between roles based on dependencies between objectives - captured also in institutional settings. When adopting a role, an agent also gets capabilities that basically correspond to possible activities or actions that the agent with that role is able/permitted to perform. For the specification of interactions the set of rules and conventions that govern agent behaviour to be specified by institutional statements is particularly interesting. There are different types of those statements for describing which behaviour can be expected by an agent and what happens if the agent does not fulfil the expectations: not following a rule results in sanctions, a norm is behaviour without sanctions if not followed. The weakest notion of an institutional statement is shared strategy. In Table 3 we give a few examples of institutional statements of a customer actor in the supermarket scenario.

These elements set up the constitutional structure of the model. In addition, the modeller needs to specify the physical context (environmental model) and the operational environment, which describes how an agent influences the overall system state. A simulation has an *action arena* which contains so called *action situations*. The latter basically describes some kind of plan structure organizing atomic actions in an institutional context for an agent exhibiting a

Table 3. Institutional Statements defining the expectations on how customer agents behave

Type of Statement	Statement
Rule	A customer always has to pay the goods before leaving
Norm	A customer has to wait in line behind earlier customers
Shared Strategy	Customers start to pack directly after the cashier accounted for a good
...	...

particular role. Interactions between different agents takes place within an entity actions. So, actually what happens during interaction is hidden quite deeply in the overall model specification.

In Sect. 2 we mentioned that UML can be used to formulate interactions. How this could look like at a rather high abstraction level is shown in Fig. 2.

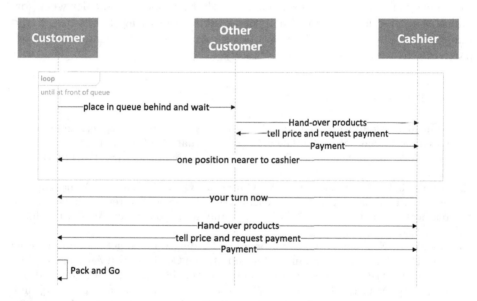

Fig. 2. Protocol-like definition of interactions between waiting customers and the cashier handling one after the other.

One can see that the different frameworks and approaches actually focus on different problems. Our affordance/affordance schema concepts actually concentrate on the selection of the interaction partner in a more flexible, yet less predictable way as in more organization-oriented approaches. One may interpret it as more specific and apt for agents that actually reason about their next action than in the IODA methododology.

6   Discussion and Conclusion

In this contribution we clarified the notion of affordances and introduced affordance schemata showing that such a distinction is necessary when distinguishing between what happens during simulation runtime and what a modeller explicitly formulates. We put those concepts into an interaction modelling context. The questions remain whether these concepts can be really useful, what to do such that they become useful and how to evaluate their usefulness? The current stage of our research is quite preliminary, as we first wanted to clearly agree on what we actually model when specifying affordances. The current contribution thus cannot be more than a discussion paper. For creating a methodology we would need to make assumptions on meta-models formulating a context such that we can formalize every detail necessary to fully support the complete modelling and simulation process. Based on such a meta-model we could then create tools that directly support modelling - and as we explicitly approach interactions hopefully support model analysis in an improved way. However, we are not sure whether yet another methodology provides a good idea. What we actually want to achieve is to propose a suitable language that supports a modeller when formulating interactions. It should help the modeller to stay aware of when, if and under which circumstances interactions happen and which agents with which particular features participate in the interaction.

References

1. Afoutni, Z., Courdier, R., Guerrin, F.: A multiagent system to model human action based on the concept of affordance. In: 4th International Conference on Simulation and Modeling Methodologies, Technologies And Applications (SIMULTECH), 8 p. (2014)
2. Amblard, F., Bouadjio-Boulic, A., Gutiérrez, C.S., Gaudou, B.: Which models are used in social simulation to generate social networks? a review of 17 years of publications in JASSS. In: 2015 Winter Simulation Conference (WSC), pp. 4021–4032 (2015)
3. Awaad, I., Kretschmar, G., Hertzberg, J.: Finding ways to get the job done: an affordance-based approach. In: 24th International Conference on Automated Planning and Scheduling (ICAPS 2014), Portsmouth, USA, pp. 499–503, June 2014
4. Bauer, B., Müller, J.P., Odell, J.: Agent UML: a formalism for specifying multiagent software systems. In: Ciancarini, P., Wooldridge, M.J. (eds.) AOSE 2000. LNCS, vol. 1957, pp. 91–103. Springer, Heidelberg (2001). https://doi.org/10.1007/3-540-44564-1_6
5. Bauer, B., Odell, J.: Uml 2.0 and agents: how to build agent-based systems with the new UML standard. Eng. Appl. Artif. Intell. **18**(2), 141–157 (2005)
6. Bersini, H.: UML for ABM. J. Artif. Soc. Soc. Simul. **15**(1), 9 (2012)
7. Chella, A., Cossentino, M., Faso, U.L.: Applying UML use case diagrams to agents. In: Proceedings of AI*IA 2000 Conference, pp. 13–15 (2000)
8. Ferber, J.: Multi-Agent System: An Introduction to Distributed Artificial Intelligence. Addison Wesley Longman, Harlow (1999)

9. Ferber, J., Gutknecht, O.: A meta-model for the analysis and design of organizations in multi-agent systems. In: Proceedings of the 3rd International Conference on Multi Agent Systems (ICMAS 1998), pp. 128–135. IEEE, Washington, DC (1998)

10. Ferber, J., Müller, J.-P.: Influences and reactions: a model of situated multiagent systems. In: Proceedings of the ICMAS 1996. AAAI Press (1996)

11. Ferdinandus, G.R., Peeters, M., van den Bosch, K., Meyer, J.-J.C.: Automated scenario generation - coupling planning techniques with smart objects. In: Proceedings of the 5th International Conference on Computer Supported Education, Aachen, Germany, pp. 76–81 (2013)

12. Ghorbani, A., Bots, P., Dignum, V., Dijkema, G.: MAIA: a framework for developing agent-based social simulations. J. Artif. Soc. Soc. Simul. **16**(2), 9 (2013)

13. Gibson, J.J.: The Ecological Approach to Visual Perception. Houghton Mifflin, Boston (1979)

14. Hübner, J., Sichman, J.F., Boissier, B.: Developing organised multi-agent systems using the MOISE + model: programming issues at the system and agent levels. Int. J. Agent-Oriented Softw. Eng. **1**(3/4), 370–395 (2007)

15. Janowicz, K., Scheider, S., Pehle, T., Hart, G.: Geospatial semantics and linked spatiotemporal data -past, present, and future. Semantic Web **3**(4), 321–332 (2012)

16. Jonietz, D.: From Space to Place - A Computational Model of Functional Place. Thesis, University of Augsburg, Geographical Information Science (2016)

17. Jonietz, D., Timpf, S.: An affordance-based simulation framework for assessing spatial suitability. In: Tenbrink, T., Stell, J., Galton, A., Wood, Z. (eds.) COSIT 2013. LNCS, vol. 8116, pp. 169–184. Springer, Cham (2013). https://doi.org/10.1007/978-3-319-01790-7_10

18. Jonietz, D., Timpf, S.: On the relevance of Gibson's affordance concept for geographical information science. Cogn. Process. Int. Quaterly Cogn. Sci. **16**(suppl. 1), 265–269 (2015)

19. Joo, J., Kim, N., Wysk, R.A., Rothrock, L., Son, Y.-J., Oh, Y.-G., Lee, S.: Agent-based simulation of affordance-based human behaviors in emergency evacuation. Simul. Model. Pract. Theory **13**, 99–115 (2013)

20. Jordan, T., Raubal, M., Gartrell, B., Egenhöfer, M.J.: An affordance-based model of place in GIS. In: Poiker, T., Chrisman, N. (eds.) Proceedings of 8th International Symposium on Spatial Data Handling, Vancouver, CA, pp. 98–109 (1998)

21. Kapadia, M., Singh, S., Hewlett, W., Faloutsos, P.: Egocentric affordance fields in pedestrian steering. In: Proceedings of the 2009 symposium on Interactive 3D graphics and games (I3D 2009), pp. 215–223. ACM, New York (2009)

22. Klügl, F.: Using the affordance concept for model design in agent-based simulation. Ann. Math. Artif. Intell. **78**, 21–44 (2016)

23. Ksontini, F., Mandiau, R., Guessoum, Z., Espié, S.: Affordance-based agent model for traffic simulation. J. Auton. Agents Multiagent Syst. **29**(5), 821–849 (2015)

24. Kubera, Y., Mathieu, P., Picault, S.: Everything can be agent! In: van der Hoek, W., Kaminka, G., Lespérance, Y., Luck, M., Sen, S. (eds.) Proceedings of 9th International Joint Conference on Autonomous Agents and Multi-Agent Systems, AAMAS 2010, Toronto, Canada, p. 1547f (2010)

25. Kubera, Y., Mathieu, P., Picault, S.: IODA: an interaction-oriented approach for multi-agent based simulations. Auton. Agent. Multi-Agent Syst. **23**(3), 303–343 (2011)

26. Kuhn, W.: An image-schematic account of spatial categories. In: Winter, S., Duckham, M., Kulik, L., Kuipers, B. (eds.) COSIT 2007. LNCS, vol. 4736, pp. 152–168. Springer, Heidelberg (2007). https://doi.org/10.1007/978-3-540-74788-8_10

27. Norman, D.A.: The Invisible Computer. MIT Press, Cambridge (1999)
28. Paris, S., Donikian, S.: Activity-driven populace: a cognitive approach to crowd simulation. IEEE Comput. Graphics Appl. **29**(4), 34–43 (2009)
29. Raubal, M.: Ontology and epistemology for agent-based wayfinding simulation. Int. J. Geogr. Inf. Sci. **15**, 653–665 (2001)
30. Raubal, M., Moratz, R.: A functional model for affordance-based agents. In: Rome, E., Hertzberg, J., Dorffner, G. (eds.) Towards Affordance-Based Robot Control. LNCS (LNAI), vol. 4760, pp. 91–105. Springer, Heidelberg (2008). https://doi.org/10.1007/978-3-540-77915-5_7
31. Şahin, E., Çakmak, M., Doğar, M.R., Uğur, E., Üçoluk, G.: To afford or not to afford: a new formalism of affordances towards affordance-based robot control. Adapt. Behav. **15**(4), 447–472 (2007)
32. Stoffregen, T.: Affordances as properties of the animal environment system. Ecol. Psychol. **15**(2), 115–134 (2003)
33. Timpf, S.: Simulating place selection in urban public parks. In: International Workshop on Social Space and Geographic Space, SGS 2007, Melbourne (2007)

Towards a Fast Detection of Opponents in Repeated Stochastic Games

Pablo Hernandez-Leal[(✉)] and Michael Kaisers

Intelligent and Autonomous Systems Group, Centrum Wiskunde & Informatica,
Amsterdam, The Netherlands
{Pablo.Hernandez,Michael.Kaisers}@cwi.nl

Abstract. Multi-agent algorithms aim to find the best response in
strategic interactions. While many state-of-the-art algorithms assume
repeated interaction with a fixed set of opponents (or even *self-play*), a
learner in the real world is more likely to encounter the same strategic
situation with changing counter-parties. This article presents a formal
model of such *sequential interactions*, and a corresponding algorithm
that combines the two established frameworks *Pepper* and *Bayesian policy reuse*. For each interaction, the algorithm faces a repeated stochastic
game with an unknown (small) number of repetitions against a random
opponent from a population, without observing the opponent's identity.
Our algorithm is composed of two main steps: first it draws inspiration from multiagent algorithms to obtain acting policies in stochastic
games, and second it computes a belief over the possible opponents that is
updated as the interaction occurs. This allows the agent to quickly select
the appropriate policy against the opponent. Our results show fast detection of the opponent from its behavior, obtaining higher average rewards
than the state-of-the-art baseline Pepper in repeated stochastic games.

Keywords: Stochastic games · Reinforcement learning
Multi-agent learning · Policy reuse

1 Introduction

Learning to act in multiagent systems has received attention mainly from game
theory and reinforcement learning (RL). The former has proposed algorithms
that converge under different scenarios [14] and the latter has focused on acting
optimally in stochastic scenarios [12], typically with limited a priori information
about the interaction. Interactions among several agents are usually modelled
as a normal-form or stochastic game, and a wide variety of learning algorithms
targets this setting [7,9,13]. However, results are typically based on the assumption of self-play (i.e., all participants use the same algorithm) and a long period
of repeated interactions. In contrast, we focus on *sequential interactions*, i.e.,
the agent is paired with stochastically drawn opponents, with whom the agent

© Springer International Publishing AG 2017
G. Sukthankar and J. A. Rodriguez-Aguilar (Eds.): AAMAS 2017 Best Papers,
LNAI 10642, pp. 239–257, 2017.
https://doi.org/10.1007/978-3-319-71682-4_15

interacts in short periods while observing joint actions, but without observing the opponent's identity.

Recent works have proposed algorithms for learning in repeated stochastic games [17,20], however, it is an open problem how to act quickly and optimally when facing different opponents [20]. Recent work on Stochastic Bayesian Games has compared several ways to incorporate observations into beliefs over opponent types when those types are re-drawn after every state transition [1]. In contrast, we assume that opponents are redrawn for interactions over several repeated stochastic games. The learning algorithm needs to optimally reuse previously learned information from distinct but similar interactions – a challenge that has been largely studied by transfer learning (TL). TL has been applied mostly in single-agent domains where information from learned *source* tasks can be reused in a new target task [32]. Determining how two tasks are similar, what information to be transferred and when it should be transferred are open problems in TL. Related to TL there are different areas that also share a connection with the problem of how to efficiently reuse previously learned information, e.g., policy reuse [21], to avoid long learning times; ad-hoc coordination [4], to collaborate with unknown agents in multi-agent teams; and learning in non-stationary environments [18] to adapt to changing conditions.

We contribute to the state of the art in two ways: First, by providing a more natural formal model of *sequential interactions* and second with an algorithm for quick detection of opponents in that setting. Our proposed algorithm Bayes-Pepper builds on top of two previously successful frameworks:

- Pepper [15], a learning algorithm for repeated stochastic games, is used to obtain policies on how to act against the possible opponents. Pepper uses the paradigm of optimism in face uncertainty together with a joint action learner to learn a policy in stochastic games.
- Bayesian Policy Reuse (BPR) [30] is used as a fast detection process to identify the opponent and select the appropriate acting policy. While previously BPR has been evaluated in single-agent tasks [30] and repeated normal-form games [25], this is the first time it is extended to stochastic games.

Our setting assumes a population of opponents that can be divided into different groups. First, Bayes-Pepper needs to compute policies and models of the opponents, which we assume happens at an offline phase. Second, in an online phase a random process pairs the learning agent against the opponent for a stochastic game.[1] The learning agent has no control over this process and does not observe the opponent identity. When the game finishes the learning agent receives an observation (reward) and updates the belief accordingly. Subsequently, the agent is paired with a new opponent. A formal definition of the game is given in Sect. 3.3.

This paper is presented as follows: Sect. 2 presents the related work in transfer learning, policy reuse and learning in non-stationary environments. Section 3

---

[1] We present experiments with one opponent, however, our approach could be generalized to more opponents taking the Cartesian product of all opponents as a single one.

describe the formal models of reinforcement learning and game theory. Section 4 presents the proposed Bayes-Pepper algorithm. Section 5 presents experimental results in repeated stochastic games. Section 6 provides a discussion considering the results and provides directions for future research. Finally, Sect. 7 summarizes the conclusions of this work.

2   Related Work

This article tackles the problem of finding a best response when being repeatedly paired with unknown opponents from an unknown population with known types. We propose an algorithm that aims to identify opponents while best-responding in face of residual uncertainty. Our setting and approach shares similarities to transfer learning, policy reuse and ad-hoc coordination which we review in this section.

Transfer learning was first used in machine learning to transfer between learning tasks in a supervised learning scenario. Recently, TL has gained attention in the RL community in particular in single-agent scenarios. An ideal fully autonomous RL transfer agent needs to complete three phases [32]:

- Given a target task, select an appropriate set of source tasks from which to transfer.
- Learn how the source task(s) and target task are related.
- Transfer knowledge from the source task(s) to the target task.

There are different evaluation metrics for TL algorithms (e.g., jumpstart, asymptotic performance, total rewards, among others) and even though these three steps are usually connected, TL has focused on them independently. For example, for the transfer step different ideas have been evaluated, e.g., models, instances and policies.

One approach that transfers instances from similar tasks was proposed by Lazaric et al. [29]. They proposed a measure to identify which source tasks are more likely to have samples similar to those in the target task, namely *task compliance*. Moreover, to select which instances to transfer from a task they propose the *relevance* measure. However, the approach was proposed for single-agent domains with continuous state and action spaces, and does not naturally transfer to our setting. Closer to our approach, Boutsioukis et al. [8] proposed TL by extending the Q-learning reuse algorithm [31] to multiagent scenarios. In contrast to our ambition of transfer from interactions against different opponents, their goal is to transfer information learned from a task with n agents to a different task with $m \neq n$ agents. In particular, they propose an inter-task transfer approach (i.e., the state and action spaces are not the same in the target and source tasks) and the evaluation was performed on the predator-prey domain transferring information tasks learned with different number of predators (agents). Policy reuse techniques are another area with a similar spirit since these approaches assume to start with a set of policies to use, and the problem is to select among them when facing a new task. Fernandez and Veloso [21] use policy reuse as a probabilistic bias when learning new, similar tasks in single agent domains. Bayesian Policy Reuse (BPR) [30] assumes prior knowledge of the performance of different policies over different tasks. BPR computes a belief over the possible tasks which is updated at every interaction and is used to select the policy that maximises the expected reward given the current belief. Bayesian reasoning has also been used in RL to learn when there is a group of related tasks with similar structure. Lazaric and Ghavamzadeh [28] proposed an algorithm assuming tasks have common state and action spaces and their value functions are sampled from a common prior. In contrast, our approach extends the BPR algorithm (see Sect. 3.2) to identify opponents rather than tasks, and combines it with a multiagent learning algorithm.

Ad-hoc coordination is another related problem where an agent needs to coordinate with an unknown agent but when a set of previous models is known. In this setting, Barrett et al. [4] proposed the PLASTIC algorithm that learns how to cooperate with other teammates based on a collection of policies to select from, which is similar to our approach. The algorithm selects at each interaction the most likely teammate type and acts following the corresponding policy. However, this approach does not consider changing agents over the course of interactions as we do.

Learning in non-stationary environments is another related area since these approaches explicitly model changes in the environment. Their goal is to learn an optimal policy and at the same time detect when the environment has changed to a different one, updating the acting policy accordingly. One algorithm designed for single agent tasks with a changing environment is the Reinforcement Learning with Context detection (RL-CD) [18]. RL-CD learns a model of the specific task and assumes an environment that changes infrequently among different *contexts*. To detect a new context RL-CD computes a quality measure of the learned models. Hernandez-Leal et al. [23,26] addressed a similar problem in two-player repeated normal-form games. In this case, the opponent has different stationary strategies to select from and the learning agent needs to learn online how to act optimally against each strategy while detecting when the opponent changes to a different one. Since the opponent might reuse one previous strategy at a later stage of the interaction the learning agent should keep previous models and policies in order to quickly detect them [25]. While this might be the closest state of the art, these approaches do not consider repeated *stochastic* games.

Experimental evidence suggests that people learn heuristics which later are transferred across different games [5]. Based on these results there is another category of algorithms that aims to learn in one game and to generalize how to act in a different game (known as general game playing). Banerjee and Stone [3] proposed a transfer approach in two-player, alternate move, complete information games facing stationary opponents. The idea is to learn general features that can be reused across games, for example, they learned from played games on Tic-tac-toe and transferred information to a more complex game (Othello).

In contrast with previous approaches we focus on two-player repeated stochastic games. An agent faces an opponent whose identity is unknown to the agent and where every few interactions a random process selects a new opponent from the population (the agent does not know when these changes happen).

3 Preliminaries

In this section, first we review the formal model of reinforcement learning. Then, we describe the Bayesian policy reuse framework [30]. Later, we present our sequential interactions model in the context of stochastic games. Finally, we describe Pepper [15] which inspires our proposed Bayes-Pepper algorithm.

3.1 Reinforcement Learning

Reinforcement learning (RL) is one important area of machine learning that formalizes the interaction of an agent with its environment, e.g., using a Markov decision process (MDP). An MDP is defined by the tuple $\langle S, A, R, T \rangle$ represents the world divided up into a finite set of possible states. A represents a finite set of available actions. The transition function $T : S \times A \rightarrow \Delta(S)$ maps each state-action pair to a probability distribution over the possible successor states, where $\Delta(S)$ denotes the set of all probability distributions over S. Thus, for each $s, s' \in S$ and $a \in A$, the function T determines the probability of a transition from state s to state s' after executing action a. The reward function $R : S \times A \times S \rightarrow \mathbb{R}$ defines the immediate and possibly stochastic reward that an agent would receive for being in state s, executing action a and transitioning to state s'.

MDPs are adequate models to obtain optimal decisions in *single* agent environments. Solving an MDP will yield a policy $\pi : S \rightarrow A$, which is a mapping from states to actions. An optimal policy π^* is the one that maximises the expected discounted reward. There are different techniques for solving MDPs assuming a complete description of all its elements. One of the most common techniques is the value iteration algorithm [6] which is based on the Bellman equation:

$$V^\pi(s) = \sum_{a \in A} \pi(s, a) \sum_{s' \in S} T(s, a, s')[R(s, a, s') + \gamma V^\pi(s')],$$

with $\gamma \in [0, 1)$. This equation expresses the *value* of a state which can be used to obtain the optimal policy $\pi^* = \arg\max_\pi V^\pi(s)$, i.e., the one that maximises that value function, and the optimal value function $V^*(s)$.

$$V^*(s) = \max_\pi V^\pi(s) \quad \forall s \in S.$$

Value iteration requires complete and accurate representation of states, actions, rewards and transitions. However, this may be difficult to obtain in many domains. For this reason, RL algorithms learn from experience without having a complete description of the MDP a priori. In contrast, an RL agent interacts with the environment in discrete time-steps. At each time, the agent chooses an action from the set of actions available, which is subsequently executed in the environment. The environment moves to a new state and the reward associated with the transition is emitted. The goal of a RL agent is to maximise the expected reward. In this type of learning the learner is not told which actions to take, but instead must discover which actions yield the best reward by trial and error.

Q-learning [33] is one well known algorithm for RL. It has been devised for stationary, single-agent, fully observable environments with discrete actions. In its general form, a Q-learning agent can be in any state $s \in S$ and can choose an action $a \in A$. It keeps a data structure $\hat{Q}(s, a)$ that represents the estimate of its expected payoff starting in state s, taking action a. Each entry $\hat{Q}(s, a)$ is an estimate of the corresponding optimal Q^* function that maps state-action pairs to the discounted sum of future rewards when starting with the given action and following the optimal policy thereafter. Each time the agent makes a transition from a state s to a state s' via action a receiving payoff r, the Q table is updated as follows:

$$\hat{Q}(s, a) = \hat{Q}(s, a) + \alpha[(r + \gamma \max_b \hat{Q}(s', b)) - \hat{Q}(s, a)]$$

with the learning rate α and the discount factor $\gamma \in [0, 1]$ being parameters of the algorithm, with α typically decreasing over the course of many iterations. Q-learning is proved to converge towards Q^* if each state-action pair is visited infinitely often under specific parameters [33].

3.2 Bayesian Policy Reuse

Bayesian policy reuse is a framework to quickly determine the best policy to select when faced with an unknown task. Formally, a task is defined as an MDP. A *policy* is a function $\pi(s)$ that specifies an appropriate action a for each state s. The return, or utility, generated from running the policy π in an interaction of a task instance is the accumulated reward, $U^\pi = \sum_{i=0}^k r_i$, with k being the length of the interaction and r_i being the reward received at step i.

Let an agent be equipped with a policy library Π for tasks in a domain. The agent is presented with an unknown task which must be solved within a limited and small number of trials. At the beginning of each trial episode, the agent can select one policy from $\pi \in \Pi$ to execute. The goal of the agent is thus to select policies to minimize the total regret incurred in the limited task duration with respect to the performance of the best alternative from Π in hindsight.

BPR assumes knowledge of *performance* models describing how policies behave on different tasks. A performance model, $P(U|\tau, \pi)$, is a probability distribution over the utility using π on a task τ. A signal σ is any information that is correlated with the performance of a policy and that is provided to the agent in an online execution of the policy on a task (e.g., immediate rewards). For a set of tasks $\mathcal{T}$ and a new instance τ^* the *belief* β is a probability distribution over $\mathcal{T}$ that measures to what extent τ^* matches the known tasks in their observation signals σ. The belief is initialized with a prior probability. After each execution on the unknown task the environment provides an observation signal to the agent, which is used to update beliefs according to Bayes' rule:

$$\beta^k(\tau) = \frac{\Pr(\sigma^k|\tau, \pi^k)\beta^{k-1}(\tau)}{\sum_{\tau' \in \mathcal{T}} \Pr(\sigma^k|\tau', \pi^k)\beta^{k-1}(\tau').} \tag{1}$$

Different mechanisms can be used to select a policy to execute. An always greedy policy selection mechanism would fail to explore, resulting in not reaching the global maximum. On the other hand a totally exploratory policy selection mechanism would not make an effort to improve performance. We thus require a balance, for which different policy selection heuristics have been proposed [30]. A policy selection heuristic $\mathcal{V}$ is a function that estimates a value for each policy through the extent to which it balances exploration with a limited degree of look-ahead for exploitation.

The *probability of improvement* heuristic for policy selection [30] considers the probability with which a specific policy can achieve a hypothesized increase in performance over the current best estimate. Assume that $U^+ \in \mathbb{R}$ is some utility which is larger than the best estimate under the current belief,

$$\hat{U} = \max_{\pi \in \Pi} \sum_{\tau \in \mathcal{T}} \beta(\tau) E[U|\tau, \pi].$$

The heuristic thus chooses the policy

$$\arg\max_{\pi \in \Pi} \sum_{\tau \in \mathcal{T}} \beta(\tau) \Pr(U^+|\tau, \pi),$$

where $U^+ > \hat{U}$.

3.3 Games

In contrast to classical RL, which considers one single agent in a stationary environment, Game theory studies rational decision making when several agents interact [22]. The core concept of a *Game* captures the strategic conflict of interest in a mathematical model. Note that different areas provide different terminology. Therefore, we will use the terms player and agent interchangeably; similarly for reward and payoff. Finally, we will refer to other agents in the environment as opponents irrespective of the domain's or agent's cooperative or adversarial nature.

A stochastic game with two players i and $-i$ consists of a set of stage games S (also known as states). In each state s players choose an action from the set $\mathbf{a} \in A(s)$. A game begins in a state $s_b \in S$. A joint action $\mathbf{a} = (a_i, a_{-i})$ is played at state s and player i receives an immediate reward $r_i(s, \mathbf{a})$, the world transitions into a new state s' according to the transition model $T(s, s', \mathbf{a})$. When a *goal* state $s_g \in S$ is encountered the game finishes, and the accumulated reward during the game is called an *episodic* reward.

We formalize *Sequential Interactions* (SI) as a specific variation of repeated stochastic games, where at each episode $k \in \{1, 2, \ldots, K\}$ a process draws a set of players $P_k \subset I$ from the population of individuals I to play a finite stochastic game that yields a reward (accumulated over the game) to each player. After the stochastic game terminates, the subsequent interaction commences. We specifically discuss the setting where the selection process is *stochastic* (as opposed to being a *strategic choice* by the agents), and the population comprises an unknown distribution over types of strategies. While in the general case these types may be unknown, our new algorithm assumes access to a priori interactions with each (proto-) type. We consider P_k and opponent rewards within the stochastic game to be unobservable, while the joint actions are observable.

3.4 Pepper

Pepper [15] (potential exploration with pseudo stationary restarts) was proposed as a framework to extend algorithms for learning in repeated normal-form games to repeated stochastic games. Pepper assumes it can observe its own immediate reward but not the opponents', and also assumes the maximum possible reward R_{max} known for each episode. It uses the principle of optimism in face of uncertainty [10] and combines it with a learning algorithm. Pepper computes the expected future rewards for a joint action **a** being in state s as:

$$R(s, \mathbf{a}) = r(s, \mathbf{a}) + \sum_{s' \in S} T(s, s', \mathbf{a})V(s') \tag{2}$$

where $V(s')$ is the expected future rewards of being in state s'. Note that given $r(\cdot), T(\cdot)$ and $V(\cdot)$, value iteration can be used to compute Eq. 2, and r and T can be learned from observations. Moreover, Pepper is initialized under the assumption that all states result in maximal reward. However, there is still the problem of updating $V(s')$ throughout the interaction. Pepper proposes a mechanism for estimating future rewards combining off-policy (e.g., Q-learning) and on-policy methods for estimating $V(s)$, i.e., an on-policy estimation based on the observed distribution of joint actions, using $n(s), n(s, \mathbf{a})$ for the number of visits to state s and the number of times joint action **a** was chosen in that state respectively:

$$V^{on}(s) = \sum_{\mathbf{a} \in A(s)} \frac{n(s, \mathbf{a})}{n(s)} R(s, \mathbf{a})$$

and a combined estimation

$$V(s) = \lambda(s)\hat{V}(s) + (1 - \lambda(s))V^{on}(s).$$

Where $\hat{V}(s)$ represents an optimistic approximation given by

$$\hat{V}(s) = \max(V^{off}(s), V^{on}(s)),$$

and where $\lambda \in [0,1]$ represents a stationarity measure initialized to one but approaching zero when the agent gets more experience.[2] Pepper uses the concept of non-pseudo stationary restarts, i.e., when $R(s)$ is observed to not be pseudo stationary $\lambda(s)$ resets to one. Let $n'(s)$ be the number of visits to state s since $R(s)$ was last observed to not be pseudo-stationary, then:

$$\lambda(s) = \max\left(0, \frac{C - n'(s)}{C}\right)$$

with $C \in \mathbb{N}^+$.

Algorithm 1. Pepper algorithmic framework

Input: States S, maximum possible reward R_{max}
1 Initialize $V(\cdot)$ with R_{max}
2 Random initial policy π
3 **for** *each episode of the stochastic game* **do**
4   Update $R(\cdot)$; *Eq.* 2
5   Update policy π
6   Observe state
7   **while** *state is not goal* **do**
8   Select action a
9   Observe state
10   Receive observation r
11   **if** *enough visits to* (s, a) **then**
12   Update rewards, $V(\cdot)$, transitions
13   Update $R(\cdot)$; *Eq.* 2
14   Update policy π

The Pepper framework is described in Algorithm 1 where different policy selection approaches can be plugged in to compute π. For example, using

$$\pi = (\arg\max_{\mathbf{a}} R(s, \mathbf{a}))_i$$

seems suitable for a friendly opponent, while

$$\pi = \arg\max_{\mathbf{a}_i} \min_{\mathbf{a}_{-i}} R(s, \mathbf{a})$$

is a minimax approach that suits other types of opponents.

Next, we present our Bayes-Pepper approach which uses Pepper in an offline phase to obtain policies. During the online phase it computes a belief over the possible opponents to tackle the uncertainty over the opponent's identity in a sequential interaction.

---

[2] Recall that $R(s, a)$ is initialized to R_{max} so it is likely to decrease in early episodes, but eventually will become pseudo-stationary.

4 Bayes-Pepper

Bayes-Pepper is composed mainly of two phases which are depicted in Fig. 1.

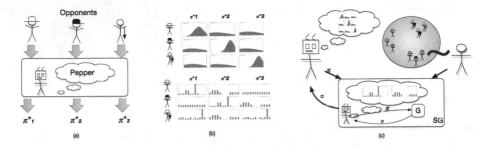

Fig. 1. Bayes-Pepper algorithm: (a) in an offline phase, Pepper is used to generate policies against each opponent and (b) performance models and transition models are generated from the learned policies. (c) In the online phase (i.e., sequential interactions) we assume a population of agents and a random process that matches the Bayes-Pepper agent and one opponent. Bayes-Pepper selects a policy π to act, in a stochastic game (SG) the agent computes an intra-belief which might override the selected policy; when the game finishes Bayes-Pepper receives an observation σ that is used to update its belief.

- An offline phase where Bayes-Pepper generates policies, transition and performance models (see Sect. 4.1). Here, the agent observes the opponent's identity.
- An online phase where a belief based approach is used to detect the opponent's identity and act with the corresponding policy. Here, the agent observes states and actions at every step of the game but only observes the accumulated reward when a stochastic game finishes. The belief is updated at every stochastic game (see Sect. 4.2) and at every state in a stochastic game (see Sect. 4.3).

Next, we describe these two phases in more detail.

4.1 Policy and Models Generation

Bayes-Pepper needs to generate policies and models for each opponent. Bayes-Pepper assumes an offline learning phase (see Algorithm 2) in which Pepper algorithm is used to obtain a policy for every opponent (lines 3–5). Then, performance models are obtained by generating list of rewards for each opponent and policy, and fitting the generated data into a distribution (in our experiments we used Gaussian distribution). The generated set of performance models can be seen as a matrix of probability distributions, see Fig. 1(b). Similarly, a list of state-action pairs is generated and fitted to a multinomial distribution to generate a transition model (lines 6–10), used in the intra-game belief, see Fig. 1(c).

Algorithm 2. Bayes-Pepper models and policy generation

---

1 $\Pi = \emptyset$
2 **for** *every opponent* $\tau \in \mathcal{T}$ **do**
3 $\quad$ Opponent τ is announced
4 $\quad$ Bayes-Pepper learns a policy π_{new} facing τ
5 $\quad$ $\Pi = \Pi \cup \pi_{new}$

6 **for** *every opponent* $\tau \in \mathcal{T}$ **do**
7 $\quad$ **for** *every* $\pi \in \Pi$ **do**
8 $\quad\quad$ Get list of rewards **r** and [s,a] pairs using π against τ
9 $\quad\quad$ Fit **r** to a distribution to obtain $\Pr(U|\tau,\pi)$
10 $\quad\quad$ Fit [s,a] to a distribution to obtain $\Pr(M|\tau,\pi)$

Algorithm 3. Bayes-Pepper detection algorithm

Input: Policy library Π, prior probabilities $\Pr(\mathcal{T})$, performance models
$\quad\quad$ $\Pr(U|\mathcal{T},\Pi)$, transition models $\Pr(M|\mathcal{T},\Pi)$, episodes K, exploration
$\quad\quad$ heuristic $\mathcal{V}$
1 Initialize beliefs $\beta^0(\mathcal{T}) = \Pr(\mathcal{T})$
2 **for** *episodes* $k = 1, \ldots, K$ **do**
3 $\quad$ Compute $v_\pi = \mathcal{V}(\pi, \beta^{k-1})$ for all $\pi \in \Pi$
4 $\quad$ $\pi^k = argmax_{\pi \in \Pi} v_\pi$
5 $\quad$ Start game with policy π_k and use intra-game belief (ζ) together with
$\quad\quad$ $\Pr(M|\mathcal{T},\Pi)$ (see Sect. 4.3)
6 $\quad$ Obtain observation signal σ^k (e.g., episodic reward)
7 $\quad$ Update belief $\beta^k(\tau) = \dfrac{\Pr(\sigma^k|\tau,\pi^k)\beta^{k-1}(\tau)}{\sum_{\tau' \in \mathcal{T}} \Pr(\sigma^k|\tau',\pi^k)\beta^{k-1}(\tau')}$

4.2 Opponent Detection Based on Rewards

Once Bayes-Pepper has a set of policies Π and its associated models it can
act in an online mode. The steps of Bayes-Pepper online detection phase are
described in Algorithm 3. Bayes-Pepper starts with a set of policies Π, prior
probabilities over the opponents $\Pr(\mathcal{T})$, performance models $\Pr(U|\mathcal{T},\Pi)$ and
transition models $\Pr(M|\mathcal{T},\Pi)$. Bayes-Pepper initializes the belief with the prior
probabilities $\Pr(\mathcal{T})$ (line 1). Then, for each episode of the sequential interaction
a loop performs the following steps:

- select a policy to execute (according to the belief β and exploration heuristic $\mathcal{V}$)
  (lines 3–4),
- use the selected policy on the stochastic game (line 5),
- receive an observation signal σ, this is, the accumulated reward of the played
  game (line 6),
- update the belief with the observation using Eq. 1 (line 7).

$\quad$ Since we assume that only the accumulated reward for the game is observed
when the game finishes, a basic approach is to select a policy to play for the entire

stochastic game. However, this might result in suboptimal results, for example when the opponent changes in the next interaction (as experimentally shown in Sect. 5.3). To overcome this issue, we propose to update the belief during a stochastic game using the state-action pairs (which are always observable). This information is also a signal of the opponent behavior and the process is similar to the one described previously.

4.3   Intra-game Belief Detection

Let $\zeta^{\ell}(\tau)$ be the *intra-game* belief which is initialized with the belief $\beta(\tau)$. The intra-game belief is updated in a similar way using Eq. 1 with two minor differences, the observation σ is the observed frequency over state-action pairs (or states) and the transition models $P(M|T, \Pi)$ are used to obtain the likelihood of a given observation.

Since the observed frequency might change more in early stages of the game we consider weighted approach, initially giving more weight to β and with each experience giving more weight to ζ as follows:

$$\zeta^{\ell}(\tau) = w\beta(\tau) + (1 - w)\zeta^{\ell-1}(\tau) \tag{3}$$

with $w = 1$ initially and $w = w \cdot \eta^t$, with $\eta = [0, 1)$ and t the number of experienced steps in the current stochastic game. Computing this updated belief might override the policy that was selected initially which is useful to avoid using a suboptimal policy for a complete stochastic game.

5   Experiments

In this section we present results on a stochastic game represented as a grid world. We performed experiments comparing our approach Bayes-Pepper with Pepper [15] and an Omniscient agent that knows the opponent's identities and plays optimal policies against them. First, we define the setting, then we present results of Bayes-Pepper against stochastic opponents, and finally, we present results in sequential interactions against switching opponents.

5.1   Setting

Figure 2 depicts a graphical representation of the stochastic game used in the experiments. There are two players, the learning agent (A) and the opponent (O). The starting positions are marked with their initial. The learning agent receives a reward when it reaches a goal state r_A or R_A, with $r_a < R_A$. The agents can move upwards or horizontally, and the opponent has the possibility to stay in the same state; the learning agent moves always to the right and the opponent to the left; to avoid agents getting trapped the grid is a toroidal world. With every step that does not transition to a goal state the learning agent receives a penalty p_{min}. In case of collision the learning agent receives high penalty p_{max} with $p_{min} < p_{max}$.

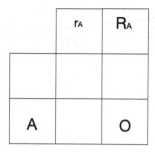

Fig. 2. A stochastic game with two players, the learning agent (A) and one opponent (O). The learning agent receives a reward when it reaches a goal state marked with r_A or R_A with $r_a < R_A$. In case of collision the opponent has priority over the state.

Note that the opponent directly influences the possible reward the learning agent can obtain. For example, since the opponent is closer to R_A it can block its way to the learning agent, in which case the best option would be to go for r_A.

For the experiments we set $r_A = 5, R_A = 17, p_{min} = -1, p_{max} = -5$. We tested against two types of stochastic opponents:

- A *defecting* opponent, Opp_D, that aims to block the way to R_A. It stays in the blocking position with probability 0.8.
- A *cooperative* opponent, Opp_C, that ignores the learning agent's actions and moves upwards with probability 0.2, and left otherwise.

The optimal policy against Opp_D is to go directly to r_A obtaining an accumulated reward of 3. In contrast, when facing Opp_C the agent should go for R_A obtaining an accumulated reward of 14.

5.2 Opponent Detection

In this experiment we evaluated how quickly Bayes-Pepper responds without knowing the opponent's identity in comparison with the learning process of Pepper. In this case, Bayes-Pepper starts with the policies against Opp_C and Opp_D and with a uniform prior over them.

Figure 3 depicts the average episodic rewards against the two types obtained over 10 independent trials facing the same type during the interaction. From the results we observe that Bayes-Pepper obtains higher rewards from the beginning of the interaction due to its fast detection. In contrast, Pepper takes more episodes to learn the appropriate policy. Table 1 shows average rewards where it can be seen that Bayes-Pepper obtained similar rewards to those of the Omniscient agent.

5.3 Switching Opponents

Now, we compare against switching opponents in sequential interactions, this is, during a repeated interaction of 150 games where the opponent changes

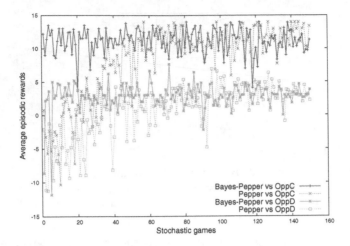

Fig. 3. Comparison of Bayes-Pepper and Pepper against two stochastic opponents in a repeated SG for 150 episodes. Bayes-Pepper obtains rewards close to the best response faster than Pepper.

Table 1. Comparison of average rewards with std. dev. ($\pm$) obtained in 10 trials.

	Bayes-Pepper	Pepper	Omniscient
Opp$_C$	11.19 ± 5.30	8.26 ± 8.80	12.40 ± 2.30
Opp$_D$	2.87 ± 4.05	0.87 ± 8.87	3.02 ± 2.97
Switching	5.44 ± 6.74	2.33 ± 8.01	8.48 ± 5.21

frequently and the learning agent does not know when the switches happen. To model switching opponents an opponent is selected randomly and is paired with the learning agent for a random number of games (uniformly from 5 to 10 repetitions).

Figure 4 depicts the average (a) episodic and (b) cumulative rewards of the compared approaches, and Table 1 shows the average episodic rewards for the 150 games. From the results we note that Bayes-Pepper obtained higher cumulative rewards than Pepper. This happens because Bayes-Pepper knows how to act optimally against every opponent, however, it needs to identify it. In contrast, Pepper learns how to optimize against the mixed behaviour of the two types. Note that in many cases when an opponent changed Bayes-Pepper was capable of obtaining competitive scores, this happens mainly due to the intra-game detection, since it triggers a change to a different policy when the transitions are not consistent with the learned model. In contrast, even when Pepper is learning within stochastic games and is able to update its policy it obtains suboptimal results, because it fails to obtain the reward R_A against the cooperative opponent.

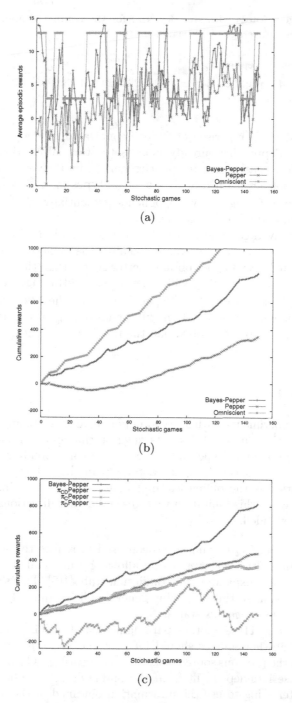

Fig. 4. Bayes-Pepper, Pepper and the Omniscient agent against switching opponents in a repeated interaction of 150 episodes. Average (a) episodic and (b) cumulative rewards. (c) Cumulative rewards of Bayes-Pepper and policies learned with Pepper.

Table 2. Average rewards with std. dev. ($\pm$) for the compared approaches against switching opponents (average of 10 trials).

Bayes-Pepper	π_C^{Pepper}	π_D^{Pepper}	π_{CD}^{Pepper}
5.44 ± 6.74	0.46 ± 18.01	2.38 ± 4.21	3.02 ± 2.75

We note that the comparison of Bayes-Pepper against Pepper might not be fair since Bayes-Pepper has already policies to start. With this in mind, we also evaluated three baselines using policies learned by Pepper against switching opponents: the policies learned against a single types π_C^{Pepper}, π_D^{Pepper} and one policy learned after facing the two opponents sequentially π_{CD}^{Pepper}, in this case there is no learning during the interaction.

Table 2 shows average rewards over the 150 games and Fig. 4(c) depicts cumulative rewards of the compared approaches against the same switching opponents. Results show that Bayes-Pepper obtains better scores than the rest. On the one side π_D^{Pepper} is a cautious policy which never takes advantage of a cooperative opponent, on the other side π_C^{Pepper} quickly obtains the best scores against a cooperative opponent but also gets highly penalized against a defecting opponent. π_{CD}^{Pepper} obtains better scores than the previous two but is not as good as Bayes-Pepper.

6   Discussion

We presented experiments with Bayes-Pepper against switching opponents in repeated stochastic games. The results suggest that our approach is capable of detecting the opponent type and act with the corresponding policy. Bayes-Pepper's main advantage is its quick detection in the online phase. However, its main limitation is the offline learning phase to obtain acting policies and models. To overcome this limitation we foresee different directions in which this work could be extended:

– State abstraction. CQ-learning has been used to reduce the state space representation in multi-agent systems by allowing a minimal state space representation and only expanding for conflicting states [19]. This same idea could be extended to our setting by having general policies and only update partial policies for some *dangerous* states.
– Lifelong learning [11] is another paradigm related to TL where information obtained from other sources should increase the performance on the target tasks and on the previous source tasks (reverse transfer). Currently, the offline learning phase is independent from the online detection phase, however, it would be interesting to use the information obtained in the online phase to update the policies and models learned in the offline phase.
– Multi-armed bandits [2] are a common formalism for selecting among different actions and this approach has been extended to select among experts [16].

Contextual multi-armed bandits are an extension in which the player also observes context information which can be used to determine the selection process [27]. How to incorporate contextual information in our setting is another challenge to address in future work.

7   Conclusions

Many learning algorithms for multiagent systems assume self-play or stationary opponents. We focus on the scenario of repeated stochastic games but with the difference of assuming a population of opponents and a stochastic process that at every game matches the learning agent with an opponent. Our first contribution is to provide a formal model of this setting, namely, sequential interactions. Our second contribution is an algorithm for quick detection of opponents in repeated stochastic games. Our proposed Bayes-Pepper algorithm draws inspiration from multiagent learning algorithms and policy reuse approaches to detect opponents. One advantage is that Bayes-Pepper is capable of detecting the opponent and responding with the appropriate policy faster than other learning algorithms for repeated stochastic games. The main limitation is the need of an offline learning phase where the policies and models can be obtained. As future work we propose to not only reuse policies but also transfer information from models and policies when facing unknown opponents, and eventually learning the set of opponent strategies in the population online, as existing preliminary work in this direction [24].

Acknowledgments. This research has received funding through the ERA-Net Smart Grids Plus project Grid-Friends, with support from the European Union's Horizon 2020 research and innovation programme.

References

1. Albrecht, S.V., Crandall, J.W., Ramamoorthy, S.: Belief and truth in hypothesised behaviours. Artif. Intell. **235**, 63–94 (2016)
2. Auer, P., Cesa-Bianchi, N., Fischer, P.: Finite-time analysis of the multiarmed bandit problem. Mach. Learn. **47**(2/3), 235–256 (2002)
3. Banerjee, B., Stone, P.: General game learning using knowledge transfer. In: International Joint Conference on Artificial Intelligence, pp. 672–677 (2007)
4. Barrett, S., Stone, P.: Cooperating with unknown teammates in complex domains: a robot soccer case study of ad hoc teamwork. In: Proceedings of the 29th Conference on Artificial Intelligence, pp. 2010–2016. Austin, Texas, USA (2014)
5. Bednar, J., Chen, Y., Liu, T.X., Page, S.: Behavioral spillovers and cognitive load in multiple games: an experimental study. Games Econ. Behav. **74**(1), 12–31 (2012)
6. Bellman, R.: A Markovian decision process. J. Math. Mech. **6**(5), 679–684 (1957)
7. Bloembergen, D., Tuyls, K., Hennes, D., Kaisers, M.: Evolutionary dynamics of multi-agent learning: a survey. J. Artif. Intell. Res. **53**, 659–697 (2015)
8. Boutsioukis, G., Partalas, I., Vlahavas, I.: Transfer learning in multi-agent reinforcement learning domains. In: Sanner, S., Hutter, M. (eds.) EWRL 2011. LNCS (LNAI), vol. 7188, pp. 249–260. Springer, Heidelberg (2012). https://doi.org/10.1007/978-3-642-29946-9_25

9. Bowling, M., Veloso, M.: Multiagent learning using a variable learning rate. Artif. Intell. **136**(2), 215–250 (2002)
10. Brafman, R.I., Tennenholtz, M.: R-MAX a general polynomial time algorithm for near-optimal reinforcement learning. J. Mach. Learn. Res. **3**, 213–231 (2003)
11. Brunskill, E., Li, L.: PAC-inspired option discovery in lifelong reinforcement learning. In: Proceedings of the 22nd Conference on Artificial Intelligence, pp. 1599–1610 (2014)
12. Busoniu, L., Babuska, R., De Schutter, B.: A comprehensive survey of multiagent reinforcement learning. IEEE Trans. Syst. Man Cybern. Part C (Appl. Rev.) **38**(2), 156–172 (2008)
13. Chakraborty, D., Stone, P.: Multiagent learning in the presence of memory-bounded agents. Auton. Agents Multi-Agent Syst. **28**(2), 182–213 (2013)
14. Conitzer, V., Sandholm, T.: AWESOME: a general multiagent learning algorithm that converges in self-play and learns a best response against stationary opponents. Mach. Learn. **67**(1–2), 23–43 (2006)
15. Crandall, J.W.: Just add pepper: extending learning algorithms for repeated matrix games to repeated markov games. In: Proceedings of the 11th International Conference on Autonomous Agents and Multiagent Systems, pp. 399–406. Valencia, Spain (2012)
16. Crandall, J.W.: Towards minimizing disappointment in repeated games. J. Artif. Intell. Res. **49**(1), 111–142 (2014)
17. Crandall, J.W.: Robust learning for repeated stochastic games via meta-gaming. In: Proceedings of the Twenty-Fourth International Joint Conference on Artificial Intelligence, pp. 3416–3422. Buenos Aires, Argentina (2015)
18. Da Silva, B.C., Basso, E.W., Bazzan, A.L., Engel, P.M.: Dealing with non-stationary environments using context detection. In: Proceedings of the 23rd International Conference on Machine Learnig, pp. 217–224. Pittsburgh, Pennsylvania (2006)
19. De Hauwere, Y.M., Vrancx, P., Nowe, A.: Learning multi-agent state space representations. In: Proceedings of the 9th International Conference on Autonomous Agents and Multiagent Systems, pp. 715–722. Toronto, Canada (2010)
20. Elidrisi, M., Johnson, N., Gini, M., Crandall, J.W.: Fast adaptive learning in repeated stochastic games by game abstraction. In: Proceedings of the 13th International Conference on Autonomous Agents and Multiagent Systems, pp. 1141–1148. Paris, France (2014)
21. Fernández, F., Veloso, M.: Probabilistic policy reuse in a reinforcement learning agent. In: Proceedings of the 5th International Conference on Autonomous Agents and Multiagent Systems, pp. 720–727. ACM, Hakodata, Hokkaido, Japan (2006)
22. Fudenberg, D., Tirole, J.: Game Theory. The MIT Press, Cambridge (1991)
23. Hernandez-Leal, P., Munoz de Cote, E., Sucar, L.E.: A framework for learning and planning against switching strategies in repeated games. Connect. Sci. **26**(2), 103–122 (2014)
24. Hernandez-Leal, P., Kaisers, M.: Learning against sequential opponents in repeated stochastic games. In: The 3rd Multi-disciplinary Conference on Reinforcement Learning and Decision Making, Ann Arbor (2017)
25. Hernandez-Leal, P., Taylor, M.E., Rosman, B., Sucar, L.E., Munoz de Cote, E.: Identifying and tracking switching, non-stationary opponents: a bayesian approach. In: Multiagent Interaction without Prior Coordination Workshop at AAAI, Phoenix, AZ, USA (2016)

26. Hernandez-Leal, P., Zhan, Y., Taylor, M.E., Sucar, L.E., Munoz de Cote, E.: Efficiently detecting switches against non-stationary opponents. Auton. Agents Multi-Agent Syst. **31**(4), 767–789 (2017)

27. Langford, J., Zhang, T.: The epoch-greedy algorithm for multi-armed bandits with side information. In: Advances in Neural Information Processing Systems, pp. 817–824 (2008)

28. Lazaric, A., Ghavamzadeh, M.: Bayesian multi-task reinforcement learning. In: Proceedings of the 27th International Conference on Machine Learning, Haifa, Israel (2010)

29. Lazaric, A., Restelli, M., Bonarini, A.: Transfer of samples in batch reinforcement learning. In: International Conference on Machine Learning, pp. 544–551. ACM, Helsinki, Finland (2008)

30. Rosman, B., Hawasly, M., Ramamoorthy, S.: Bayesian policy reuse. Mach. Learn. **104**(1), 99–127 (2016)

31. Taylor, M.E., Stone, P., Liu, Y.: Transfer learning via inter-task mappings for temporal difference learning. J. Mach. Learn. Res. **8**, 2125–2167 (2007)

32. Taylor, M.E., Stone, P.: Transfer learning for reinforcement learning domains: a survey. J. Mach. Learn. Res. **10**, 1633–1685 (2009)

33. Watkins, J.: Learning from delayed rewards. Ph.D. thesis, King's College, Cambridge, UK (1989)

Event Calculus Agent Minds Applied to Diabetes Monitoring

Nicola Falcionelli[1](✉), Paolo Sernani[1], Albert Brugués[2],
Dagmawi Neway Mekuria[1], Davide Calvaresi[2,3], Michael Schumacher[2],
Aldo Franco Dragoni[1], and Stefano Bromuri[4]

[1] Università Politecnica delle Marche, Ancona, Italy
{n.falcionelli,d.n.mekuria}@pm.univpm.it,
{p.sernani,a.f.dragoni}@univpm.it
[2] University of Applied Sciences Western Switzerland, Sierre, Switzerland
{albert.brugues,michael.schumacher}@hevs.ch
[3] Scuola Superiore Sant'Anna, Pisa, Italy
d.calvaresi@sssup.it
[4] Open University of the Netherlands, Heerlen, The Netherlands
stefano.bromuri@ou.nl

Abstract. The increasing incidence of chronic diseases is a major challenge for the healthcare sector. Personal Health Systems (PHSs) address the self-management of chronic diseases, by decentralizing the health monitoring outside hospitalized environments. Rule based agents allow bringing domain experts' knowledge into PHSs. However, agents must meet the requirements of real monitoring scenarios, characterized by massive streams of events. Hence, with the aim to monitor the health status of diabetic patients, two logic-based agent minds for an agent-oriented PHS are presented. One agent mind is based on the standard version of jREC, a Prolog-based implementation of Cached Event Calculus, while the other is a customization of the standard jREC mind that exploits an event-indexing technique. Both of them are as well integrated into MAGPIE, a Java agent platform. The paper then compares and analyzes the performances of the proposed agent minds, by computing the time needed to trigger different type of alerts, when the number of recorded events (e.g. values of physiological parameters) increases. The results show that the customized jREC mind performs much better when an high number of events need to be checked, making its use advisable in monitoring scenarios.

1 Introduction

The incidence of chronic diseases in the population is recognized as a major challenge for the healthcare sector [2]. For instance, the number of people affected by diabetes has doubled in the last 20 years [33]. Statistics from WHO report that

This paper has already been published in:
© Springer International Publishing AG 2017
S. Montagna et al. (Eds.): A2HC/A-HEALTH 2017, LNAI 10685, pp. 40–56, 2017.
https://doi.org/10.1007/978-3-319-70887-4_3

© Springer International Publishing AG 2017
G. Sukthankar and J. A. Rodriguez-Aguilar (Eds.): AAMAS 2017 Best Papers,
LNAI 10642, pp. 258–274, 2017.
https://doi.org/10.1007/978-3-319-71682-4_16

more than 400 million individuals live with diabetes, and losses in the GDP for diabetes-related costs from 2011 to 2030 are estimated at 1.7 trillion USD [32].

Personal Health Systems (PHSs) aim at supporting the self-management of chronic diseases and reducing the healthcare costs, supporting medical doctors in following the patients' disease evolution [10]. PHSs implement the *"healthcare to anyone, anytime, and anywhere"* paradigm, by increasing both the coverage and the quality of healthcare [31]. In fact, PHSs bring the health technology to domestic environments, by localizing healthcare services to the specific needs, practices, and situations of people and their social contexts [24]. PHSs ensure the continuity of care, focusing on a knowledge-based approach integrating past and current data of each patient together with statistical evidence [29]. A PHS is composed of three tiers [30]: Tier 1 is the Body Area Network (BAN), i.e. the set of sensors on the patient's body to monitor her health parameters; Tier 2 is the personal server, usually a mobile device, which collects and aggregates the parameters and events produced by the BAN; Tier 3 is the remote server which processes and stores the data from the personal server and supports doctors in following the treatment of patients at home.

Beyond the modeling capabilities of agent-based frameworks [28] and their still opened challenges [12,13], Multi-Agent Systems have been proved useful in the healthcare sector implementing modularity, distribution, and personalization for data management, decision support systems, planning and resource allocation, and remote care [18], being ideal for PHSs. In [8], an agent-based platform called MAGPIE implements a programmable expert PHS to monitor patients suffering from diabetes. In particular, that agent platform adds scalability to the PHS by shifting from Tier-3 to Tier-2 the computation needed for the patient monitoring. To obtain such scalability, the agents, composed by an agent body and an agent mind, run directly on the personal server. The agent body is the part of the agent that collects the data acting as an interface between the BAN in Tier-1 and the agent mind. The agent mind, based on an Event Calculus (EC) engine, is the part of the agent that checks the data collected from the body to perform the monitoring task and trigger alerts for the medical doctors logging in Tier-3. The approach of MAGPIE allows improving the scalability of the PHS when the number of patients increases, compared to a centralized PHS where the computation is performed in Tier-3. However, another aspect has to be taken into account: the scalability of the agent mind when the number of events increases. In fact, the use of rule engines based on EC usually restricts the number of events and rules to be applied in a real monitoring scenario, where short time delays are needed to apply corrective actions. Thus, the next step to apply the agent-based PHS in real scenarios requiring long-term monitoring is to develop agent minds capable of caching and retrieving events efficiently.

This paper addresses such issue by proposing two agent minds for the MAGPIE agent platform presented in [8]. The agent minds have been implemented using jREC, a Cached Event Calculus (CEC) reasoner based on Java and tuProlog [5], to move the computational complexity from query to update time by caching the maximum validity intervals for fluents. Even if both based on jREC and integrated into the MAGPIE agent platform, the proposed agent minds differ on the way in which they handle event streams. One is a straightforward integration of the jREC engine, and the other is based on an indexing technique that gives to jREC the ability to process event streams more efficiently.

In addition, as the main contribution of the paper, the performances of the jREC-based agent minds are evaluated on the time required to trigger an alert, when the number of events generated by the agent body increases. Diabetes has been adopted as the use case for the monitoring rules to be checked.

The rest of the paper is organized as follows. Section 2 presents the paper background on EC, CEC, jREC and red-black trees. Section 3 describes an overview of the entire PHS in which the agent minds runs, shows the encoding of the monitoring rules to check alerts based on glucose and blood pressure levels in diabetic patients, as well as problems and solutions that arise when massive streams of events have to be handled by reasoning engines. Section 4 presents the experimental results to evaluate the two agent minds. Section 5 describes the work related to the presented research. Section 6 draws the conclusions of the paper and outlines the future work.

2   Background

This section introduces the concepts on which the proposed agent minds are based on.

2.1   Event Calculus

EC is a logic formalism for reasoning about actions and their effects in time [20]. Therefore, it is a suitable tool for modeling expert systems representing the evolution in time of an entity by means of the production of events. EC is based on many-sorted first-order predicate calculus, known as domain-independent axioms, which are represented as normal logic programs that are executable in Prolog. The underlying time model of EC is linear. EC manipulates fluents, where a fluent represents a property that can have different values over time. The term $F = V$ denotes that a fluent F has value V as a consequence of an action that took place at some earlier time-point and not terminated by another action in the meantime. Table 1 summarizes the main EC predicates. Predicates, functions, symbols and constants start with lowercase letter, while variables start with uppercase letter. Predicates in the text are referenced as predicate/N, where predicate is the name of the predicate and N its arity (e.g. number of arguments).

The domain independent axioms of EC are the following:

$$\text{holdsAt}(F = V, 0) \leftarrow \text{initially}(F = V). \tag{1}$$

$$\begin{aligned}
\text{holdsAt}(F = V, T) \leftarrow \\
\text{initiatesAt}(F = V, T_s), T_s < T, \\
\text{not broken}(F = V, [T_s, T]).
\end{aligned} \tag{2}$$

Predicate (1) states that a fluent F holds value V at time 0, if it has been initially set to this value. For any other time $T > 0$, the Predicate (2) states

Table 1. Main Event Calculus predicates

Predicate	Meaning
initially(F = V)	The value of fluent F is V at time 0
holdsAt(F = V,T)	The value of fluent F is V at time T
holdsFor(F = V,[T_{min},T_{max}])	The value of fluent F is V between T_{min} and T_{max}
initiatesAt(F = V,T)	At time T the fluent F is initiated to have value V
terminatesAt(F = V,T)	At time T the fluent F is terminated from having value V
broken(F = V,[T_{min},T_{max}])	The value of fluent F is either terminated at T_{max}, or initiated to a different value than V between T_{min} and T_{max}
happensAt(E,T)	An event E takes place at time T updating the state of the fluents

that the fluent holds at time T if it has been initiated to value V at some earlier time point Ts, and it has not been broken on the meanwhile.

$$broken(F = V, [Tmin, Tmax]) \leftarrow$$
$$terminatesAt(F = V, T), Tmin < T, Tmax > T. \tag{3}$$

$$broken(F = V_1, [Tmin, Tmax]) \leftarrow$$
$$initiatesAt(F = V_2, T_i), V_1 \neq V_2,$$
$$Tmin < Ti, Tmax > Ti. \tag{4}$$

Predicates (3) and (4) specify the conditions that break a fluent. Predicate (3) states that a fluent is broken between two time points $Tmin$ and $Tmax$ if within this interval it has been terminated to have value V. Alternatively, Predicate (4) states that a fluent is broken within a time interval if it has been initiated to hold a different value.

$$holdsFor(F = V, [Tmin, Tmax]) \leftarrow$$
$$initiatesAt(F = V, Tmin),$$
$$terminiatesAt(F = V, Tmax),$$
$$not\ broken(F = V, [Tmin, Tmax]). \tag{5}$$

$$holdsFor(F = V, [Tmin, infPlus]) \leftarrow$$
$$initiatesAt(F = V, Tmin), \qquad (6)$$
$$not \ broken(F = V, [Tmin, +\infty]).$$

$$holdsFor(F = V, [infMin, Tmax]) \leftarrow$$
$$terminatesAt(F = V, Tmax), \qquad (7)$$
$$not \ broken(F = V, [-\infty, Tmax]).$$

Predicates (5), (6) and (7) deal with the validity intervals of fluents. In particular, Predicate (5) specifies that a fluent F keeps value V for a time interval going from $Tmin$ to $Tmax$ if nothing happens in the middle that breaks such an interval. Predicates (6) and (7) behave in the same way, but deal with open intervals.

The domain dependent predicates in EC are typically expressed in terms of the initiatesAt/2 and terminatesAt/2 predicates. One example of a common rule for initiatesAt/2 is

$$initatesAt(F = V, T) \leftarrow$$
$$happensAt(Ev, T), \qquad (8)$$
$$Conditions[T].$$

The above definition states that a fluent is initiated to value V at time T if an event Ev happens at this time point, and some optional conditions depending on the domain are satisfied. In relation with MAGPIE, the agent platform in which the proposed agent mind has been integrated, these events that must happen are physiological measurements from the patient.

2.2   Cached Event Calculus and jREC

Straightforward implementations of EC [20] have time and memory complexity which are not practical for developing real applications. This is due to the fact that every time the EC engine is queried, the computation starts from scratch, and all fluents validity intervals are calculated again. Cached Event Calculus (CEC), proposed by Chittaro and Montanari [14], tries instead to overcome this inefficiency by giving EC a memory mechanism, and moving computation from query time to update time.

CEC formalizes the concept of Maximal Validity Interval (MVI), that represents a time interval in which a particular fluent holds without being terminated by any event. A fluent is also associated to a list of MVIs, in order to express all the time intervals in which that fluent holds continuously.

Whenever the rule engine is updated (e.g. by inserting a new event occurrence), the fluents' MVIs are calculated, and then stored for further use, allowing incremental computation for following updates. Also, every time a new event is added to the database, CEC manages to compute MVIs only for the fluents that can vary with that event, and does not check the MVIs of those fluents that cannot possibly change, thus avoiding unnecessary computation.

jREC is a reasoning tool based on Java and tuProlog that implements a lightweight version of CEC [5]. Since MAGPIE is also written in Java, it has been chosen to implement the proposed agent minds, in order to ensure seamless integration with the agent platform.

jREC consists of three main components:

- The Prolog theory, which represents the actual CEC axiomatization that is loaded into tuProlog;
- The Java engine, which allows to query and update the database without having to interact directly with tuProlog, as well as adding specific domain-dependent theories;
- The Tester, which is a GUI based stand-alone tool for editing theories, visualizing fluents' MVIs and event occurrences, mainly used for prototyping and developing domain-dependent theories.

2.3 Red-Black Trees

A red-black tree (RBT) is a well known data structure proposed by Rudolf Bayer in 1972 [3]. It is a binary search tree which provides $O(log(n))$ Worst Case time complexity for operations such as node searching, insertion and deletion, as well as $O(n)$ Worst Case space complexity [3]. This is made possible thanks to node coloration: every node of the tree is augmented with an extra bit, and based on the value of such bit, the node is considered to be red or black.

The aforementioned operations rely on such coloration feature to achieve Worst Case logarithmic time complexity and linear space complexity. In fact, every operation that modifies the RBT has to comply with very precise policies which constrain how the nodes should be moved or re-painted. These policies guarantees that the nodes in an RBT are always balanced after every operation, giving such data structure the epiteth of self-balancing. Even though the obtained balance is not perfect, it is proven to be good enough to provide the declared performances [3].

Red-black trees can be effectively exploited as indexing data structures. As it will be also explained in Sect. 3, one of the agent minds that are proposed in this work relies on such RBT-based indexing in order to efficiently process event streams.

3 System Overview

The implemented agent minds run in Tier-2 of the MAGPIE agent-based PHS for self monitoring of diabetes. The entire PHS is depicted in Fig. 1. Each patient has its own agent composed by a body (Tier-1) and a mind (Tier-2) running on the personal server: in Tier-1 data are collected from the patients through a BAN; in Tier-2, the agent minds are responsible to trigger possible alerts based on the patients' physiological values, running domain dependent rules which could be customized for each patient. The triggered alerts have to be sent as a notification to medical doctors connected to Tier-3.

Fig. 1. The agentified PHS. The agent mind runs in Tier-2, to monitor the patient's physiological values.

3.1 MAGPIE Agent Platform

MAGPIE is an agent platform integrated with the Android OS. It plays the role of Tier-2 in a PHS by connecting the patient and the medical doctor, with the aim of improving the management of chronic diseases. From the side of the patient it collects physiological values, whereas from the medical side it models the medical knowledge in terms of monitoring rules expressed as domain dependent axioms of EC. Interested readers can find in [9] a description of the MAGPIE architecture and its integration with Android. In relation to this work, a monitoring rule is defined as a combination of events that trigger an alert to be notified to a medical doctor, where an event is considered as the measurement of a physiological parameter. Therefore, the following two types of monitoring rules are specified:

– Complex rules: consist of the combination of two or more events in a specific time window, where the order in which the events happen is not considered.
– Sequential rules: consist of the sequence of two or more events in a specific time window, where the particular order in which the events occur matters.

3.2 Diabetes Monitoring Rules

In order to detect alert conditions related to diabetes, a sequential and a complex rule patterns are proposed. These rule patterns are based on the literature available for glucose and blood pressure monitoring [6,16] and checks physiological values collected by the patient's BAN. The patterns identify alert conditions in the patient's health status by modeling the sensor inputs as events that are evaluated in the body of the rules. The two patterns are:

Pattern 1: Brittle diabetes, defined as a glucose rebound going from less than 3.8 mmol/l to more than 8.0 mmol/l in a period of six hours. This pattern can be expressed with a sequential rule.

Pattern 2: Pre-hypertension, defined as two events of high blood pressure in a period of one week. This pattern can be expressed with a complex rule.

Pattern 1 is implemented as follows:

$$\text{initiatesAt}(F = A, T) : -$$
$$\text{happensAt}(\text{ev}(2, A, W), T),$$
$$\text{happensAt}(\text{ev}(1, A, \_), T_1),$$
$$T_s \text{ is } (T - W), \tag{9a}$$
$$T > T_1,$$
$$T_1 >= T_s,$$
$$\text{no\_alert}(A, T_s).$$

$$\text{terminatesAt}(F = A, T) : -$$
$$\text{happensAt}(\text{ev}(1, A, \_), T). \tag{9b}$$

$$\text{happensAt}(\text{ev}(1, \text{'brittle diabetes'}, E), T) : -$$
$$\text{hours\_to\_epoch}(6, E),$$
$$\text{happensAt}(\text{glucose}(G), T), \tag{9c}$$
$$G =< 3.8.$$

$$\text{happensAt}(\text{ev}(2, \text{'brittle diabetes'}, E), T) : -$$
$$\text{hours\_to\_epoch}(6, E),$$
$$\text{happensAt}(\text{glucose}(G), T), \tag{9d}$$
$$G >= 8.$$

Rules (9a) and (9b) represent a generic sequential rule template with two events. In particular, the fluent F (i.e. the alert) is initiated with value A when: (i) two temporal ordered events occur inside a certain time window and (ii) when the fluent does not hold anywhere else inside the time window (no_alert/2). The fluent F is instead terminated when the first event of the ordering happens.

Rules (9c) and (9d) customize the template for the glucose monitoring use case. They instantiate the variables of the ev/3 term, specifying the time window width (W), the alert name (A) and the threshold values for G.

Pattern 2 is expressed in the following way:

$$\text{initiatesAt}(F = A, T) : -$$
$$\text{happensAt}(\text{alertcheck}(A, W, NMax_1), T),$$
$$T_s \text{ is } (T - W),$$
$$\text{count\_events\_tw}(N_1, \text{evc}(1, A), T_s, T), \tag{10a}$$
$$N_1 >= NMax_1,$$
$$\text{no\_alert}(A, T_s).$$

$$\text{terminatesAt}(F = A, T) : -$$
$$\text{happensAt}(\text{alertcheck}(A, W, \_), T), \tag{10b}$$
$$\text{holdsAt}(F = A, T).$$

$$\text{happensAt}(\text{evc}(1, \text{'pre-hypertension'}), T) : -$$
$$\text{happensAt}(\text{blood\_pressure}(S, D), T),$$
$$S >= 130, \tag{10c}$$
$$D >= 80.$$

$$\text{happensAt}(\text{alertcheck}(\text{'pre-hypertension'}, E, 2), T) : -$$
$$\text{weeks\_to\_epoch}(1, E), \tag{10d}$$
$$\text{happensAt}(\text{evc}(1, \text{'pre-hypertension'}), T).$$

Rules (10a) and (10b) represent a generic complex rule template with one event type. In particular, the fluent F (i.e. the alert) is initiated with value A when: (i) there are least $NMax_1$ occurrences of the alertcheck/3 event inside the time window and (ii) when the fluent does not hold anywhere else inside the time window (no_alert/2). Also, the count_events_tw/4 predicate is necessary to handle different event temporal orderings without having to duplicate the rule body for every permutation. Rules (10c) and (10d) customize the template for the hypertension monitoring use case. They instantiate the variables of the evc/2 and the alertcheck/3 terms specifying the time window width (W), the alert name (A) and the threshold values for S and D.

3.3   Event Handling with jREC

Efficient handling of massive event streams, while preserving the philosophy of Event Calculus, and in broader terms, of Logic Programming, is a non-trivial task. Techniques such as (i) event windowing/forgetting [1], (ii) theory pre-compilation [1] and (iii) a priori assumptions on event temporal ordering, can help to ease the burden of this process, but at the same time their adoption will cause the reasoning approach to be less general and less flexible. Therefore, since in real case monitoring scenarios these techniques and assumptions might simply not be applicable, finding alternatives ways to tackle the problem in a more general case becomes mandatory.

For example, jREC does not apply any simplifying assumption or technique to the event streams: this forces the reasoner to spend a very high amount of resources every time the engine's knowledge base (KB) is updated with new events. Whenever a list of new events has to be asserted into the KB, jREC must perform the following steps:

- Sort the list of new events chronologically;
- Read all the events already present in the KB and put them in a list;
- Retract all the events from the KB;
- Sort the list of KB's events chronologically;
- Merge the list of new events with the list of events read from the KB;
- Sort the newly obtained list chronologically and remove duplicates;
- Assert the events from the newly obtained list back into the KB;
- Calculate the effects on the fluents.

This procedure indeed maintains the reasoning as general and flexible as possible, but it is also the main source of jREC inefficiency, since every new event(s) insertion causes the engine to sort the event lists multiple times.

To tackle such issue, this paper proposes the integration of jREC with an indexing data structure, i.e. the previously mentioned red-black trees. RBTs will take the duty of maintaining the events temporal ordering by avoiding unnecessary sorting operations, and ensuring fast execution times.

The introduction of event indexing with RBTs allows to more precisely define the proposed agent minds:

- An agent mind based on the standard jREC implementation (standard jREC);
- An agent mind based on a custom jREC implementation, which has been augmented with RBT event indexing (RBT-index jREC).

It should be noticed that, since (i) an event normally contains multi-dimensional data (i.e. timestamp and phyisiological values), (ii) an RBT only allows single-dimensional indexing, and (iii) jREC needs the events to be ordered chronologically, the only choice is to consider the events timestamp as the key on which the indexing will be performed.

4   Test Setup and Results

The performances of the two jREC agent minds have been evaluated using the sequential and complex rule patterns described above. To accomplish that, synthetic datasets containing glucose and blood pressure measurements have been created. Each measurement is a tuple containing the value(s) and its timestamp.

4.1   Testing Protocol

To see how the performances of the agent minds evolve when the number of events increases, a series of random dataset has been created, each one containing a different number of events.

The events of each dataset are fed into the agent minds one by one, and the time needed by each agent to trigger the alert is recorded. Every experiment is repeated one-hundred times to obtain the mean and standard deviation values.

The biggest assumption of the experiment is that real datasets do not add any value to the performance evaluation. In fact, the use of synthetic datasets allowed to stress the agent minds on very specific and critical tasks.

4.2   Results and Discussion

The tests have been executed on an i7-6700K@4.20 GHz CPU with 16 GB@2400 MHz DDR4 RAM, running Ubuntu/Linux 16.04 and Java Runtime Environment 8u121.

It should be noticed that the main results of these tests are the execution time trends, rather than the absolute values themselves (since they vary with different machines).

From the plots in Fig. 2a and b, it is clear that the two agent minds show a very different behaviour: the execution time of the standard jREC agent mind grows in a polynomial fashion, and is considerably higher than the counterpart's. In fact, from the plots in Fig. 3a and b it can be observed that the execution time of the RBT-index jREC agent mind follows a logarithmic-like curve.

The polynomial trend exhibited by the standard jREC agent mind can be explained in terms of nested sortings: in fact, every time the engine knowledge base is updated with one or more events, the logic machinery of the engine launches multiple nested sorting clauses (see also Sect. 3.3).

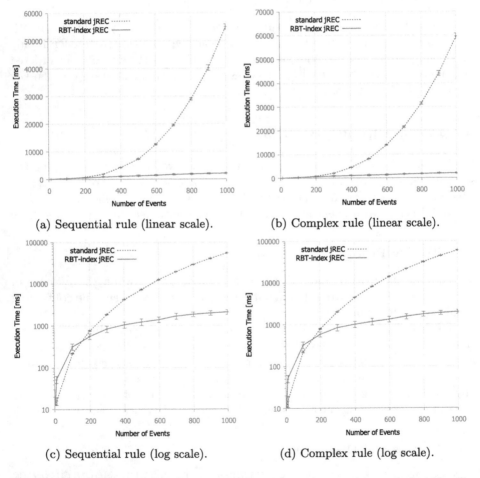

(a) Sequential rule (linear scale).   (b) Complex rule (linear scale).

(c) Sequential rule (log scale).   (d) Complex rule (log scale).

Fig. 2. Milliseconds needed by the two jREC agent minds to compute an alert, for the different rules.

On the other hand, the logarithmic-like trend of the RBT-index jREC agent mind highlight a direct correlation with the expected performances of the RBT-based indexing. It also demonstrates that, as the number of event grows, the execution time introduced by the reasoning on such events plays a minor role on the overall execution time. This can be explained by considering that the average number of events falling inside a rule timewindow is constant, since (i) the average event inter-arrival time and (ii) rules time-windows duration are fixed.

As a last remark on the performance gap, the logarithmic plots in Fig. 2c and d clearly highlight that, after 1000 events, the improvement almost reaches 2 orders of magnitude.

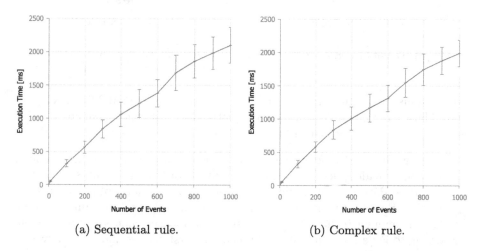

(a) Sequential rule.   (b) Complex rule.

Fig. 3. Detail of Fig. 2a and b, showing only RBT-index jREC agent mind's trends.

The execution times of the two agent minds in a scenario with a small number of events highlights some peculiar behaviours. As can be clearly seen from the plots (Fig. 4a and b), when the number of events is smaller than 100, the standard jREC agent mind shows better performances than the other. When the number of events reaches 200, the situation is the opposite, with the RBT-index agent mind being on top. This effect is due to the additional overhead in RBT-index agent mind: to be more precise, with very few events, the event-handling time gain obtained with the exploitation of the RBT-based indexing is not enough to compensate the additional overhead introduced by such data structure. By increasing the number of events, the event-handling time gain becomes increasingly more predominant over the data structure overhead, allowing the RBT-index jREC agent mind to perform better on massive event streams.

With the machine used for the tests, it is shown that the RBT-index jREC agent mind exhibits a clear performance improvement over the standard jREC agent mind. The execution time trends suggest that even scenarios with more

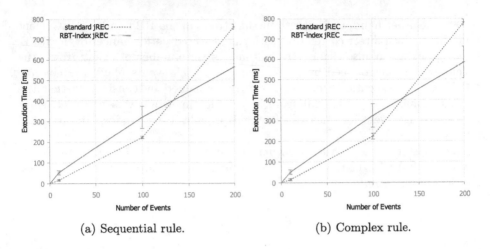

(a) Sequential rule.   (b) Complex rule.

Fig. 4. Detail of Fig. 2a and b, with number of events going from 0 to 200.

than a thousand events are reasonably manageable by the RBT-index agent mind. Thus, some possible real-case applications for the said agent mind, with the proposed rule patterns, can be:

- Detecting Brittle Diabetes with Continuous Glucose Monitoring devices. They can provide glucose measurements up to one minute [16], so referring to the Rule Pattern 1, it would mean a worst case scenario of 360 events.
- Detecting Pre-Hypertension conditions with digital arm sphygmomanometers. It is enough to have two blood pressure measurements per day [6], so referring to Rule Pattern 2, it would mean a worst case scenario of 14 events.

Even though the standard jREC agent mind performs slightly better with a low number of events, this is not enough to justify its usage only in such scenario.

5   Related Work

Multi-Agent Systems (MASs) meet the requirements of the healthcare sector: context awareness, reliability, data abstraction and interoperability, unobtrusiveness [4]. From a requirements engineering perspective, goal-oriented and agent-based design methodologies are useful to tailor pervasive systems to end-users and stakeholders' needs [11]. When appied to PHSs, agent-based modeling has the potential to bring the decision making at the level of self-management of chronic diseases [21]. In the implementation phase, MASs in PHSs pursue the enhancement of home-based self-care by using networks of sensors and remote assistance, to increase the satisfaction of the patient and make an efficient use of resources [17].

Reasoning agents in PHSs allow to transfer part of the knowledge from domain experts to the handheld devices used to perform the self-management of chronic diseases. Beyond PHSs, other applications include energy management [25], to control energy demand and production, home automation [26], to coordinate the available appliances, and ambient assisted living [23,27], with monitoring purposes. In the context of PHSs, EC and MASs have been successfully applied to the self-management of diabetes [7,19]. However, such works do not take into account the scalability of the PHS. In fact, a clear advantage of reasoning agents in the Tier-2 of PHSs is the system scalability with increasing number of patients, as showed in [8]. Nevertheless, in such research, the scalability of the agent minds with high streams of events is not considered. Thus, there is the need to find the suitable tools to implement agent minds, which are supposed to run in portable devices, even with high numbers of events and large datasets. This is especially true for EC given its complexity. Indeed, non-logic based pattern recognition has been proved, overall, more efficient than traditional EC when performing predictions. However, it lacks the potential of coding domain experts' knowledge into logic rules and needs to train on large amounts of data. Hence, caching and windowing techniques to make EC efficient and applicable with large scale dataset have been investigated [1,5]. There are some already available tools that allow logic programming in terms of EC. One of such is DECReasoner [22], a Discrete Event Calculus Reasoner: it implements EC without any caching mechanism and, thus, it is not usable for this research, due to its computation time with the datasets used for the performance tests. A more efficient EC implementation is RTEC, which adds to EC support for handling event streams [1]. However, RTEC techniques such as event windowing and theory pre-compilation do not match the flexibility requirements of the proposed PHS. In addition, it is not compatible with the platform, as the agent-oriented PHS is based on tuProlog and Java [8]. Thus, jREC has been used to implement a prototype for the proposed agent mind, since it implements Cached Event Calculus [5] with tuProlog, a Java-based Prolog engine [15]. Moreover, being Java-based, jREC and tuProlog can run on Android devices, allowing to run the proposed agent mind on handheld devices.

6   Conclusions

In this work, two rule-based minds for monitoring agents running on Tier-2 of a PHS have been presented and tested. Being both integrated into the MAGPIE agent platform, one is based on the plain jREC reasoner and the other is a customization of the standard jREC reasoner augmented with an RBT-based indexing technique. In order to be used in real monitoring scenarios, the agent minds have to be able to process massive event streams, represented by the patient's physiological values. Therefore, in addition to the customization of the jREC reasoning engine, the main contribution of this paper is the performance evaluation of proposed agent minds on the time needed to trigger alerts based on glucose and blood pressure levels, in a diabetes monitoring scenario. Two real application scenarios for the proposed agent minds are the detection of brittle diabetes, with Continuous Glucose Monitoring, and the detection of Pre-Hypertension conditions, with devices such as digital arm sphygmomanometers.

As future work, since PHSs are intended for the self-management of diabetes with handheld devices, the tests should be performed on mobile phones, to obtain more realistic figures. In this direction, the performance of the RBT-index jREC agent mind can be further enhanced by improving the sophistication of the current indexing solution. Furthermore, in order to validate the usefulness of the rules, the tests should run on real datasets. Lastly, the system can be applied to other use cases, in order to model rules for other diseases.

References

1. Artikis, A., Sergot, M., Paliouras, G.: An event calculus for event recognition. IEEE Trans. Knowl. Data Eng. **27**(4), 895–908 (2015)
2. Bauer, U.E., Briss, P.A., Goodman, R.A., Bowman, B.A.: Prevention of chronic disease in the 21st century: elimination of the leading preventable causes of premature death and disability in the USA. Lancet **384**(9937), 45–52 (2014)
3. Bayer, R.: Symmetric binary B-trees: data structure and maintenance algorithms. Acta Informatica **1**(4), 290–306 (1972). https://doi.org/10.1007/BF00289509
4. Bergenti, F., Poggi, A.: Multi-agent systems for e-health: recent projects and initiatives. In: 10th Workshop on Objects and Agents, WOA 2009 (2009)
5. Bragaglia, S., Chesani, F., Mello, P., Montali, M., Torroni, P.: Reactive event calculus for monitoring global computing applications. In: Artikis, A., Craven, R., Kesim Çiçekli, N., Sadighi, B., Stathis, K. (eds.) Logic Programs, Norms and Action. LNCS (LNAI), vol. 7360, pp. 123–146. Springer, Heidelberg (2012). https://doi.org/10.1007/978-3-642-29414-3_8
6. British Hypertension Society: Home blood pressure monitoring protocol, 17 February 2017. http://www.bhsoc.org/files/4414/1088/8031/Protocol.pdf
7. Bromuri, S., Puricel, S., Schumann, R., Krampf, J., Ruiz, J., Schumacher, M.: An expert personal health system to monitor patients affected by gestational diabetes mellitus: a feasibility study. J. Ambient Intell. Smart Environ. **8**(2), 219–237 (2016)
8. Brugués, A., Bromuri, S., Barry, M., del Toro, O.J., Mazurkiewicz, M.R., Kardas, P., Pegueroles, J., Schumacher, M.: Processing diabetes mellitus composite events in MAGPIE. J. Med. Syst. **40**(2), 44 (2016)
9. Brugués, A., Bromuri, S., Pegueroles-Valles, J., Schumacher, M.I.: MAGPIE: an agent platform for the development of mobile applications for pervasive healthcare. In: Proceedings of the 3rd International Workshop on Artificial Intelligence and Assistive Medicine (AI-AM/NetMed), pp. 6–10 (2014)
10. Calvaresi, D., Cesarini, D., Marinoni, M., Buonocunto, P., Bandinelli, S., Buttazzo, G.: Non-intrusive patient monitoring for supporting general practitioners in following diseases evolution. In: Ortuño, F., Rojas, I. (eds.) IWBBIO 2015. LNCS, vol. 9044, pp. 491–501. Springer, Cham (2015). https://doi.org/10.1007/978-3-319-16480-9_48
11. Calvaresi, D., Cesarini, D., Sernani, P., Marinoni, M., Dragoni, A.F., Sturm, A.: Exploring the ambient assisted living domain: a systematic review. J. Ambient Intell. Humaniz. Comput. pp. 1–19 (2016)
12. Calvaresi, D., Marinoni, M., Sturm, A., Schumacher, M., Buttazzo, G.: The challenge of real-time multi-agent systems for enabling IoT and CPS. In: Proceedings of IEEE/WIC/ACM International Conference on Web Intelligence (WI 2017) (2017)

13. Calvaresi, D., Schumacher, M., Marinoni, M., Hilfiker, R., Dragoni, A.F., Buttazzo, G.: Agent-based systems for telerehabilitation: strengths, limitations and future challenges. In: Proceedings of the 10th Workshop on Agents Applied in Health Care (A2HC 2017) (2017)
14. Chittaro, L., Montanari, A.: Efficient temporal reasoning in the cached event calculus. Comput. Intell. **12**(3), 359–382 (1996). https://doi.org/10.1111/j.1467-8640.1996.tb00267.x
15. Denti, E., Omicini, A., Ricci, A.: Multi-paradigm Java-Prolog integration in tuProlog. Sci. Comput. Program. **57**(2), 217–250 (2005)
16. Dungan, K.: Monitoring technologies - continuous glucose monitoring, mobile technology, biomarkers of glycemic control. In: De Groot, L.J., Beck-Peccoz, P., Chrousos, G., Dungan, K., Grossman, A., Hershman, J.M., Singer, F. (eds.) Endotext [Internet] (2014)
17. Isern, D., Moreno, A.: A systematic literature review of agents applied in healthcare. J. Med. Syst. **40**(2), 43 (2015)
18. Isern, D., Sánchez, D., Moreno, A.: Agents applied in health care: a review. Int. J. Med. Inform. **79**(3), 145–166 (2010)
19. Kafalı, Ö., Bromuri, S., Sindlar, M., van der Weide, T., Aguilar Pelaez, E., Schaechtle, U., Alves, B., Zufferey, D., Rodriguez-Villegas, E., Schumacher, M.I., et al.: Commodity12: a smart e-health environment for diabetes management. J. Ambient Intell. Smart Environ. **5**(5), 479–502 (2013)
20. Kowalski, R., Sergot, M.: A logic-based calculus of events. New Gener. Comput. **4**(1), 67–95 (1986)
21. Montagna, S., Omicini, A., Angeli, F.D., Donati, M.: Towards the adoption of agent-based modelling and simulation in mobile health systems for the self-management of chronic diseases. In: Proceedings of the 17th Workshop "From Objects to Agents", Catania, Italy, 29–30 July 2016, pp. 100–105 (2016)
22. Mueller, E.T.: Commonsense Reasoning: An Event Calculus Based Approach, 2nd edn. Morgan Kaufmann Publishers Inc., San Francisco (2015)
23. Nefti, S., Manzoor, U., Manzoor, S.: Cognitive agent based intelligent warning system to monitor patients suffering from dementia using ambient assisted living. In: 2010 International Conference on Information Society, pp. 92–97 (2010)
24. Peine, A., Moors, E.H.: Valuing health technology - habilitating and prosthetic strategies in personal health systems. Technol. Forecast. Soci. Change **93**, 68–81 (2015). Science, Technology and the "Grand Challenge" of Ageing
25. Ramchurn, S.D., Vytelingum, P., Rogers, A., Jennings, N.: Agent-based control for decentralised demand side management in the smart grid. In: The 10th International Conference on Autonomous Agents and Multiagent Systems, vol. 1, pp. 5–12 (2011)
26. Ruta, M., Scioscia, F., Loseto, G., Sciascio, E.D.: Semantic-based resource discovery and orchestration in home and building automation: a multi-agent approach. IEEE Trans. Indus. Inform. **10**(1), 730–741 (2014)
27. Sernani, P., Claudi, A., Dragoni, A.: Combining artificial intelligence and netmedicine for ambient assisted living: a distributed BDI-based expert system. Int. J. E-Health Med. Commun. **6**(4), 62–76 (2015)
28. Silverman, B.G., Hanrahan, N., Bharathy, G., Gordon, K., Johnson, D.: A systems approach to healthcare: agent-based modeling, community mental health, and population well-being. Artif. Intell. Med. **63**(2), 61–71 (2015)
29. Tartarisco, G., Baldus, G., Corda, D., Raso, R., Arnao, A., Ferro, M., Gaggioli, A., Pioggia, G.: Personal health system architecture for stress monitoring and support to clinical decisions. Comput. Commun. **35**(11), 1296–1305 (2012)

30. Touati, F., Tabish, R.: U-healthcare system: state-of-the-art review and challenges. J. Med. Syst. **37**(3), 9949 (2013)
31. Varshney, U.: Pervasive healthcare and wireless health monitoring. Mob. Netw. Appl. **12**(2–3), 113–127 (2007)
32. World Health Organization: Global Report on Diabetes. World Health Organization, Geneva (2016)
33. Zimmet, P., Alberti, K.G., Magliano, D.J., Bennett, P.H.: Diabetes mellitus statistics on prevalence and mortality: facts and fallacies. Nature Rev. Endocrinol. **12**(10), 616–622 (2016)

Multiple-Profile Prediction-of-Use Games

Andrew Perrault[(⊠)] and Craig Boutilier

University of Toronto, Toronto, ON M5S 3H8, Canada
{perrault,cebly}@cs.toronto.edu

Abstract. *Prediction-of-use (POU) games* [14] address the mismatch between energy supplier costs and the incentives imposed on consumers by fixed-rate electricity tariffs. However, the framework does not address how consumers should coordinate to maximize social welfare. To address this, we develop multiple-profile prediction-of-use (MPOU) games, an extension of POU games in which agents report *multiple acceptable electricity use profiles*. We show that MPOU games share many attractive properties with POU games attractive (e.g., convexity). However, MPOU games introduce new incentive issues that limit our ability to exploit convexity effectively, a problem we analyze and resolve. We validate our approach with experimental results using utility models learned from real electricity use data.

1 Introduction

Prediction-of-use games were developed by Robu et al. [14], hereafter RVRJ, to address the mismatch between the cost structure of energy suppliers and the incentive structure induced by traditional fixed-rate tariffs faced by consumers. In most countries, energy suppliers face a two-stage market, where they purchase energy at lower rates in anticipation of future consumer demand and then reconcile supply and demand exactly at a higher rate at the time of realization through a balancing market [20]. The cost to energy suppliers is thus highly dependent on their ability to predict future consumption. Since consumers typically have little incentive to consume predictably, suppliers generally use past behavior to predict consumption. The uncertainty in these predictions incurs some additional cost for suppliers.

One way to improve supplier predictions is to incentivize consumers to report *predictions of their own consumption*, thus offering access to their private information about the future. RVRJ analyze mechanisms where flat tariffs are replaced with *prediction-of-use (POU) tariffs*, in which consumers make a payment based on both their actual consumption and the accuracy of their prediction. Similar tariffs have, in fact, been deployed in practice [3]. RVRJ analyze the *cooperative game* induced by POU tariffs, in which consumers form *buying coalitions* that reduce (aggregate) consumption uncertainty, and find that, under

---

C. Boutilier—Now at Google Research, Mountain View, CA.

© Springer International Publishing AG 2017
G. Sukthankar and J. A. Rodriguez-Aguilar (Eds.): AAMAS 2017 Best Papers,
LNAI 10642, pp. 275–295, 2017.
https://doi.org/10.1007/978-3-319-71682-4_17

normally-distributed prediction error, the game is *convex*. Convexity is a powerful property that significantly reduces the complexity of important problems in cooperative games, both analytically and computationally.

While attractive, the POU model has a significant shortcoming. Though the POU model could be adapted to model how consumers change their consumption in reaction to price changes, consumers cannot *coordinate* their consumption choices. A consumer's optimal *consumption profile*—a random variable representing the individual's possible behaviors or patterns of energy consumption—depends on the profiles others use. In POU games, the only consumer choice is what coalition to join—a consumer's demand is represented by a *single* prediction, reflecting just one selected (or average) consumption profile for each individual. In essence, consumers predict their behavior without knowing anything about others in the game. While the POU model can offer social welfare gains when the profiles are selected optimally, we show they can result in significant welfare loss when profile selection is uncoordinated.

We introduce *multiple-profile POU (MPOU) games*, which extend POU games to admit *multiple* consumer profiles. This allows consumers to coordinate the behaviors that change their predictions, facilitating the full realization of the benefits of the POU model. We show that MPOU games have many of the same properties that make the POU model tractable, e.g., convexity, which makes the stable distribution of the benefits of cooperation easy to compute. In addition, we show that MPOU games are individually rational and that consumer utility is monotone increasing as the number of truthfully-reported profiles increases. However, MPOU games also present a new challenge in coalitional allocation: since one can only observe an agent's (stochastic) consumption—not their underlying behavior—determining stabilizing payments for coalitional coordination requires novel techniques. We introduce *separating functions*, which incentivize agents to take a specific action in settings where actions are only *partially observable*.

We experimentally validate our techniques, using household utility functions that we learn (via structured prediction) from publicly-available electricity use data. We find that the MPOU model provides a gain of 3–5% over a fixed-rate tariff across several test scenarios, while a POU tariff *without* consumer coordination can result in losses of up to 30%. These experiments represent the first end-to-end study of the welfare consequences of POU tariffs.

The remainder of the paper is organized as follows. Section 2 reviews cooperative games, the POU model and related work. Section 3 introduces MPOU games and Sect. 4 proves their convexity. Section 5 outlines the new class of incentive problems that arises when the mechanism designer cannot (directly) observe an agent's selected profile, and develops a general solution to that problem. Section 6 briefly discusses manipulation. In Sect. 7, we describe an approach for learning consumer utility models from real-world electricity usage data, and experimentally validate the value of MPOU games using these learned models in Sect. 8.

2   Background

We begin with basic background on cooperative games, POU games, and their related work.

2.1   Cooperative Games

A prediction-of-use game is an instance of a *cooperative game with transferrable utility* [11], where agents can make arbitrary monetary payments to each other. In a cooperative game, the set N of agents divides into a set of *coalitions*, i.e., a disjoint partitioning of the agents. In a *profit game*, the *characteristic function* $v : 2^N \to \mathbb{R}$ represents the value that any subset of agents can achieve by cooperating. A profit game is a tuple $\langle N, v \rangle$.

The agents in a coalition $C \subseteq N$ distribute the benefits of cooperation however they choose. An *allocation* is a payment function $t : N \to \mathbb{R}$ that assigns some payment (which may be negative) to each agent. An allocation is *efficient* if it distributes the entire value, i.e., $\sum_{i \in N} t(i) = v(N)$. Agents receive no "individual" value under this model—all value is redistributed via coalitional payments. In practice, the individual value accrued by an agent may be deducted from its payment in order to reduce total transfers.

A major goal of cooperative game theory is to find allocations that prevent agents from *defecting* from their coalition, thus achieving *stability*. An allocation that stabilizes the *grand coalition* of all agents is in the *core*:

Definition 1. *Allocation t is in the* core *of profit game $\langle N, v \rangle$ if it is efficient and $\sum_{i \in S} t(i) \geq v(S)$ for all $S \subseteq N$.*

The core is a strong stability concept, so much so that certain profit games have an empty core (i.e., there are no core allocations). Another central solution concept is the *Shapley value* $s_C(i)$ of an agent i in coalition $C \subseteq N$, which emphasizes fairness and always exists. It values each agent according to the marginal value they contribute to the coalition when averaged over all *join orders* (i.e., the order in which agents are added to C):

$$s_C(i) = \sum_{S \subseteq C \setminus \{i\}} \frac{|S|!(|C| - |S| - 1)!}{|N|!} (v(S \cup \{i\}) - v(S)) \qquad (1)$$

A *convex game* is one where the value contributed by an agent to a coalition never decreases as more agents are added to that coalition:

Definition 2. *Profit game $\langle N, v \rangle$ is* convex *if $v(T \cup \{i\}) - v(T) \geq v(S \cup \{i\}) - v(S)$, for all $i \in N$, $S \subseteq T \subseteq N \setminus \{i\}$.*

Convex games have several important properties [17]. First, the grand coalition maximizes social welfare. Second, the Shapley value is in the core. Finally, a core allocation must exist and is computable in polynomial time in the number of agents.

2.2   Prediction-of-Use Games

A *prediction-of-use (POU) game* is a tuple $\langle N, \Pi, \tau \rangle$, where N is a set of agents, Π is a set of consumption profiles, and τ is a POU tariff. Each $i \in N$ uses electricity according to a *consumption profile* in Π, a normal random variable with mean μ_i and standard deviation σ_i, say, in kilowatt-hours (kWh). Let x_i denote i's realized consumption, $x_i \sim \mathcal{N}(\mu_i, \sigma_i)$. Agents are assumed to truthfully report their profiles to the coalition. We do not address elicitation or estimation of consumption here, but see below.

A *POU tariff* has the form $\tau = \langle p, \underline{p}, \bar{p} \rangle$, and is intended to better align the incentives of the consumer and electricity supplier, whose costs are greatly influenced by how predictable demands are. Each agent i is asked to predict a *baseline consumption* b_i, and is charged p for each unit of x_i, plus a penalty that depends on the accuracy of their prediction: $\bar{p}$ for each unit their realized x_i exceeds the baseline, and $\underline{p}$ for each unit it falls short:

$$\psi(x_i, b_i, \tau) = \begin{cases} p_j \cdot x_i + \bar{p} \cdot (x_i - b_i) & \text{if } b_i \leqslant x_i \\ p_j \cdot x_i + \underline{p} \cdot (b_i - x_i) & \text{if } b_i > x_i \end{cases} \qquad (2)$$

To ensure agents have no incentive to artificially inflate consumption, we require $0 \leqslant \bar{p}$ and $0 \leqslant \underline{p} \leqslant p$ [14]. An agent i should report a baseline that minimizes her expected payment. RVRJ show that i does this by predicting $b^* = \mu_i + \sigma_i \Phi^{-1}(\frac{\bar{p}}{\bar{p}+\underline{p}})$, where Φ^{-1} is the inverse normal CDF. They also show that i's expected payment under the optimal baseline is $\mu_i p + \sigma_i L(\underline{p}, \bar{p})$ where $L(\underline{p}, \bar{p}) = \int_0^{\frac{\bar{p}}{\bar{p}+\underline{p}}} \Phi^{-1}(y) dy$.

To be more predictable in *aggregate*, agents may form a coalition C, where C reports its aggregate demand and is charged as if it were a single agent. C's aggregate consumption is the sum of the normal random variables corresponding to the members' profiles, itself normal with mean $\mu(C) = \sum_{i \in C} \mu_i$ and std. dev. $\sigma(C) = \sqrt{\sum_{i \in C} \sigma_i^2}$. This aggregate prediction generally has lower variance w.r.t. the mean, thus reducing total penalty payments facing C under POU tariffs (compared to members acting individually).

RVRJ analyze *ex-ante* POU games. In the ex-ante game, all agent decisions, as well as any internal transfers, or payments, are based on *expected* consumption (realized consumption plays no role). This approach is justified when agents are risk-neutral, expected-utility maximizers and coalitions form at the time of consumption prediction, not at the time of consumption. The characteristic value of coalition C is

$$v(C) = -\mu(C)p - \sigma(C)L(\underline{p}, \bar{p}) \qquad (3)$$

and they show that the ex-ante POU game is convex.[1]

---

[1] Technically, they define the game as a *cost game* and show that the game is concave, while we use a profit game, but results from the two perspectives translate directly.

2.3   Related Work

POU games are closely related to *newsvendor games* [10], where a supplier must purchase inventory in advance of demand and faces a penalty for oversupply (storage costs) and undersupply (lost profit). Unlike POU games, the players are the suppliers, the demand distribution is known, and the primary object of study is the value that suppliers can gain by pooling their inventory.

In addition to POU games, others have proposed the formation of cooperatives or coalitions among electricity consumers. Rose et al. [15] develop a similar mechanism for truthfully eliciting consumer demand. Kota et al. [7] and Akasiadas and Chalkiadakis [2] propose using coalitions to improve reliability and shift peak power loads. Perrault and Boutilier [12] focus on the formation of groups of consumers with multiple profiles to reduce peak loads. None of this work offers the theoretical guarantees of RVRJ.

Beyond electricity markets, several authors have studied the problem of group purchasing in an AI context. Lu and Boutilier [8] study a restrictive class of buyer preferences (unit demand, only the supplier affects utility) and seller price functions (volume discounts), which has strong theoretical guarantees. Similarly, optimally matching a group of cooperative buyers to sellers has been studied [9,16].

3   Multiple-Profile POU Games

We extend POU games by allowing agents to report *multiple profiles*, each reflecting different behaviors or consumption patterns, and each with an inherent utility or value reflecting comfort, convenience, flexibility or other factors. These profiles correspond to different discrete choices the consumer makes, e.g., what temperature to set the air conditioner at or when to do laundry or dishes. This will allow an agent, when joining or bargaining with a coalition, to trade off cost—especially the cost of predictability—with her inherent utility. A *multiple-profile POU (MPOU) game* is a tuple $\langle N, \{\Pi_i\}, V, \tau \rangle$. Given set of agents N, each agent $i \in N$ has a non-empty *set of demand profiles* Π_i, where each profile $\pi_{i,k} = \langle \mu_{i,k}, \sigma_{i,k} \rangle \in \Pi_i$ reflects a consumption pattern (as in a POU model). Agent i's *valuation function* $V_i : \Pi_i \to \mathbb{R}$ indicates her value or relative preference (in dollars) for her demand profiles.[2] Admitting multiple profiles allows us to reason about an agent's response to the incentives that emerge with POU tariffs and in coalitional bargaining. Finally, τ is a POU tariff. We use the same definition of POU tariffs and agent baselines as in POU games above. Notice that the optimal baseline report for an agent is now defined relative to the profile they use.

As in POU games, agents are motivated to form coalitions to reduce the relative variance in their predictions. However, for a coalition C to accurately report its aggregate demand, its members must select and commit to a specific

---

[2] Such profiles and values may be explicitly elicited or estimated using past consumption data (see Sect. 6).

usage profile. We denote an *assignment of profiles to agents* as $A : N \rightarrow \times_{i \in N} \Pi_i$. Under such an assignment, C's consumption is normal, with mean $\mu(C, A) = \sum_{i \in C} \mu(A(i))$ and std. dev. $\sigma(C, A) = \sqrt{\sum_{i \in C} \sigma^2(A(i))}$. The aggregate value accrued by the coalition (prior to supplier payments) is the sum of its members' values: $V(C, A) = \sum_{i \in C} V_i(A(i))$.

As in RVRJ, we begin by analyzing ex-ante MPOU games, where agents make decisions and payments before consumption is realized. The characteristic value v of a coalition C is the maximum value that coalition can achieve in expectation under full cooperation, that is, assuming an optimal profile assignment and baseline report. We thus define $v(C) = \max_A v(C, A)$, where

$$v(C, A) = V(C, A) - \mu(C, A)p - \sigma(C, A)L(\underline{p}, \bar{p}) \tag{4}$$

Notice that profile selection does not arise in the POU setting.

In the following sections, we present a mechanism for MPOU games with which the grand coalition organizes the individual consumption behavior of its members (all agents in N) and the payments that flow among them. The mechanism proceeds as follows:

1. Agents report their consumption profiles to the mechanism (we assume this report is truthful).
2. The mechanism calculates an assignment A of agents to profiles that maximizes social welfare. We elaborate on this assignment optimization at the end of this section.
3. The mechanism calculates an ex-ante core stable payment $t(i)$ for each agent i that is based on all agents using their assigned profiles. We address payment computation in Sect. 4.
4. In Sect. 5, we find that some agents have an incentive to defect from the assigned profile, and we design *separating functions* to prevent these defections. The mechanism calculates a separating function D_i for each agent with an incentive to defect from their assigned profile.
5. At realization time, each agent i receives $t(i)$. Each agent i that has a separating function receives $D_i(x_i)$, where x_i is his/her realized consumption.

In the MPOU model, calculating a social welfare-maximizing assignment of agents to profiles requires solving a non-convex optimization problem. We do this using a mixed integer program with objective function given by (4), a binary assignment variable for each agent-profile pair, and a constraint that each agent is assigned exactly one profile. The last term of the objective is non-convex: $\sigma(C, A) = \sqrt{\sum_{i \in C} \sigma^2(A(i))}$. We replace the negative square root with a piecewise linear upper bound, which requires two binary variables per segment. As in other assignment problems, we can relax the assignment variables: in practice, relaxed solutions that are very close to integral.

4 Properties of MPOU Games

It is natural to ask whether, like POU games, ex-ante MPOU games are convex, since convexity simplifies the analysis of stability and fairness. We show that this is, in fact, the case. We begin with a technical lemma.

Lemma 1. *Let $\langle N, \Pi, \tau, V \rangle$ be an MPOU game. Let $i \in N$ and $S \subset T \subseteq N \setminus \{i\}$ and $j \in T \setminus S$. Then we have:*

$$v(S \cup \{i\}) - v(S) \leqslant v(S \cup \{i, j\}) - v(S \cup \{j\}) \tag{5}$$

Proof. We let $A^*(S)$ denote the assignment of profiles that maximizes the social welfare of S. In the case where there are multiple social welfare-maximizing configurations of S, we use the one with highest aggregate variance. We observe that $v(T, A^*(S)) \leqslant v(T)$ because $A^*(S)$ imposes a constraint on the behavior of S. For technical reasons, we break the proof into two cases based on whether it is more beneficial for i) i to join coalition S when S is configured to maximize $v(S \cup \{i\})$ or ii) i to join coalition $S \cup \{j\}$ when $S \cup \{j\}$ is configured to maximize $v(S \cup \{j\})$.

Case 1. $v(S \cup \{i\}) - v(S, A^*(S \cup \{i\})) > v(S \cup \{i, j\}, A^*(S \cup \{j\})) - v(S \cup \{j\})$

On both sides of the inequality, we are adding $\{i\}$ to a set of agents without changing the configuration of that set of agents. Thus, the inequality implies that $\{i\}$ contributes more value on the left side than on the right side. Since the amount of value that $\{i\}$ contributes depends only on the variance of the coalition that it is joining, the inequality implies that $\sigma(S \cup \{j\}, A^*(S \cup \{j\})) < \sigma(S, A^*(S \cup \{i\}))$.

Since j contributes a non-negative amount of variance, $\sigma(S, A^*(S \cup \{j\})) \leqslant \sigma(S \cup \{j\}, A^*(S \cup \{j\}))$, and likewise, $\sigma(S, A^*(S \cup \{i\})) \leqslant \sigma(S \cup \{i\}, A^*(S \cup \{i\}))$. Applying these inequalities yields $\sigma(S, A^*(S \cup \{j\})) < \sigma(S \cup \{i\}, A^*(S \cup \{i\}))$, implying:

$$v(S \cup \{j\}) - v(S, A^*(S \cup \{j\})) < v(S \cup \{i, j\}, A^*(S \cup \{i\})) - v(S \cup \{i\}) \tag{6}$$

Then, applying the inequalities $v(S, A^*(S \cup \{j\})) \leqslant v(S)$ and $v(S \cup \{i, j\}, A^*(S \cup \{i\})) \leqslant v(S \cup \{i, j\})$, and rearranging terms:

$$v(S \cup \{i\}) - v(S) < v(S \cup \{i, j\}) - v(S \cup \{j\}) \tag{7}$$

which is a stronger version of the lemma.

Case 2. $v(S \cup \{i\}) - v(S, A^*(S \cup \{i\})) \leqslant v(S \cup \{i, j\}, A^*(S \cup \{j\})) - v(S \cup \{j\})$

Applying the inequality $v(S, A^*(S \cup \{i\})) \leqslant v(S)$ on the left side yields:

$$v(S \cup \{i\}) - v(S) \leqslant v(S \cup \{i, j\}, A^*(S \cup \{j\})) - v(S \cup \{j\}) \tag{8}$$

Applying on the right side $v(S \cup \{i, j\}, A^*(S \cup \{j\})) \leqslant v(S \cup \{i, j\})$ yields the lemma:

$$v(S \cup \{i\}) - v(S) \leqslant v(S \cup \{i, j\}) - v(S \cup \{j\}) \tag{9}$$

From Lemma 1, we immediately obtain:

Theorem 1. *The ex-ante MPOU game is convex.*

Proof. If $S = T$, then $v(S \cup \{i\}) - v(S) = v(T \cup \{i\}) - v(T)$ since the welfare-maximizing configurations of S and T are the same. If $S \subset T$, we repeatedly apply Lemma 1 to "grow" S one agent a time, creating a series of inequalities, until we relate S and T.

Since the ex-ante MPOU game is convex, the Shapley value is in the core, hence we can compute a core allocation by averaging the payments from any number of join orders. In our experiments, we approximate the Shapley value by sampling [4].

It is important that agents are incentivized to participate in the mechanism. We show that MPOU games are *individually-rational*—no agent receives less utility than her best outside option, i.e., what she would receive if she chose not to participate in the mechanism. To achieve this, we augment an instance of the game by adding a dummy profile to each agent with value equal to that of their (best) outside option.

Theorem 2. *Let G be an MPOU game where each agent has a profile $\pi_{out}^{(i)}$ with $V(\pi_{out}^{(i)}) = \theta_i$, $\sigma(\pi_{out}^{(i)}) = \mu(\pi_{out}^{(i)}) = 0$, where θ_i is the value of i's outside option. Then, G is ex-ante individually rational if core payments are used.*

Proof. Core payments exist because G is an MPOU game, hence convex. Suppose, by way of contradiction, agent i receives an expected payment less than θ_i. The stability condition of core payments requires that $t(i) \geq v(\{i\})$. However, this contradicts the fact that $v(\{i\}) \geq \theta_i$.

5 Incentives in MPOU Games

MPOU games introduce a new coordination problem for coalitions that do not arise in POU games. In a fully-cooperative MPOU game, a coalition C agrees on a joint consumption profile prior to reporting its (aggregate) predicted demand. Despite this agreement, an agent $i \in C$ may have incentive to actually use a profile that differs from the one agreed to. For instance, suppose agent i has two profiles, π_0 and π_1, with $V_i(\pi_0) > V_i(\pi_1)$, and that to maximize the social welfare of C, i should use π_1 (and receive coalitional payment $t(i)$). By deviating from her agreed upon profile, i can increase her net utility (from $t(i)$ to $V_i(\pi_0) - V_i(\pi_1) + t(i)$).

Typically, a penalty should be imposed for such a deviation to ensure that C's welfare in maximized. Unfortunately, i's profile cannot be directly observed. Only her realized consumption x_i is observable, and it is related only *stochastically* to her underlying behavior (adopted profile). As such, any such transfer or penalty in the coalitional allocation must depend on x_i, showing that an ex-ante analysis is insufficient for MPOU games (in stark contrast to POU games). Furthermore, since x_i is stochastic, it could have arisen from i using either profile (i.e., we have no direct signal of the i's chosen profile), which makes the design of such transfers even more difficult. Finally, the poor choice of a transfer function may compromise the convexity of the ex-ante game, undermining our ability to compute core payments.

To address these challenges, we use a *separating function* $D_i(x_i)$. For each agent i, D_i maps i's realized consumption to an additional *ex-post separating payment*.

Definition 3. D_i *is a* separating function *(SF) for* i *under assignment* A *if it satisfies the* incentive *and* zero-expectation *conditions.*

- ***Incentive:*** $\mathbb{E}_{x_i \sim A(i)}[D_i(x_i)] > \mathbb{E}_{x_i \sim \pi}[D_i(x_i)] + V_i(\pi) - V_i(A(i))$ *for any* $\pi \in \Pi_i$ *such that* $\pi \neq A(i)$.
- ***Zero-expectation:*** $\mathbb{E}_{x_i \sim A(i)}[D_i(x_i)] = 0$.

Intuitively, the incentive condition ensures that the agent is incentivized to use the assigned profile, and the zero-expectation condition requires that the payments introduced by the incentive condition do not affect the agent's expected payment if she uses the assigned profile. Since agents are assumed to be risk neutral, each agent's payoffs are unaffected by addition of a SF as long as the agent uses the profile assigned by the coalition. Thus, payments remain in the core after the addition of an SF.[3]

The rest of this section describes how to find SFs. We begin by showing that a weaker form of separating function can trivially be transformed into a SF.

Definition 4. D_i *is a* weak separating function *(WSF) for* i *under assignment* A *if* $\mathbb{E}_{x_i \sim A(i)}[D_i(x_i)] > \mathbb{E}_{x_i \sim \pi}[D_i(x_i)]$ *for any* $\pi \in \Pi_i$ *such that* $\pi \neq A(i)$.

Remark 1. Let D_i be a WSF for i under assignment A. Then, $D_i' = w_0 D_i + w_1$ is an SF, where $w_0 = \max_{\pi \in \Pi_i, \pi \neq A(i)} \frac{V_i(\pi) - V_i(A(i))}{\mathbb{E}_{x_i \sim A(i)}[D_i(x_i)] - \mathbb{E}_{x_i \sim \pi}[D_i(x_i)]}$ and $w_1 = -\mathbb{E}_{x_i \sim A(i)}[w_0 D_i(x_i)]$.

Thus, it is sufficient to find a WSF. When an agent has only two profiles, this is straightforward: we let D_i be the PDF of the assigned profile minus the PDF of the unassigned profile. The proof for this statement is algebraic, using the fact that $\mathcal{N}(x; \mu_0, \sigma_0)\mathcal{N}(x; \mu_1, \sigma_1)$ has a closed form that is proportional to a normal PDF in x.

Theorem 3. *Let* i *be an agent with two profiles* π_0 *and* π_1 *and let* $A(i) = \pi_0$. *Then, w.l.o.g.,* $D_i(x_i) = \mathcal{N}(x_i; \mu_0, \sigma_0) - \mathcal{N}(x_i; \mu_1, \sigma_1)$ *is a WSF for* i *under* A.

Proof. We show that the minimum of $\mathbb{E}_{x \sim \mathcal{N}(\mu_0, \sigma_0)}[\mathcal{N}(x; \mu_0, \sigma_0) - \mathcal{N}(x; \mu_1, \sigma_1)] - \mathbb{E}_{x \sim \mathcal{N}(\mu_1, \sigma_1)}[\mathcal{N}(x; \mu_0, \sigma_0) - \mathcal{N}(x; \mu_1, \sigma_1)]$ occurs when $\mu_1 = \mu_0$ and $\sigma_1 = \sigma_0$, and that the value of the expression at that point is positive.

We make use of the fact that $\mathcal{N}(x; \mu_1, \sigma_1)\mathcal{N}(x; \mu_2, \sigma_2)$ is a function proportional to the PDF of a normal distribution. Specifically,

$$\mathcal{N}(x; \mu_0, \sigma_0)\mathcal{N}(x; \mu_1, \sigma_1)$$
$$= \mathcal{N}\left(\mu_0; \mu_1, \sqrt{\sigma_0^2 + \sigma_1^2}\right) \mathcal{N}\left(x; \frac{\sigma_0^{-2}\mu_0 + \sigma_1^{-2}\mu_1}{\sigma_0^{-2} + \sigma_1^{-2}}, \frac{\sigma_0^2\sigma_1^2}{\sigma_0^2 + \sigma_1^2}\right) \quad (10)$$

---

[3] Our use of zero-expectation payments for risk-neutral agents is mechanically similar to Cremer and McClean's [5] revenue-optimal auction for bidders with correlated valuations.

Then, by expanding terms and applying (10):

$$\mathbb{E}_{x \sim \mathcal{N}(\mu_0, \sigma_0)}[\mathcal{N}(x; \mu_0, \sigma_0) - \mathcal{N}(x; \mu_1, \sigma_1)]$$
$$- \mathbb{E}_{x \sim \mathcal{N}(\mu_1, \sigma_1)}[\mathcal{N}(x; \mu_0, \sigma_0) - \mathcal{N}(x; \mu_1, \sigma_1)]$$
$$= \frac{1}{2\sigma_0 \sqrt{\pi}} - 2\mathcal{N}\left(\mu_1; \mu_0, \sqrt{\sigma_0^2 + \sigma_1^2}\right) + \frac{1}{2\sigma_1 \sqrt{\pi}} \qquad (11)$$

We then minimize with respect to μ_1 and σ_1. Since the middle term is the only one that contains μ_1, we can minimize it separately:

$$- \frac{2}{\sqrt{2\pi(\sigma_0^2 + \sigma_1^2)}} \exp\left(-\frac{(\mu_0 - \mu_1)^2}{2(\sigma_0^2 + \sigma_1^2)}\right) \qquad (12)$$

Since the argument of the exponent is always non-positive, it is maximized when it is zero, i.e., $\mu_1 = \mu_0$. Making this substitution yields:

$$\frac{1}{2\sigma_0 \sqrt{\pi}} - \frac{2}{\sqrt{2\pi(\sigma_0^2 + \sigma_1^2)}} + \frac{1}{2\sigma_1 \sqrt{\pi}} \qquad (13)$$

Setting the derivative with respect to σ_1^2 to zero yields two real roots of $\sigma_0 = \pm \sigma_1$. The second derivative at these points is positive. Thus, it is a minimum. The value of the original expression at this point is 0 and positive otherwise.

With more than two profiles, this approach does not always work. Instead, we can use a linear program (LP) to find coefficients of a linear combination of the profile PDFs. Formally, denote the PDFs of the profiles as $\mathcal{N}_i(x_i) = \langle \mathcal{N}(x_i; \mu_0, \sigma_0), \ldots, \mathcal{N}(x_i; \mu_{|\Pi_i|-1}, \sigma_{|\Pi_i|-1}) \rangle$, their weights as $\boldsymbol{y}_i$, and search over $\boldsymbol{y}_i \in \mathbb{R}^{|\Pi_i|}$ for a separating function of the form $D_i(x_i, \boldsymbol{y}_i) = \boldsymbol{y}_i \cdot \mathcal{N}_i(x_i)$. We use an LP that minimizes the L_1-norm of $\boldsymbol{y}_i$ subject to $\mathbb{E}_{x_i \sim A(i)}[D_i(x_i, \boldsymbol{y}_i)] > \mathbb{E}_{x_i \sim \pi}[D_i(x_i, \boldsymbol{y}_i)]$ for all $\pi \in \Pi_i, \pi \neq A(i)$. Ideally, we would also like to minimize the variance of the separating payment, giving agents maximal certainty w.r.t. this payment; however, this objective is not tractable in an LP (we leave this question to future work). In our experiments below, we do, however, assess the variance of the separating payment.

A feasible $\boldsymbol{y}_i$ corresponds to a linear combination of vectors whose sum has only positive entries. We call these the *difference vectors* of D_i. While we cannot prove that a feasible $\boldsymbol{y}_i$ always exists, viewing the problem in terms of difference vectors suggests why they exist in practice:

Definition 5. *Let $A(i)$ be π_0 (w.l.o.g.). For each profile $\pi_k \in \Pi_i$ the difference vector $\boldsymbol{d}_k = \mathbb{E}_{x \sim \pi_k}[\mathcal{N}(x; \pi_0, \sigma_0] - \langle \mathbb{E}_{x \sim \pi_k}[\mathcal{N}(x; \mu_1, \sigma_1)], \ldots, \mathbb{E}_{x \sim \pi_k}[\mathcal{N}(x; \mu_{|\Pi_i|-1}, \sigma_{|\Pi_i|-1})] \rangle$.*

Note that these vectors do not depend on $\boldsymbol{y}_i$. We can restate the LP constraints using difference vectors:

Theorem 4. *Let i have profiles Π_i and let A assign a profile to i. There exists $\boldsymbol{y}_i \in \mathbb{R}^{|\Pi_i|}$ that makes $D_i(x_i, \boldsymbol{y}_i)$ a WSF if and only if there is a linear combination of the difference vectors of $D_i(x_i, \boldsymbol{y}_i)$ that has only positive entries.*

Proof. First, we prove the forward direction. Let c be the coefficients of the linear combination of the difference vectors that has only positive entries, i.e., $\sum_{k \in |\Pi_i|} c_k d_k = b$ where b is element-wise positive. Then, $\mathbb{E}_{x_i \sim A(i)}[D_i(x_i, c)] - \mathbb{E}_{x_i \sim \pi}[D_i(x_i, c)] = c d_k = b_{k-1}$. Since b is element-wise positive, letting $y_i = c$ makes $D_i(x_i, y_i)$ a separating function.

The reverse direction is also straightforward. Suppose $D_i(x_i, y_i)$ is a separating function. Then, let $b_{k-1} = \mathbb{E}_{x_i \sim A(i)}[D_i(x_i, c)] - \mathbb{E}_{x_i \sim \pi}[D_i(x_i, c)] = y_i \cdot d_k$. Thus, taking y_i as the coefficients of the linear combination of difference vectors equals b, which has only positive entries.

Corollary 1. *Let d_k be the difference vectors for agent i. If the difference vectors are linearly independent, a setting of y_i exists that makes $D_i(x_i, y_i)$ a WSF.*

Proof. If the difference vectors are linearly independent, there exists a coefficient vector c that makes $\sum_{k \in |\Pi_i|} c_k d_k$ elementwise positive. We can take $y_i = c$ to satisfy the corollary.

We generally expect a random set of vectors to be linearly independent as the set of matrices drawn from the reals with non-independent rows has Lebesgue measure zero. We have yet to encounter an instance where a separating function does not exist in our experiments. It is an open question as to whether a separating function of this form always exists.

6   Manipulation in MPOU Games

While we defer a thorough discussion of manipulation of MPOU games to future work, we briefly discuss a simple form of manipulation: *adding profiles to, or removing profiles from, an agent's report.* Formally, we say that an agent can *manipulate* an MPOU game if they gain expected utility by misreporting their true set of profiles. Here, we simplify the discussion by assuming that agents have a true underlying set of profiles, and we rely on the results of the previous section by assuming that each agent can be incentivized to use their assigned profile without changing their expected payoff.

Agents are not incentivized to strategically withhold information if they otherwise report truthfully. However, reporting additional untruthful profiles will benefit the agent, as long as those profiles are not assigned by the mechanism.

Theorem 5. *Let G be an MPOU game, let G' be identical to G except agent i reports an additional profile $\pi^{(i)}_{extra}$. Let all of i's reported profiles be truthful except $\pi^{(i)}_{extra}$ and let at least one of these conditions hold: (i) $\pi^{(i)}_{extra}$ is truthful or (ii) $\pi^{(i)}_{extra}$ is not the assigned profile. Then, agent i's payoff in G' is greater than or equal to its payoff in G if payments are used that average marginal contributions over the same join orders.*

Proof. First, we establish that i's Shapley value is greater with the additional profile. Each time agent i is added to a coalition S in a join order, agent i's marginal contribution to $v(S \cup \{i\})$ with the extra profile is greater than or equal to its contribution with its original profiles. Thus, $t_{G'}(i) \geq t_G(i)$.

This condition is not sufficient to ensure that i increases her payoff, which is equal to her coalitional payment minus the reported value of the assigned profile plus the true value of the assigned profile. In condition (i), the Shapley value equals the payoff value and in condition (ii), the assigned profile is the same in G and G'. Thus, i's payoff is greater or equal in G' in either case.

Note that the theorem applies both to the Shapley value, which can be expressed as an average over marginal contributions over join orders, and to sampling-based approximations, such as the ones used in our experiments.

We outline two ways of combatting manipulation by reporting additional profiles. The first is to simply limit the number of reported profiles, either by creating a cap or by charging agents per profile they report, limiting the amount agents can gain by manipulating. This approach leads to a non-truthful equilibrium, and it penalizes agents who have more complicated utility functions.

The second approach emerges from an approximation to the Shapley value that happens to remove the incentive to add additional profiles that are not selected. Recall that i's Shapley value in coalition C can be interpreted as the average marginal value that i contributes over all orders that agents join C. Computing this requires recalculating the optimal assignment of profiles before and after i joins since the addition of i may cause change the optimal assignment for the other agents. Because this is computationally expensive, we approximate it by fixing agents to the profile they are assigned in the grand coalition. Formally, we let i's *Shapley value with fixed profiles* be

$$s_C(i, N) = \sum_{S \subseteq C \setminus \{i\}} \frac{|S|!(|C| - |S| - 1)!}{|N|!} (v(S \cup \{i\}, A^*(N)) - v(S, A^*(N))) \quad (14)$$

Recall that $v(S, A^*(N))$ is the value of coalition S under the assignment that maximizes the value of coalition N, i.e., the grand coalition. We find the approximation is quite close to the true Shapley value in our setting. The approximation sacrifices exact convexity because it does not discriminate between agents based on how attractive their unassigned profiles are, which has the additional consequence that, as long as agents report their true profiles, they have no incentive to add additional false ones.

Theorem 6. *Let G be an MPOU game, let G' be identical to G except agent i reports an additional profile $\pi_{extra}^{(i)}$. Let all of i's reported profiles be truthful except $\pi_{extra}^{(i)}$. Then, agent i's payoff in G' is less than or equal to its payoff in G, if payments are used that average marginal contributions over the same join orders and fix i's profile to its assigned profile.*

Proof. Since we assume that i reports all of its profiles truthfully, the true value of $\pi_{extra}^{(i)}$ is 0. Then, either the mechanism selects $\pi_{extra}^{(i)}$ or it does not. If it does, i's payoff will be negative since it receives 0 value from $\pi_{extra}^{(i)}$, and thus, its payoff decreased because the mechanism is individually rational according to Theorem 2. If it does not, i's payoff is unchanged because $\pi_{extra}^{(i)}$ does not affect its payoff.

7   Learning Utility Models

To empirically test the MPOU framework and our separating functions, we require consumer utility functions. As we know of no data set with such utility functions, we learn household (agent) utility models from real electricity usage data from Pecan Street Inc. [13].[4] We define our prediction period as 4–7 pm each day, when electricity usage typically peaks in Austin, Texas, where the data was collected. We decompose utility into two parts: $V_i^{(\mu)}(w, \mu)$ describes the value an agent i derives from her mean consumption given a vector w of weather conditions; and $V_i^{(\sigma)}(\sigma, \mu)$ represents utility derived from variance in consumption behavior. Agent i's utility is $V_i(w, \mu, \sigma) = V_i^{(\mu)}(w, \mu)V_i^{(\sigma)}(\sigma, \mu)$.

Estimating $V_i^{(\mu)}$ is difficult, since we lack data for some aspects of the problem. Thus, we make some simplifying assumptions: (i) consuming $0 \, \text{kWh}$ yields value \$0; and (ii) $V_i^{(\mu)}(w, \mu)$ is concave and increasing. We learn a model for each of 25 households that have complete data from 2013–15 (about 1100 data points per household), using select weather conditions w and mean consumption between 4–7 pm as input, and outputting value (in dollars). We use this valuation function to predict consumption by maximizing an agent's net utility under the observed price:

$$V_i^{(\mu)}(w, \mu) = z_i^{(0)}(w) \left(\mu - z_i^{(1)}(w) \right)^{z_i^{(2)}(w)} + z_i^{(3)}(w) \tag{15}$$

constraining $z_i^{(0)} > 0$, $z_i^{(1)} > 0$, $0 < z_i^{(2)} < 1$, $z_i^{(3)}(w) \geq 0$ (Fig. 1 depicts the utility model). We use a homogenous function to represent utility [18]. The term $z_i^{(3)}(w)$ has no influence on predictions: it can be viewed as inherent value due to weather, and accounts for the flexibility provided by the $z_i^{(1)}$ term, which may create valuations where consumption 0 yields negative value (violating our assumptions). To prevent this, we set $z_i^{(3)}(w)$ to ensure the tangent at the predicted consumption for \$0.64 (the largest price in the data set) passes through (0, 0) (see Fig. 2). When this tangent crosses the y-axis above 0, we set $z_i^{(3)}(w) = 0$ and splice in an exponential ax^b that passes through (0, 0) and matches the derivative at the splice point.

For training, we use the model to predict consumption by solving the net utility maximization problem, $\max_\mu (V_i(w, \mu) - \mu p)$, yielding:

$$\hat{\mu}(w, p) = \frac{p}{z_i^{(0)}(w)z_i^{(2)}(w)}^{\frac{1}{z_i^{(2)}(w)-1}} + z_i^{(1)}(w) \tag{16}$$

---

[4] Publicly available at pecanstreet.org.

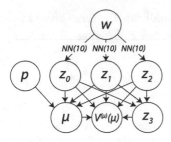

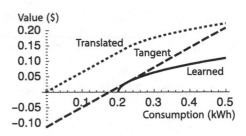

Fig. 1. The learned valuation model. $NN(10)$ denotes a neural network with 10 hidden units.

Fig. 2. Translating the valuation function to pass through the origin

We represent $z_i^{(0)}, z_i^{(1)}$ and $z_i^{(2)}$ in fully-connected single-layer neural networks, each with 10 hidden units and ReLU activations, and train the model with backpropagation. We implement the model in TensorFlow [1] using the squared error loss function and the Adam optimizer [6]. We use Dropout [19] with a probability of 0.7 on each hidden unit.

We split the data into 80% train and 20% test for each household. Table 1 compares the prediction accuracy of our model ("valuation") to (i) an unstructured neural network, and (ii) the best constant prediction for each household. The unstructured net learns a mapping from $\langle w, p \rangle$ to μ directly using 10 hidden units, without an intervening utility model.[5] The best constant prediction disregards weather and price data, and simply predicts average consumption for that household. Table 1 shows that the valuation model overfits somewhat, but that predictive accuracy is on par with the unstructured model. This shows that our constraints on the form of the valuation function are not unduly restrictive and validates the value predictions produced by these learned models. However, we believe these value functions significantly underestimate value because we lack consumption observations when the price is higher is than $0.64.

Table 1. Comparison of model prediction accuracy by root-mean-square error (RMSE). We divide each household's consumption amounts by their largest observed consumption.

Model	Mean train RMSE	Std. dev. train RMSE	Mean test RMSE	Std. dev. test RMSE
Valuation	0.137	0.0168	0.148	0.0194
Unstructured	0.142	0.0226	0.144	0.0284
Constant	0.204	0.0345	0.205	0.0411

---

[5] Our other implementation choices are the same as the valuation model, except we use Dropout of 0.5.

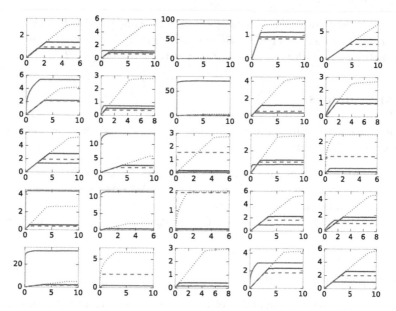

Fig. 3. Learned value models for the 25 households with consumption mean (kwh) on the x-axis and value (\$) on the y-axis. The red line represents the median weather conditions. The dotted line represents the median day with 90th percentile or higher temperature. The dashed and green lines are the same for sunshine and humidity, respectively.

Figure 3 shows the learned valuation for the 25 households. Each line represents a household's response to different weather conditions. While temperature is the most significant predictor of power usage, different households appear to exhibit sensitivity to different factors (e.g., the household on the right is highly sensitive to humidity).

Modeling Unpredictable Consumption. Unfortunately, we do not have access to electricity usage data where consumers are charged differently depending on the accuracy of their predictions. Our model of the value of unpredictable consumption is thus speculative, but uses the Pecan Street data as a starting point. We assume that each household chooses the σ that maximizes its utility (since they are not being charged for σ), and that it has an optimal fraction β_i of σ/μ that does not depend on other conditions. We estimate β_i from the data by treating each data point as having an observed σ equal to the absolute error in consumption prediction made by the learned valuation model. We assume no value is gained by increasing σ above the optimal ratio, and use an exponential to represent the loss in value when σ is reduced,

$$V_i^{(\sigma)}(\sigma, \mu) = \max\left(\frac{\mu/\sigma}{\beta_i}, 1\right)^{\gamma_i}, \tag{17}$$

where γ_i is a constant representing i's cost for being predictable. A higher γ_i means that consumer i values variance more highly. In our experiments, we sample γ_i from the uniform distribution over the interval $[0.1, 2]$.

8   Experiments

We experimentally evaluate our mechanism for MPOU games. The questions we study experimentally are:

1. How important is consumer coordination under POU tariffs?
2. What is the social welfare gain from using an MPOU model vs. a flat tariff?
3. How important is an agent's choice of reported profiles?
4. What are the variances of the payments introduced by the separating functions?

8.1   Experimental Setup

We first describe the experimental setup: how we select agents, profiles and tariffs. For each trial, we select weather conditions w uniformly at random from the Pecan Street data. To generate agents, we sample from our 25 learned household utility models, using w as input and adding a small amount of zero mean noise to the model parameters. We sample γ_i from the uniform distribution $[0.1, 2]$ for each agent i. Each data point is an average of 100 trials with 5000 agents, unless otherwise noted. One of the goals of our experiments is to study the consequences of different choices of reported profile. To do this, we vary the way profiles are generated. Each agent has four profiles: a *base profile* (predicted to be optimal under a flat rate tariff with rate equal to the fixed-rate p of the POU tariff), and three others reflecting reduced consumption mean or variance. The first reduces the base profile *mean* by the amount required to reduce value by $u\%$, which we call the *profile spacing*. The second reduces *variance* to reduce value by $u\%$. The third reduces both. We vary u throughout the experiments.

To generate tariffs, we vary the amount of emphasis each puts on accurate predictions vs. the amount consumed. We let the *predictivity emphasis* (PE) of a tariff w.r.t. a group of agents be the fraction of the expected total cost paid for prediction penalties when each uses her base profile. In practice, PE should be set to match the properties of the reserve power generation capacity that is available: a higher PE corresponds to more expensive reserves. A tariff is *revenue-equivalent* to another with respect to a specific set of profiles if the revenue of the two is the same for that set. All of our tariffs will be revenue-equivalent with respect to the set of base profiles. To find a revenue-equivalent tariff with a certain PE, we use a numerical solver to find a tariff of the form $\langle p, r, r \rangle$ with the appropriate total cost. Intuitively, a higher PE should result in larger benefits from POU tariffs, and we find that to be the case in our experiments.

To generate Shapley values, we ample a number of join orders equal to the logarithm of the number of agents in the instance. Shapley values were very close to linear in the std. dev. of the assigned profile. The average Shapley payment for prediction was $0.41 per kWh of uncertainty across trials with PE 10%, and $0.82 per kWh with PE 20%.[6] Within a single trial, the std. dev. of this ratio was less than 0.01 on average, suggesting that it is not necessary to optimize the choice of profiles every time an agent added in a join order—it is sufficient to fix each agent's profile to the assigned one. We exploit this fact to run larger experiments.

8.2   Results

We first address the question of how important it is for agents to coordinate their consumption under a POU tariff. We define the *uncoordinated POU setting* as the scenario where agents are subject to a POU tariff, but do not coordinate their consumption behavior, i.e., each agent uses the profile that individually maximizes her net utility relative to that POU tariff. Then, as is standard in that setting, the grand coalition forms and makes the optimal baseline prediction. Figure 4 shows the social welfare derived by agents in the uncoordinated POU setting as a percentage of their social welfare under a revenue-equivalent fixed-rate tariff. We see that the average social welfare achieved in the uncoordinated POU setting is less than that of the fixed rate setting for all profile spacings. Individual agents react to the POU tariff by increasing their predictivity, and thus decreasing their realized value, but they do not account for the predictivity discount that results from being part of a coalition. As profile spacing increases, more agents shift away from their base profile and social welfare decreases, reaching 70% when spacing is 25%. These results underscore the need for a way for agents to coordinate their profile choices under POU tariffs and highlight one of the main challenges of successfully implementing a POU tariff in practice.

Next, we study the social welfare gain that can be achieved by a POU tariff when agents coordinate optimally under the MPOU framework. Figure 5 shows the effect of profile spacing (u) on the welfare gained by switching from a fixed-rate tariff to a revenue-equivalent POU tariff.[7] Overall welfare gains are moderate, around 3.13% for PE of 10% and 4.4-4.9% for PE of 20%. A higher PE results in a larger social welfare gain because agents only benefit from cooperating when trading off predictivity for inherent utility. Profile spacing appears to have limited impact on social welfare gain, suggesting that most of the gain is achieved by the effective reduction in fixed-rate price under a POU tariff. We note that these experiments are the first to study end-to-end social welfare gain from a POU tariff.

Figure 5 appears to indicate that personalizing profile spacing based on each agent's value for predictivity would increase social welfare further. We can see this because increasing profile spacing increases welfare up to a spacing of 15% for both PE levels, but the number of agents that shift profiles decreases as

---

[6] This and other tariffs in this section have $0.2 \leqslant \underline{p} = \bar{p} \leqslant 1.5$.

[7] Each instance took around 3 min on a single thread of 2.6 Ghz Intel i7, 8 GB RAM.

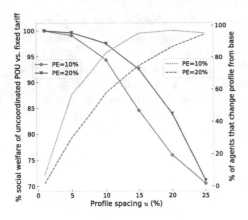
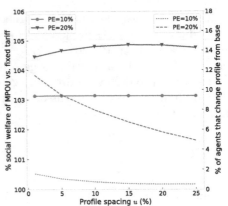

Fig. 4. Profile spacing vs. % of social welfare of fixed-rate tariff for uncoordinated POU setting and % of agents that change profile

Fig. 5. Profile spacing vs. social welfare % gain from fixed-rate tariff and % of agents that change profile

spacing is increased (shown on the right-side axis). Thus, we hypothesize that welfare could be further increased if agents with higher γ spaced their profiles farther apart than those with lower.

Next, we address the question of uncertainty introduced by separating payments. Recall that while separating payments have expectation zero, they introduce additional uncertainty to agent payments. We find that the amount of uncertainty introduced is, in fact, minimal, and decreases with instance size and increased PE. Figure 6 shows the same of the standard deviation of the separating payment to the Shapley payment for predictivity. The std. dev. of the separating payment is on average 15–20% of predictivity payment for PE of 10% and 7.5–10% for PE of 20%, and increases slightly as profile spacing increases. Note that only agents that actually require a separating function are taken into account, around 1–2% of all agents for PE of 10% and 5–10% for PE of 20%, on average. More agents require separating payments as PE increases, but the uncertainty introduced by each decreases. Note that these are uncertainties for a single instance of the game, and if the game is played repeatedly (e.g., every day), the aggregate uncertainty will decrease as the independent random variables are added.

Figure 7 shows the same uncertainty ratio for a single large instance versus the predictivity flexibility (γ) of each agent. This instance has PE of 20%, 100,000 agents, profile spacing of 15% and takes 90 min to solve. The ratio is shown for the 4876 agents that require separating functions. The magnitude of the introduced uncertainty is smaller in this larger instance with an average of 2.07% (and not exceeding 3% for any agent). In addition, predictivity flexibility has little affect on the introduced uncertainty: the linear least-squares fit (red line) has slope of less than 10^{-4}.

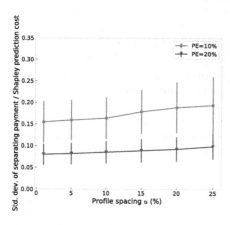

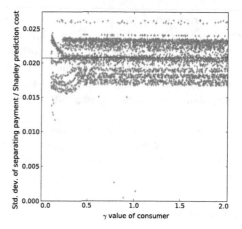

Fig. 6. Comparison of the standard deviation of the separating function payment to the ex-ante payment for prediction accuracy. Bars show one standard deviation. 5000 agents, 100 trials

Fig. 7. Comparison of the standard deviation of the separating function payment to the ex-ante payment for prediction accuracy

9   Conclusion

We have introduced *multiple-profile POU (MPOU) games*, a framework for coordinating agent behavior under POU tariffs. MPOU games allow agents to express their consumption utility functions, while maintaining convexity of the basic POU model. MPOU games introduce a new class of incentive problems due to agent actions being partially observable: we introduce *separating payments* to restore proper incentives. Our experimental utility models are learned from historical electricity usage data in a novel way. Our experiments show that, while social welfare gained by introducing the MPOU model (w.r.t. a fixed-rate tariff) appear moderate, the gains relative to a POU tariff are substantial. The gains over a fixed-rate tariff may be worthwhile in a large system and may be further enhanced by more sophisticated agent utility and behavior profile models. They depend both on the predictivity emphasis (PE) of reserve generation and on consumers' value for consuming unpredictably, which are both areas where more real-world data is needed. We find that the uncertainty introduced by separating payments decreases as instance size increases, and decreases in aggregate as more iterations of the game are played. Increased PE increases the number of agents that need separating functions, but the uncertainty introduced decreases.

Interesting future directions for POU/MPOU games remain. Following up on our approach, we could more precisely test social welfare gain with better access to household utility data, especially for variance of consumption, and data about the PE of generation mixes. Other critical aspects of the system are the ability of agents to manipulate, which we only briefly touch on, and how to elicit household utility functions. Thinking more broadly, it would be desirable to allow agents to make predictions contingent on intermediate predictions (e.g., of weather) thus reducing the need for agents to make accurate weather forecasts.

While our discussion of POU and MPOU games has focused on electricity markets, we believe the approach may be more widely applicable in other cases where agents are contending with a scarce resource, e.g., internal allocation of computing resources across groups in a company or university.

Acknowledgments. Perrault was supported by an Ontario Graduate Scholarship. We gratefully acknowledge the support of NSERC. We thank Valentin Robu, Meritxell Vinyals, Marek Janicki, Jake Snell, and the anonymous reviewers for their helpful suggestions.

References

1. Abadi, M., Agarwal, A., Barham, P., Brevdo, E., Chen, Z., Citro, C., Corrado, G.S., Davis, A., Dean, J., Devin, M., Ghemawat, S., Goodfellow, I., Harp, A., Irving, G., Isard, M., Jia, Y., Jozefowicz, R., Kaiser, L., Kudlur, M., Levenberg, J., Mané, D., Monga, R., Moore, S., Murray, D., Olah, C., Schuster, M., Shlens, J., Steiner, B., Sutskever, I., Talwar, K., Tucker, P., Vanhoucke, V., Vasudevan, V., Viégas, F., Vinyals, O., Warden, P., Wattenberg, M., Wicke, M., Yu, Y., Zheng, X.: TensorFlow: Large-scale machine learning on heterogeneous systems (2015). http://tensorflow.org/, Software available
2. Akasiadis, C., Chalkiadakis, G.: Agent cooperatives for effective power consumption shifting. In: Proceedings of the Twenty-seventh AAAI Conference on Artificial Intelligence (AAAI-13), pp. 1263–1269. Bellevue, WA (2013)
3. Braithwait, S., Hansen, D., O'Sheasy, M.: Retail electricity pricing and rate design in evolving markets, pp. 1–57. Edison Electric Institute, Washington, D.C. (2007)
4. Castro, J., Gómez, D., Tejada, J.: Polynomial calculation of the Shapley value based on sampling. Comput. Oper. Res. **36**(5), 1726–1730 (2009)
5. Cremer, J., McLean, R.P.: Full extraction of the surplus in Bayesian and dominant strategy auctions. Econometrica **56**(6), 1247–1257 (1988)
6. Kingma, D.P., Ba, J.L.: Adam: a method for stochastic optimization. In: Proceedings of the 3rd ACM SIGKDD International Conference on Learning Representations (ICLR-15), San Diego (2015)
7. Kota, R., Chalkiadakis, G., Robu, V., Rogers, A., Jennings, N.R.: Cooperatives for demand side management. In: Proceedings of the Twenty-First European Conference on Artificial Intelligence (ECAI-12), pp. 969–974. Montpellier, France (2012)
8. Lu, T., Boutilier, C.: Matching models for preference-sensitive group purchasing. In: Proceedings of the Thirteenth ACM Conference on Electronic Commerce (EC'12), pp. 723–740. Valencia, Spain (2012)
9. Manisterski, E., Sarne, D., Kraus, S.: Enhancing cooperative search with concurrent interactions. J. Artif. Intell. Res. **32**(1), 1–36 (2008)
10. Müller, A., Scarsini, M., Shaked, M.: The newsvendor game has a nonempty core. Games Econ. Behav. **38**(1), 118–126 (2002)
11. Osborne, M.J., Rubinstein, A.: A Course in Game Theory. MIT Press, Cambridge (1994)
12. Perrault, A., Boutilier, C.: Approximately stable pricing for coordinated purchasing of electricity. In: Proceedings of the Twenty-fourth International Joint Conference on Artificial Intelligence (IJCAI-15), Buenos Aires (2015)

13. Rhodes, J.D., Upshaw, C.R., Harris, C.B., Meehan, C.M., Walling, D.A., Navrátil, P.A., Beck, A.L., Nagasawa, K., Fares, R.L., Cole, W.J., et al.: Experimental and data collection methods for a large-scale smart grid deployment: methods and first results. Energy **65**, 462–471 (2014)
14. Robu, V., Vinyals, M., Rogers, A., Jennings, N.: Efficient buyer groups with prediction-of-use electricity tariffs. IEEE Trans. Smart Grid (2017)
15. Rose, H., Rogers, A., Gerding, E.H.: A scoring rule-based mechanism for aggregate demand prediction in the smart grid. In: Proceedings of the Eleventh International Joint Conference on Autonomous Agents and Multiagent Systems (AAMAS-12), pp. 661–668. Valencia, Spain (2012)
16. Sarne, D., Kraus, S.: Cooperative exploration in the electronic marketplace. In: Proceedings of the Twentieth National Conference on Artificial Intelligence (AAAI-05), pp. 158–163. Pittsburgh (2005)
17. Shapley, L.S.: Cores of convex games. Int. J. Game Theory **1**, 11–26 (1971)
18. Simon, C.P., Blume, L.: Mathematics for Economists, vol. 7. Norton, New York (1994)
19. Srivastava, N., Hinton, G.E., Krizhevsky, A., Sutskever, I., Salakhutdinov, R.: Dropout: a simple way to prevent neural networks from overfitting. J. Mach. Learn. Res. **15**(1), 1929–1958 (2014)
20. Team, G.M.: Electricity and gas supply market report. Technical report. 176/11, The Office of Gas and Electricity Markets (Ofgem), December 2011

Author Index